Communications in Computer and Information Science

2689

Series Editors

Gang Li , *School of Information Technology, Deakin University, Burwood, VIC, Australia*

Joaquim Filipe, *Polytechnic Institute of Setúbal, Setúbal, Portugal*

Zhiwei Xu, *Chinese Academy of Sciences, Beijing, China*

Rationale

The CCIS series is devoted to the publication of proceedings of computer science conferences. Its aim is to efficiently disseminate original research results in informatics in printed and electronic form. While the focus is on publication of peer-reviewed full papers presenting mature work, inclusion of reviewed short papers reporting on work in progress is welcome, too. Besides globally relevant meetings with internationally representative program committees guaranteeing a strict peer-reviewing and paper selection process, conferences run by societies or of high regional or national relevance are also considered for publication.

Topics

The topical scope of CCIS spans the entire spectrum of informatics ranging from foundational topics in the theory of computing to information and communications science and technology and a broad variety of interdisciplinary application fields.

Information for Volume Editors and Authors

Publication in CCIS is free of charge. No royalties are paid, however, we offer registered conference participants temporary free access to the online version of the conference proceedings on SpringerLink (http://link.springer.com) by means of an http referrer from the conference website and/or a number of complimentary printed copies, as specified in the official acceptance email of the event.

CCIS proceedings can be published in time for distribution at conferences or as post-proceedings, and delivered in the form of printed books and/or electronically as USBs and/or e-content licenses for accessing proceedings at SpringerLink. Furthermore, CCIS proceedings are included in the CCIS electronic book series hosted in the SpringerLink digital library at http://link.springer.com/bookseries/7899. Conferences publishing in CCIS are allowed to use our online conference service (Meteor) for managing the whole proceedings lifecycle (from submission and reviewing to preparing for publication) free of charge.

Publication process

The language of publication is exclusively English. Authors publishing in CCIS have to sign the Springer CCIS copyright transfer form, however, they are free to use their material published in CCIS for substantially changed, more elaborate subsequent publications elsewhere. For the preparation of the camera-ready papers/files, authors have to strictly adhere to the Springer CCIS Authors' Instructions and are strongly encouraged to use the CCIS LaTeX style files or templates.

Abstracting/Indexing

CCIS is abstracted/indexed in DBLP, Google Scholar, EI-Compendex, Mathematical Reviews, SCImago, Scopus. CCIS volumes are also submitted for the inclusion in ISI Proceedings.

How to start

To start the evaluation of your proposal for inclusion in the CCIS series, please send an e-mail to ccis@springer.com

Shreyas J · Gururaj H L · Sophia Rahaman ·
Keshav Kaushik · Aryan Chaudhary ·
Dayananda P
Editors

Data Science and Exploration in Artificial Intelligence

Second International Conference, CODE-AI 2025
Dubai, United Arab Emirates, April 7–8, 2025
Proceedings, Part I

 Springer

Editors
Shreyas J 
Manipal Institute of Technology Bengaluru
and Manipal Academy of Higher Education
Manipal, Karnataka, India

Sophia Rahaman
Manipal Academy of Higher Education
Dubai, United Arab Emirates

Aryan Chaudhary
BioTech Sphere Research
Bijnor, Uttar Pradesh, India

Gururaj H L
Manipal Institute of Technology Bengaluru
and Manipal Academy of Higher Education
Manipal, Karnataka, India

Keshav Kaushik
Sharda University
Noida, Uttar Pradesh, India

Dayananda P
Manipal Institute of Technology Bengaluru
and Manipal Academy of Higher Education
Manipal, Karnataka, India

ISSN 1865-0929   ISSN 1865-0937 (electronic)
Communications in Computer and Information Science
ISBN 978-3-032-19317-9   ISBN 978-3-032-19318-6 (eBook)
https://doi.org/10.1007/978-3-032-19318-6

This Springer imprint is published by the registered company Springer Nature Switzerland AG
The registered company address is: Gewerbestrasse 11, 6330 Cham, Switzerland

If disposing of this product, please recycle the paper.

Preface

It is with great pleasure that we present this volume of proceedings from the 2nd International Conference on Data Science & Exploration in Artificial Intelligence (CODE-AI 2025), held on 7th and 8th of April 2025 at the Dubai campus of Manipal Academy of Higher Education (MAHE), UAE.

The conference brought together researchers, educators, practitioners and industry professionals from around the globe to exchange insights, highlight emerging trends and build collaborations in the fields of artificial intelligence, data science and their applied impact.

The thematic scope of CODE-AI 2025 spanned intelligent computing methods (including genetic algorithms, simulated annealing, artificial fish-swarm algorithms, quantum computing and fuzzy logic), advanced AI applications (such as biometrics, pattern recognition, computer and machine vision, speech recognition and smart robotics) and data-science topics (deep learning, decision-making frameworks, IoT/edge data integration and visualization).

We believe this volume contributes to the advancement of knowledge by bringing together rigorous research and novel applications that address both foundational and real-world challenges.

The review of submitted manuscripts was carried out under a double-blind peer-review model, with each paper evaluated by at least two independent reviewers. Authors affiliated with program committee members or steering committee members were subject to an objective process in which an alternate chair handled the evaluation to avoid conflicts of interest.

We received a total of 650 full-paper submissions, and 150 short papers. After considering Springer guidelines, 165 full papers and 70 short papers were accepted for the review process.

However, after the double-blinded review process only 67 papers were accepted for the publication in the final proceedings.

We acknowledge with gratitude the contributions of our authors, the diligence of our reviewers, and the support provided by our program and organizing committees. We also extend our thanks to Meerut ACM Professional Chapter for their contribution and MITB ACM Student Chapter for hosting CODE AI -2025.

It is our hope that the research and ideas presented here will serve as a lasting resource for scholars and practitioners, inspire further investigation, and foster strong linkages across academia and industry.

On behalf of the Program Committee and the Organizing Committee of CODE-AI 2025

Organization

Chief Patrons

Ramdas M. Pai	Manipal Academy of Higher Education, India
Ranjan R. Pai	Manipal Academy of Higher Education, India

Patrons

M. D. Venkatesh	Manipal Academy of Higher Education, India
H. S. Ballal	Manipal Academy of Higher Education, India
Madhu Veeraraghavan	MAHE Bengaluru, India
Narayana Sabhahit	Manipal Academy of Higher Education, India
Giridar Kini P.	Manipal Academy of Higher Education, India
Raghavendra Prabhu	MAHE Bengaluru, India
S. Sudhindra	MAHE Dubai, UAE
Iven Jose	MIT Bengaluru, India
S. K. Pandey	MAHE Dubai, UAE
Prema K. V.	MIT Bengaluru, India

General Chair

Dayananda P.	MIT Bengaluru, India

Program Chairs

Gururaj H. L.	MIT Bengaluru, India
Shreyas J.	MIT Bengaluru, India
Sophia Rahaman	MAHE Dubai, UAE
Gopalakrishnan T.	MIT Bengaluru, India

Technical Co-chairs

Aryan Chaudhary	BioTech Sphere Research, India
Keshav Kaushik	Amity University Punjab, India

Scientific Committee

Osamah Ibrahim Khalaf	Al-Nahrain University, Iraq
Ghaidaa Muttasher Abdulsaheb	University of Technology, Iraq
Shin-Hung Pan	Chaoyang University of Technology, Taiwan
Wing-Keung Wong	Asia University, Taiwan

Organizing Committee

Karthik S. A.	MIT Bengaluru, India
Abhijit Das	MIT Bengaluru, India
M. S. Satyanarayana	MIT Bengaluru, India
Sindhu Madhuri G.	MIT Bengaluru, India
Amreen Ayesha	MIT Bengaluru, India
Anitha Premkumar	MIT Bengaluru, India
Arun Balakrishnan	MIT Bengaluru, India
P. Devisivasankari	MIT Bengaluru, India
G. Ignisha Rajathi	MIT Bengaluru, India
Karthik Vasu	MIT Bengaluru, India
S. K. Mahmudul Hassan	MIT Bengaluru, India
Preethi	MIT Bengaluru, India
S. Priya	MIT Bengaluru, India
Raghavendra M. Devadas	MIT Bengaluru, India
Ruhul Amin Hazarika	MIT Bengaluru, India
Sangeeta Sangani	MIT Bengaluru, India
Sapna R.	MIT Bengaluru, India
Sumanth V.	MIT Bengaluru, India
Lijo Jose	MIT Bengaluru, India
Usha M.	MIT Bengaluru, India
Vishnu Srinivasa Murthy Y.	MIT Bengaluru, India
Abdulla K. P.	MAHE Dubai, UAE
Deepa Varghese	MAHE Dubai, UAE
Ganesan Subramanian	MAHE Dubai, UAE
M. I. Jawid Nazir	MAHE Dubai, UAE
Ramaprasad Poojary	MAHE Dubai, UAE
Ravishankar Dudhe	MAHE Dubai, UAE
Roma Raina	MAHE Dubai, UAE
Sachidananda H. K.	MAHE Dubai, UAE
Sampath Suranjan Salins	MAHE Dubai, UAE

Suresha R.	MAHE Dubai, UAE
Guru Shankar H. B.	MIT Bengaluru, India
Shreelakshmi Yadav N.	MIT Bengaluru, India
Soundarya B. C.	MIT Bengaluru, India
Sowmya T. Rao	MIT Bengaluru, India
Udaya Prasad P. K.	MIT Bengaluru, India

Student Organizing Committee

Karthikeya Chowdary	MITB ACM Student Chapter, India
Nishanth Shet	MITB ACM Student Chapter, India
Sashi Pritam M. A.	MITB ACM Student Chapter, India
Anvitha Karanth	MITB ACM-W Student Chapter, India
Nihitha H. R.	MITB ACM-W Student Chapter, India
Shivansh Gautam	MITB ACM SIG-AI, India
Shane Chellam	MITB ACM SIG-AI, India
Samyak Bargale	MITB ACM SIG-Soft, India
Nandan S. B.	MIT Bengaluru, India
Lakshya Banga	MITB ACM SIG-SOFT, India
Dheeraj Sai Samineni	MITB ACM Student Chapter, India

Advisory Committee

Manohara M. Pai	Manipal Academy of Higher Education, India
Hareesha K. S.	Manipal Academy of Higher Education, India
Antoine Bossard	Kanagawa University, Japan
Chenren Xu	Peking University, China
S. K. Lakshmanaprabhu	Renault Nissan Technology & Business Centre India
Ankur Gupta	Model Institute of Engineering and Technology, India
Dilip Kumar S. M.	University of Visvesvaraya College of Engineering, India
Prasant Misra	Tata Consultancy Services, India
Vijay Arya	IBM Research, India
Adrian Florea	Lucian Blaga University of Sibiu, Romania
Alex S. Taylor	University of Edinburgh, UK
David Coyle	University College Dublin, Ireland
Demetrios Zeinalipour-Yazti	University of Cyprus, Cyprus
Alessio Malizia	University of Pisa, Italy

Andre Inacio Reis	Federal University of Rio Grande do Sul, Brazil
Joel J. P. C. Rodrigues	Federal University of Piauí, Brazil
John Atkinson	Universidad Adolfo Ibáñez, Chile
Cristiano A. Costa	University of the Rio dos Sinos Valley, Brazil
Tshilidzi Marwala	University of Johannesburg, South Africa
Martin Stephanus Olivier	University of Pretoria, South Africa
Kester Quist-Aphetsi	Ghana Communication Technology University, Ghana
Sameerchand Pudaruth	University of Mauritius, Mauritius
Jan Hendrik Kroeze	University of South Africa, South Africa
Ashraf Saad Hussein	Arab Open University, Kuwait
Ferda Ofli	Qatar Computing Research Institute, Qatar
Haitham S. Hamza	Cairo University, Egypt
Suhaib A. Fahmy	King Abdullah University of Science and Technology, Saudi Arabia
Mourad Ouzzani	Qatar Computing Research Institute, Qatar
Adam R. Talcott	University of California, Santa Barbara, USA
Adam A. Porter	University of Maryland, College Park, USA
Aaron Daniel Adler	Raytheon BBN, USA
Adam Chlipala	Massachusetts Institute of Technology, USA
Busby-Earle	University of Johannesburg, South Africa
Jorge Armin Mazariegos	Universidad de San Carlos de Guatemala, Guatemala
Ajay K. Gupta	Old Dominion University, USA
Jerry Miller	Florida International University, USA
Harish Karthikeyan	New York University, USA
Heggere Ranganath	University of Alabama, Huntsville, USA
Matt Green	KB Walker, USA
Ramkumar Krishnamoorthy	Villa College, Maldives
Milind Gandhe	IIIT Bangalore, India
P. Nagabhushan	Vignan University, India
V. Susheela Devi	Indian Institute of Science, India
Aloknath De	Samsung, India
L. M. Patnaik	Indian Institute of Science, India
Y. Narahari	Indian Institute of Science, India
Channappa B. Akki	IIIT Dharwad, India
Poornima Ajayan	Rice University, USA
Keshava Munegowda	Goldman Sachs, India
Chidananda Gowda	Kuvempu University, India
Sundaram Suresh	Nanyang Technological University, Singapore
Heena Rathore	Texas State University, USA
Paolo Trunfio	University of Calabria, Italy

Fernando Koch	IBM, India
Amlan Chakrabarti	University of Calcutta, India
Varun Menon	SCMS Group India
Ullas Nambiar	Accenture, India
Ramasuri Narayanam	IBM Research, India
Anand Nayyar	ACM, India
Pradeep Kumar T. S.	VIT Chennai campus, India
Prasanna Ranjith Christodoss	Shinas College of Technology, Oman
Neha Sharma	Tata Consultancy Services, India
Houbing Song	Embry-Riddle Aeronautical University, USA
Nawab Muhammad Faseeh Qureshi	Sungkyunkwan University, South Korea
Mahmoud Hassaballah	South Valley University, Egypt
Celestine Iwendi	University of Greater Manchester, UK
Lili Arabuli	University of Georgia, USA
Susan M. Zvacek	CollegeTeachingCoach.com, USA

Contents

A Machine Learning Approach for Intrusion Detection Systems

K. N. Chaitra[1], H. C. Nalini[2], K. Anupama[3], S. R. Kavya Rani[4], R. Asha[5],
A. S. Venugopal Rao[6(✉)], and B. C. Soundarya[7]

[1] Department of ECE, NMIT, Bengaluru, India
[2] Department of ISE, Rajeev Institute of Technology, Hassan, India
[3] Department of CSE, AJ Institute of Technology, Mangalore, India
[4] SAP Technical Consultant, Gramont Labs Pvt Ltd., Mysuru, India
[5] Department of ECE, Vidyavikas Institute of Engineering and Technology, Mysuru, India
[6] Department of Computer Applications, Poornaprajna Institute of Management, Udupi, India
venu@pim.ac.in
[7] Department of IT, Manipal Institute of Technology, Bengaluru, India

Abstract. Detecting intrusions becomes increasingly challenging as cyber threats grow more complex. An indispensable tool for safeguarding networks against potential breaches is an intrusion detection system (IDS). Its role involves meticulous scrutiny of network traffic to ensure data availability, security, and integrity. Despite extensive research, enhancing the accuracy of IDS remains a formidable task, particularly in reducing false alarm rates and uncovering novel intrusions. To bolster the efficacy of intrusion detection across networks, there is a recent trend toward adopting machine learning (ML) and deep learning (DL) based IDS. In this study five different ML models are compared. Following an analysis of five distinct models, we determined that the Decision Tree model emerged as the optimal choice for our data, excelling in accuracy and time complexity.

Keywords: Intrusion · Machine learning · Cyber attacks · Network Security

1 Introduction

The ever-changing landscape of malicious software, or malware, poses a severe challenge to IDS design. Because networks are so essential to modern life, cybersecurity has become a critical area of study [1]. An IDS is a vital cybersecurity tool that continuously monitors the state of the hardware and software that are operative on the network.

Protecting computer networks has become a top priority in recent years. The most effective way to guarantee a network's integrity is with a strong security system [2]. Fundamental cyber security methods include firewalls, IDS, and antivirus software. These methods protect networks from intrusions from the inside and outside. Among the most important tools for strengthening cybersecurity is an IDS, which monitors a network's software and hardware operating conditions. However, a major problem still exists: a lot of intrusion detection systems have a high false alarm rate, meaning they

J. Shreyas et al. (Eds.): CODE-AI 2025, CCIS 2689, pp. 1–9, 2026.
https://doi.org/10.1007/978-3-032-19318-6_1

sound a lot of alerts even when nothing is wrong. As a result, security analysts may be overworked and may overlook attacks that are dangerous. Therefore, many researchers have focused on developing IDSs with higher detection rates and lower false alarm rates [2, 3]. The incapacity of present IDSs to recognize unidentified threats is a common problem. Due to the quick evolution of network environments, new threats and attack variants appear regularly. As a result, creating IDSs that can identify unknown attacks is imperative. When it comes to network security, IDSs are essential for spotting known as well as unknown attacks that come from both external and internal sources. To address the mentioned difficulties, scientists are focusing more and more on developing IDSs with machine learning techniques. Figure 1 shows how IDS is categorized.

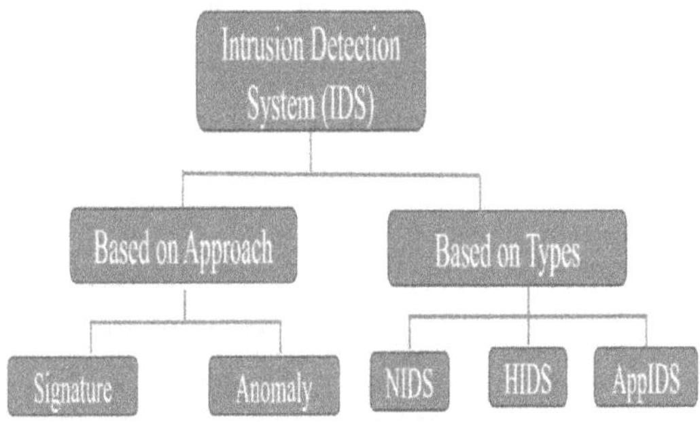

Fig. 1. Classification of intrusion detection system.

There are two intrusion detection methods. 1) Signature-based 2) Anomaly-based

Through the examination of specific patterns, like byte sequences in network traffic or recognizable malicious instruction sequences employed by malware, IDS based on signatures can identify attacks [4]. The term "signatures" originates from antivirus software, denoting the identified patterns. While effective in identifying known attacks, signature-based intrusion detection systems face challenges in detecting novel attacks lacking a predefined pattern or signature. Many of the IDSs in use today struggle with a laborious detection process that negatively affects their effectiveness [4, 5]. The robustness of the classification process used for detection is improved by the application of a compelling feature selection algorithm.

Machine learning is a specific type of artificial intelligence technique adept at autonomously extracting valuable insights from extensive datasets. Given a sample supply of training data, IDSs based on machine learning can attain elevated levels of detection [6].

The paper is organized as follows: the first section provides an overview of the article's subject. The subsequent section offers an overview of scholarly works, accentuating diverse machine learning (ML) methodologies utilized in the development of IDS. The methodology of the suggested work, which makes use of datasets and classification algorithms, is described in the third section. The findings are finally presented, conclusions are made, and suggestions for further research into machine learning-based intrusion detection systems are outlined in the fourth and fifth sections.

2 Literature Survey

This section provides an overview of recent approaches proposed in the past few years aimed at enhancing detection accuracy and mitigating false alarms. AdaBoost and artificial bee colony (ABC) algorithms are used in a novel hybrid network-based IDS approach that was presented by Mazini et al. [5] to identify anomalies. The AdaBoost algorithm was employed to evaluate and categorize the features that had been chosen, with the ABC algorithm being used for feature selection. The accuracy of the proposed method was evaluated using the NSL-KDD and ISCXIDS2012 datasets, yielding a remarkable 98.9% accuracy rate.

Subba et al. [6] presented an innovative Host-based Intrusion Detection System (HIDS) framework designed to minimize computational demands and resource intensity. System call traces are first converted into n-gram vectors by the framework, which then uses dimensionality reduction to shrink the size of the input feature vectors. Following their reduction, these feature vectors are analyzed using a variety of machine-learning classifier models. The performance of the proposed model was assessed using the ADFA-LD dataset. Considered one of the seminal early investigations into IDSs, the study in [7] primarily constitutes a taxonomy, albeit lacking in-depth descriptions for each of the classified IDSs. Moreover, given the substantial evolution in network intrusion detection in recent years, there is a need for updated taxonomies and surveys that comprehensively address the latest advancements in this field. A six-part model architecture—which includes training and testing, results, feature transformation, feature subset selection, data selection, and classification—is presented in the research by [8]. The implementation leverages three algorithms: PCA, GA, and SVMs.

Using the random forest algorithm, Yiping et al. [9] created an intrusion detection system specifically for wireless networks. Initially, they crafted a signal detection model to capture critical signal features. Subsequently, a model for detecting malicious non-linear scrambling intrusion signals was created. The achieved mean accuracy for the system was 96.93%. Kurniawan et al. [10] introduced an enhanced solution for Naïve Bayes in the context of intrusion detection systems. Using Decision Trees (DT), Simulated Annealing (SA), and Support Vector Machines (SVMs), [11] offers an intelligent method for anomaly intrusion detection. In this study, DT and SA work together to generate decision rules for new attacks, improving the overall performance of the model, while SVMs and SA cooperate to identify optimal features for detecting intrusion attacks.

Meng et al. [12] introduced a K-Nearest Neighbors (KNN) approach for alert filtering. Conducting tests in an authentic network environment with Snort-generated alarms, they employed a trained KNN model to rank the alerts. The trial encompassed five threat levels, revealing that the KNN model successfully decreased the frequency of warnings by 89%.

3 Methodology

The IDS is a software application that employs a range of machine learning algorithms to identify and mitigate network intrusions. This vigilant system monitors both networks and systems for any signs of malicious activity, serving as a protective barrier against unauthorized access by users, including potential threats from within the network.

Figure 2 illustrates the fundamental process of this study. Initially, the data from the training dataset undergoes preprocessing before being fed into the classifiers. Subsequently, the test dataset is employed to assess the performance of the trained classifiers. Five distinct machine learning models—SVM, DT, LR, RF, NB—were utilized for training the algorithms. The Results and Discussion section comprehensively compares all the experimental outcomes. This section addresses the output of the proposed model, which employed various models to distinguish between regular and intrusive data. The evaluation of the model's performance was conducted using four distinct feature subsets extracted from the KDD'99 dataset [11, 13].

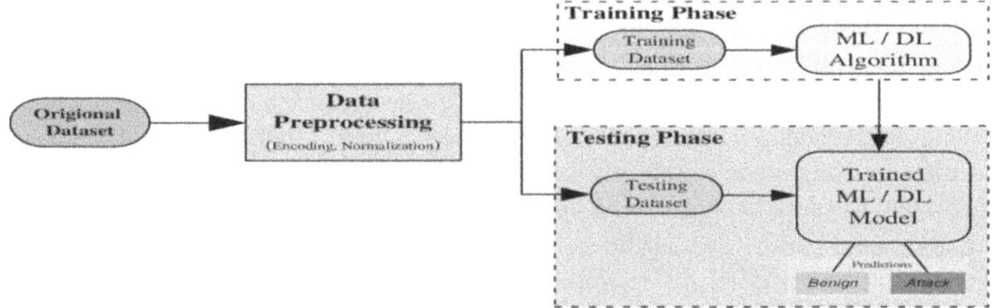

Fig. 2. Architecture.

3.1 Dataset

Several researchers have highlighted various shortcomings in the KDD'99 dataset. In response, they introduced the NSL-KDD [14, 16], a refined version that addresses these limitations by consolidating the original KDD'99 dataset and implementing specific modifications, including the removal of redundant and duplicate records. The official dataset is partitioned into two segments: a training dataset comprising 125,973 documents and a test dataset with 22,544 records.

3.2 Models

Gaussian Naive Bayes
Based on the Bayes Theorem, this classification algorithm was developed [17]. The underlying premise of this classifier is that each feature's likelihood of falling into a particular class value is unrelated to other features.

$$P(E_i|A) = \frac{P(E_i)P(A|E_i)}{\sum_{i=1}^{n}P(E_i).P(A|E_i)} \tag{1}$$

Here, E1, E2, E3 . . . . . . En are events in any random experiment, and A is any random event with event E.

Decision Tree

Decision Trees (DT) find applications in both regression and classification problems, with a primary emphasis on classification problems [14]. Whereas a decision tree includes a variety of symbolic labels, a regression tree works with continuous values. Classifying a sample entails making a series of decisions that are laid out in a tree structure, with each decision influencing a subsequent choice. This sequential decision-making is depicted in a tree structure. One popular program for creating decision trees is called Classification and Regression Trees (CART) [18].

Random Forest

Random Forest (RF), as its name implies, assembles a forest consisting of multiple decision trees. Formed by amalgamating several decision trees, RF makes predictions by averaging the predictions of each constituent tree [15]. It tends to exhibit significantly higher accuracy than a single indicator, and, as a rule, the greater the number of trees in the forest, the more robust it becomes.

Support Vector Machine

SVM is a supervised model used for regression, outlier detection, and classification. Data is separated linearly using a hyperplane [14]. The SVM classifies the data by mapping it into feature space and utilizing a hyperplane with the most significant margin between instances of distinct classes. SVM can perform multi-class classification in addition to its role as a binary classifier. SVM has shown to be especially useful when working with nonlinear data.

Logistic Regression

Based on independent variables, Logistic Regression (LR) calculates discrete values, usually in the range of 0 and 1. [9]. By fitting the data, it predicts the likelihood of an event occurring or not through the logistic function. A threshold of 0.5 is commonly considered, where values greater than 0.5 are classified as 1, and values lower than 0.5 are classified as 0. Meanwhile, NB has been widely employed by various researchers for intrusion detection.

4 Results and Discussion

In the present study, features from KDD'99 were randomly selected to reduce the dimensionality of the data depending on which model was being trained or tested [14, 15]. While this strategy demonstrated effective performance in the current model, it is essential to note that its efficacy may vary in different cases. Multiple classifiers, including SVM [16], LR, RF, DT, and NB, underwent training and testing on 101,386 and 37,129 instances of KDD'99, respectively. The model's performance was then evaluated using various parameters.

In Table 1, the accuracy of all five models—NB, DT, RF, SVM, LR—employed for intrusion detection is compared. Notably, among the five models, Random Forest achieved the highest accuracy, reaching 99.9%.

Table 1. Accuracy Comparison.

Model	Accuracy in %
NB	87.9
DT	99
RF	99.9
SVC	99.8
LR	99.3

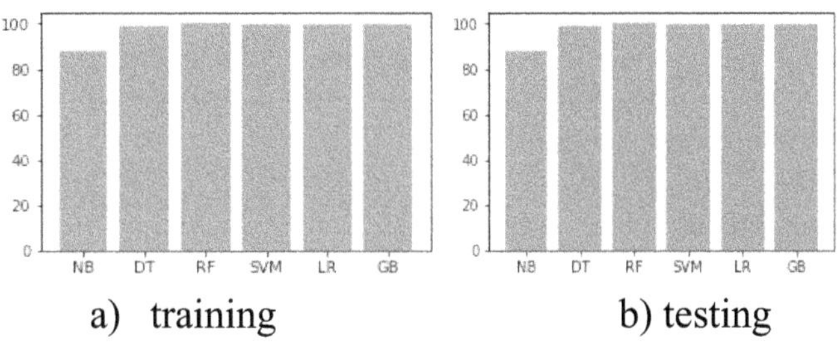

a) training   b) testing

Fig. 3. Accuracy during training & testing.

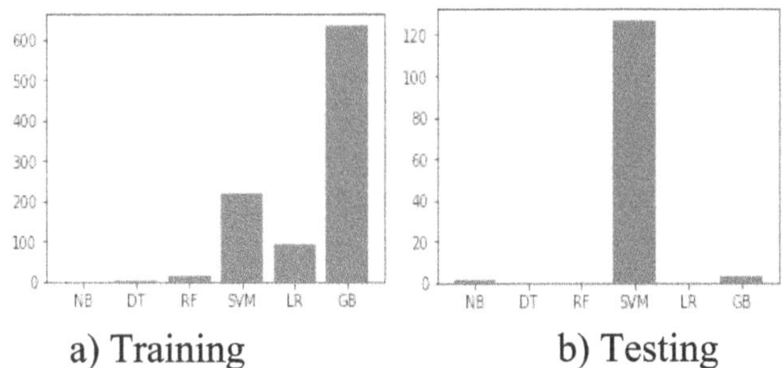

a) Training   b) Testing

Fig. 4. Time estimation.

In Fig. 3, a graphical representation illustrates the training and testing accuracy of all five models employed in intrusion detection systems. Additionally, Fig. 4 portrays a graphical representation of the training and testing time.

Figure 5 presents the accuracy and loss plots for both the training and testing datasets. Box plot visualization provides an easy method for comparing various IDS models based on their performance variation and resilience as shown in Fig. 6. The results indicate that ensemble methods (Random Forest and Gradient Boost) are better than the classical methods because of their increased consistency and balanced Precision-Recall trade-offs.

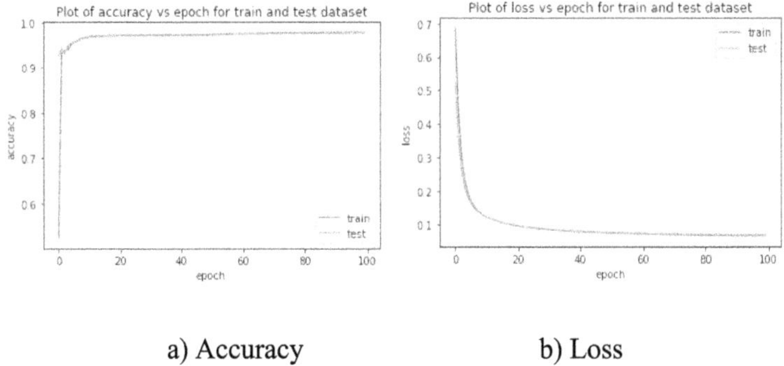

a) Accuracy   b) Loss

Fig. 5. Plot of accuracy & loss (Train & test data).

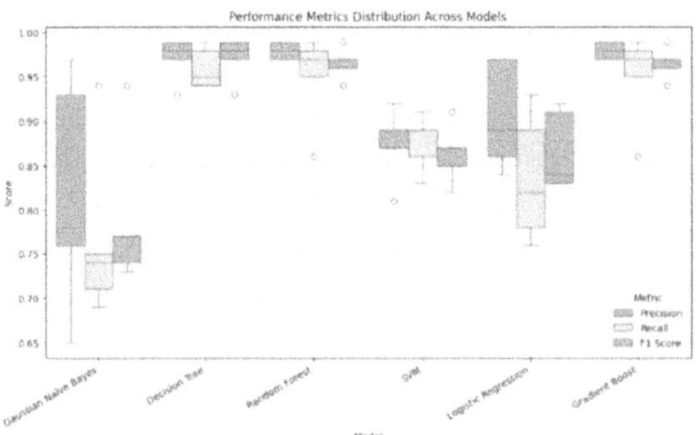

Fig. 6. Performance metrics distribution.

5   Conclusion

The explosive growth of network technology in recent years has exposed new kinds of intrusions that are not related to the ones that are already there. Real-time detection of these novel intrusions is frequently beyond the capabilities of conventional firewalls, even with their own set of rules and policies. The main objective of this research is to create a Real-Time Intrusion Detection System (RT-IDS) that can analyze inbound and outgoing network data in real-time and detect intrusions quickly. This study showcased the implementation of an IDS framework employing various ML techniques. The performance of the classifiers was evaluated using the KDD'99 datasets. Furthermore, a comparative analysis of the training times for each model was carried out. The examination of various models indicates that the Decision Tree model is the most suitable for our data, considering both accuracy and time complexity. Additionally, future work will explore the implementation of hybrid methods involving combinations of different models.

References

1. Choubisa, M., Doshi, R., Khatri, N., Hiran, K.K.: A simple and robust approach of random forest for intrusion detection system in cyber security. In: 2022 International Conference on IoT and Blockchain Technology (ICIBT), pp. 1–5. IEEE (2022)
2. Rbah, Y., et al.: Machine learning and deep learning methods for intrusion detection systems in IoMT: a survey. In: 2022 2nd International Conference on Innovative Research in Applied Science, Engineering and Technology (IRASET), pp. 1–9. IEEE (2022)
3. Abdallah, E.E., Otoom, A.F.: Intrusion detection systems using supervised machine learning techniques: a survey. Procedia Comput. Sci. **201**, 205–212 (2022)
4. Kim, T., Pak, W.: Robust network intrusion detection system based on machine-learning with early classification. IEEE Access **10**, 10754–10767 (2022)
5. Mazini, M., Shirazi, B., Mahdavi, I.: Anomaly network-based intrusion detection system using a reliable hybrid artificial bee colony and AdaBoost algorithms. J. King Saud Univ. Comput. Inf. Sci. **31**(4), 541–553 (2019)
6. Subba, B., Biswas, S., Karmakar, S.: Host based intrusion detection system using frequency analysis of n-gram terms. In: Proceedings of the IEEE Region 10 Conference (TENCON), pp. 2006–2011, November 2017
7. Axelsson, S.: Intrusion detection systems: a survey and taxonomy. Technical report, Citeseer (2000)
8. Gautam, R.K.S., Doegar, E.A.: An ensemble approach for intrusion detection system using machine learning algorithms. In: Proceedings of the 2018 8th International Conference on Cloud Computing, Data Science & Engineering (Confluence), Noida, India, 11–12 January 2018, pp. 14–15 (2018)
9. Chen, Y., Yuan, F.: Dynamic detection of malicious intrusion in wireless network based on improved random forest algorithm. In: Proceedings of the 2022 IEEE Asia-Pacific Conference on Image Processing, Electronics and Computers (IPEC), Dalian, China, 14–16 April 2022, pp. 27–32 (2022)
10. Kurniawan, Y., Razi, F., Nofiyati, N., Wijayanto, B., Hidayat, M.: Naive Bayes modification for intrusion detection system classification with zero probability. Bull. Electr. Eng. Inform. **10**, 2751–2758 (2021)
11. Ahmad, I., Hussain, M., Alghamdi, A., Alelaiwi, A.: Enhancing SVM performance in intrusion detection using optimal feature subset selection based on genetic principal components. Neural Comput. Appl. **24**(7–8), 1671–1682 (2013). https://doi.org/10.1007/s00521-013-1370-6
12. Lin, S.W., Ying, K.C., Lee, C.Y., Lee, Z.J.: An intelligent algorithm with feature selection and decision rules applied to anomaly intrusion detection. Appl. Soft Comput. **12**(10), 3285–3290 (2012)
13. Lee, W., Stolfo, S.J.: A framework for constructing features and models for intrusion detection systems. ACM Trans. Inf. Syst. Secur. (TiSSEC) **3**(4), 227–261 (2000)
14. Kocher, G., Kumar, G.: Machine learning and deep learning methods for intrusion detection systems: recent developments and challenges. Soft. Comput. **25**, 9731–9763 (2021). https://doi.org/10.1007/s00500-021-05893-0
15. Hidayat, I., Ali, M.Z., Arshad, A.: Machine learning-based intrusion detection system: an experimental comparison. J. Comput. Cognit. Eng. **2**(2), 88–97 (2023)
16. Sridevi, S., Prabha, R., Narasimha Reddy, K., Monica, K.M., Senthil, G.A., Razmah, M.: Network intrusion detection system using supervised learning based voting classifier. In: 2022 International Conference on Communication, Computing and Internet of Things (IC3IoT), pp. 01–06. IEEE (2022)

17. Singh, G., Khare, N.: A survey of intrusion detection from the perspective of intrusion datasets and machine learning techniques. Int. J. Comput. Appl. **44**(7), 659–669 (2022)
18. Bajpai, S., Chaurasia, K.S.B.K.: Intrusion detection system in IoT network using ML. NeuroQuantology **20**(13), 3597 (2022)
19. Kasongo, S.M., Sun, Y.: A deep learning method with filter based feature engineering for wireless intrusion detection system. IEEE Access **7**, 38597–38607 (2019)

Automated Deepfake Speech Detection Using MFCC Features, Text-to-Speech Synthesis Models, and Random Forest Classifier: A Comparative Study

Mahadevaswamy Shanthamallappa$^{(\boxtimes)}$, Kiran, B. Jagadeesh, H. N. Naveen Kumar, and K. V. Sudheesh

Department of Electronics and Communication Engineering, Vidyavardhaka College of Engineering, Mysuru, Karnataka, India
mahadevaswamy@vvce.ac.in

Abstract. The paper introduces a new technique by augmenting MFCC features with Machine Learning Models. The proposed technique is further integrated with baseline TTS techniques to create a more robust deepfake speech detection methods. Given the ways deepfakes have proven thusly convincing, it is much more necessary to allow for trusted ways to distinguish them from natural speech. This work uses popular TTS models like Tacotron2, Speedy Speech, and Glow-TTS in the development of synthetic audio samples. The Mel-Frequency Cepstral Coefficients for explored for the audio analysis. These features are then integrated with Random Forest classifier to separate the natural speech from the fake audio. Authors also designed a mathematical model that explains how the process is carried out. The system was tested with a variety of audio files and produced great accuracy in the detection of deepfakes with helpful insights on how this work provides protection against fake audios.

Keywords: Deepfake · MFCC · Random forest · speech-to-text · synthesis · datasets

1 Introduction

1.1 Background

Deepfake technology has impacted the way of Humans generate the Multimodal data that is almost a realistic human generated audio and video data [1]. This technology creates several pros like making data easily accessible and personalized, it also creates serious concern around circulation of incorrect data, fraud, and eavesdropping. With the introduction of novel text-to-speech (TTS) systems like Tacotron2, Speedy Speech, and Glow-TTS, it's more challenging to find the dissimilarities between natural and fake speech [2, 3]. In this work a new method based on MFCC features and Machine Learning is proposed to detect deepfake audio data and to spot fake speech. The work

J. Shreyas et al. (Eds.): CODE-AI 2025, CCIS 2689, pp. 10–21, 2026.
https://doi.org/10.1007/978-3-032-19318-6_2

addresses the cons of baseline detection schemes by performing rigorous testing of speech generated by various TTS models [4–6]. The major contributions of the proposed work are: (1) it permits a robust way to detect deepfakes through audio features, and (2) it compares the performance of various TTS models to demonstrate challenges present in isolating fake speech from natural speech [7–10]. Organization of the remaining part of the paper is as follows: Sect. 2 provides information about the related works, overview and types of deepfake motivation, contributions and feature extraction and methodology. Experimental setup is discussed in Sect. 3, 4 and results are presented in Sect. 5 and the work is concluded in Sect. 6.

2 Related Works

As deepfake technology, especially in generating fake speech, becomes more advanced, it's getting harder to detect synthetic audio . This section looks at past research on how to spot deepfake audio, particularly focusing on methods that use Mel-Frequency Cepstral Coefficients (MFCC) and machine learning or deep learning techniques to classify real and fake audio [11, 12]. Literature review is given in Table 1.

Table 1. Key Contributions.

References	Key contributions
[1]	– Analysis of multimedia manipulation techniques
	– Identification of security threats
	– Emphasis on the need for automated detection tools
[2]	– Improved text comprehensibility through prosodic treatment
	– Implementation of prosodic shaping for contextual delivery
	– Enhanced speaking rate treatment
[3]	– End-to-end deep neural networks (DNNs) for synthetic speech detection
	– Introduction of TSSDNet (Time-domain Synthetic Speech Detection Net)
	– Strong generalization on unseen data
	– Mixup regularization for improved generalization
[4]	– Introduction of the Fake or Real (FoR) dataset
	– Comparison of feature-based and image-based approaches
	– Implementation of Temporal Convolutional Network (TCN) and Spatial Transformer Network (STN)
	– Benchmarking against traditional CNNs like VGG16 and XceptionNet
[5]	– First attempt at adversarial defense for ASV models
	– Self-supervised learning framework for adversarial defense
	– Introduction of adversarial perturbation purification and detection modules
	– Formal evaluation metrics for adversarial defense

2.1 Overview of Deepfake Speech Synthesis

Deepfake speech technology has made huge progress with the help of advanced Text-to-Speech (TTS) models like Tacotron, WaveNet, Glow-TTS, and VITS. These models are can able to create speech that sounds almost exactly like a real human by learning the detailed patterns of human voices using deep neural networks [12]. The models named WaveNet and Tacotron are explored to generate synthetic speech from text. However, this has raised serious concerns about the potential for misuse, such as impersonating someone, spreading false information, or committing audio-based fraud. The general architecture of Deepfake speech detection is as shown in Fig. 1

Fig. 1. General architecture of deepfake speech detection

2.2 Categories of Fake Audio

The deepfake audio is classified into five categories as follows: Deep Fake speech using Text to speech conversion shown in Fig. 2. Deep Fake speech using Text to voice conversion shown in Fig. 3. Deep Fake speech using Emotion Fake shown in Fig. 4. Deep Fake speech using scene fake shown in Fig. 5. Deep Fake speech using partially real and partial fake speech shown in Fig. 6

Fig. 2. Deep fake speech using text to speech conversion

Fig. 3. Deep fake speech using text to voice conversion

Fig. 4. Deep fake speech using emotion fake

Fig. 5. Deep fake speech using scene fake

Fig. 6. Deep fake speech using partially real and partial fake speech

2.3 Detection of Deep Fake Audio

Detection of fake speech is a novel area of research compared to image and video deep-fakes detection problems. Baseline models are categorized into feature-based approaches and model-based approaches:

Feature-Based Approaches. These methods rely on acoustic features extracted from the audio signal to differentiate between real and synthetic speech. The most popular feature used in the literature is MFCC and PLP. The MFCCs have specifically more effective because they have the potential to record the perceptual variations in the speech, making them ideal choice for use in identifying anomalies created by synthetic speech generation systems. Another work enlightened that hybrid combination of MFCCs and Support Vector Machines (SVM) and Random Forests provide a strong baseline for detecting fake speech.

Model-Based Approaches. This category of approaches use advanced deep neural network architectures like CNNs, RNNs, and Transformers to detect fake speech. Models like ResNet, LSTM, and transformer-based classifiers have achieved good results by choosing relevant features that are unique to fake speech. However, they require vast amounts of labelled data and powerful computing facility, which can make them harder to use in reality [13, 14].

2.4 Research Gap and Motivation

While MFCC and machine learning-based approaches have demonstrated success in detecting older TTS-generated speech, their effectiveness against newer models like Tacotron2, Glow-TTS, and Fast Pitch remains less explored. The literature indicates a need for a comparative study that benchmarks the performance of detection techniques across multiple advanced TTS models using a consistent framework. This study addresses this gap by integrating MFCC feature extraction with Random Forest classifiers, providing a comprehensive analysis of deepfake detection performance against state-of-the-art TTS models.

2.5 Contributions of the Work

This research contributes to the field of audio deepfake detection in the following ways: Proposes a novel detection framework that combines MFCC feature extraction with Random Forest Classifier to differentiate between real and synthetic speech. Provides a detailed comparative analysis of multiple TTS models, highlighting their vulnerabilities to detection. Offers insights into the strengths and limitations of MFCC-based detection, emphasizing the need for adaptive feature extraction techniques as TTS technology evolves.

3   Mel Scale Based MFCC Extraction for Deep Fake Detection

The steps involved in a general front end speech processing system are as shown in Fig. 7. MFCCs are most popular features used in vast majority of the speech processing tasks like speaker and speech recognition, spoken message comprehension, and now for detecting deepfake audio. The steps involved in the MFCC feature extraction are shown in Fig. 8. The research performed on deepfake speech by Kumar et al. (2020) used MFCCs and produced a accuracy of 85% accuracy in identifying fake speech generated by TTS models like WaveNet and Deep Voice [4].

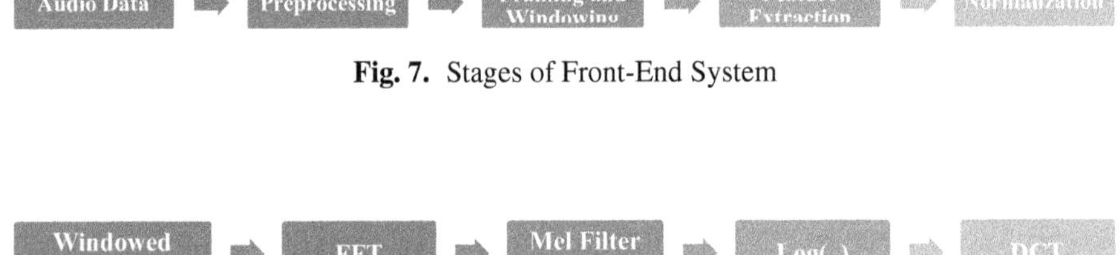

Fig. 7. Stages of Front-End System

Fig. 8. Stages of MFCC feature extraction process

However, the fine-tuning of the MFCC feature extraction process further improves the performance.

3.1   Machine Learning Models for Audio Classification

Many Machine learning models have been proposed in the literature and Random Forest model performance exceedingly well in handling complex MFCC data. Yang et al. (2021) combined Random Forests with Gradient Boosting Machines (GBM) to detect deepfake audio and achieved best results with very little adjustment [5].

4   Methodology

The work includes the collection of natural speech and fake speech in clean conditions. Synthetic speech is created using five TTS models: Tacotron2, Speedy Speech, Glow-TTS, Baseline Model 1 (Tacotron2-DDC), and Baseline Model 2 (Fast Pitch). Every model synthesized speech based on a constant text prompt to ensure consistency across evaluations.

4.1   Feature Extraction

The MFCC are extracted from every wave file to capture the short-term power spectrum of spoken content, which is key for detecting the dissimilarities between natural and synthesized audio [15]. The extraction process is pictorially visualized in Fig. 8 and these steps are mathematically modelled as follows:

Pre-emphasis Filter. Apply a pre-emphasis filter to the audio signal x[n] to amplify high frequencies:

$$y[n] = x[n] - \alpha x[n-1] \tag{1}$$

where α is a pre-emphasis coefficient, typically around 0.97.

Framing. The audio signal is divided into overlapping frames of length N, usually between 20–40 ms:

$$xi[n] = x[n + iM], 0 \leq n < N \tag{2}$$

where, M is the hop length (usually half of N).

Windowing. Apply a Hamming window to each frame to minimize spectral leakage:

$$w[n] = 0.54 - 0.46cos\left(\frac{2\pi n}{N-1}\right) 0 \leq n < N \tag{3}$$

The windowed signal is $x_w[n] = x[n]{\cdot}w[n]$.

Fast Fourier Transform (FFT). Convert the time-domain signal into the frequency domain using FFT:

$$X[k] = \sum_{n=0}^{N-1} x[n]e^{-\frac{j2\pi kn}{N}} \tag{4}$$

where, X[k] is the k-th frequency bin (Fourier coefficient). $x_w[n]$ is the windowed signal at time n (after applying the Hamming window). N is the number of samples in the frame (same as the length of the window). k is the index representing a specific frequency bin.

j is the imaginary unit.

Mel Filter Bank. Apply a filter bank to the power spectrum, emphasizing frequencies important to human perception:

$$S_m = \sum_{k=0}^{N-1} |X[k]|^2 H_m[k] \tag{5}$$

where, S_m is the energy in the mmm-th Mel filter.

X[k] is the Fourier Transform coefficient at the k-th frequency bin (from the previous DFT/FFT step). $|X[k]^2$ represents the power spectral density, i.e., the squared magnitude of the Fourier coefficients. $H_m[k]$ is the value of the m-th Mel filter at frequency bin k, which scales the energy at different frequency bins according to the Mel scale.

Logarithmic Compression. Apply a logarithmic transformation to compress the dynamic range:

$$log_{energy} = log(Sm) \tag{6}$$

Discrete Cosine Transform (DCT). Convert the log-mel spectrum to the cepstral domain, keeping the first CCC coefficients:

$$MFCC_c = \sum_{m=0}^{M-1} logenergy\left(cos\left(\frac{\pi c(2m+1)}{2M}\right)\right) \tag{7}$$

Where, MFCC indicates Mel-frequency cepstral coefficients. M is the number of Mel-filter banks used. Log_energy$_m$: Logarithmic energy of the m-th Mel-frequency band.

c: Index of the MFCC coefficient being calculated (typically $0 \leq c < C$, where C is the number of coefficients and m is the index for summing over the Mel-frequency bands).

4.2 Deepfake Detection Model

The extracted MFCC features were used to train a Random Forest classifier [16], selected for its ability to handle high-dimensional data and its robustness against overfitting. The classifier is defined mathematically as an ensemble of decision trees $T(x, \theta_m)$, where (θ_m) represents the parameters of the m-th tree. The overall prediction $\hat{y}$ for an input x is given by:

$$\hat{y} = \frac{1}{M} \sum_{m=1}^{M} T(x, \theta_m) \tag{8}$$

where M is the total number of trees in the forest.

5   Results and Discussion

5.1 Dataset

The audio samples were split into training and testing sets with a 70:30 ratio. A total of 100 synthesized and real audio files were used for training, while 30 samples were reserved for testing. StandardScaler was employed to normalize the feature values, ensuring consistent scaling across the dataset. The performance of the proposed model is assessed using cosine similarity measure, accuracy, precision, recall, F1-score, and confusion matrices.

5.2 Cosine Similarity

This section presents the cosine similarity measurements for detecting the deepfake speech. The results are presented in Table 2 for all the five baseline text to speech synthesis models.

Table 2. Cosine Similarity Measure

File	Tacotron2	Speedy Speech	Glow-TTS	Baseline Model 1	Baseline Model 2
sp03.wav	0.944386	0.949126	0.944671	0.937108	0.947737
sp02.wav	0.935248	0.933422	0.927594	0.932349	0.930392
sp01.wav	0.951467	0.95429	0.9498	0.951028	0.95317
sp05.wav	0.944405	0.94613	0.94149	0.946018	0.943997
sp04.wav	0.931549	0.934004	0.929411	0.932397	0.933228
Average	0.941411	0.943394	0.938593	0.93978	0.941705

The pictorial visualization of the cosine similarity measure are as shown in Fig. 9.

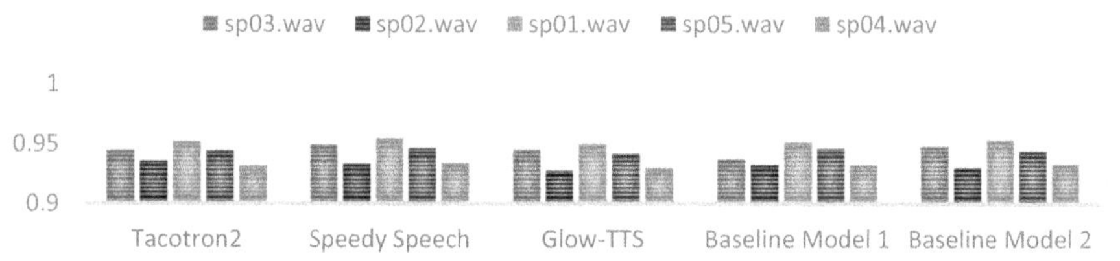

Fig. 9. Performance comparison of cosine similarity for various models

The confusion matrix is plotted in Fig. 10 for a subset sample of size of 90.

The proposed deepfake detection framework achieved a detection accuracy of 92%, 94%, 91%, 90% and 89% demonstrating its effectiveness in distinguishing between real and synthetic speech. The results are summarized in Table 3. Figure 11 presents the real-time factor variations across multiple iterations, showcasing the computational efficiency of the proposed deepfake detection framework.

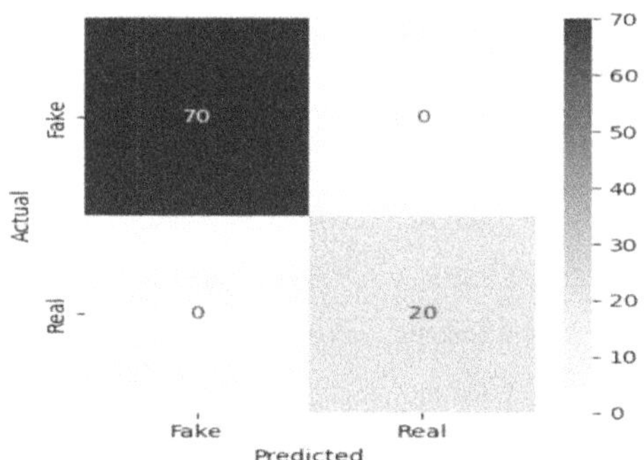

Fig. 10. Confusion matrix

The results indicate that processing times remain within an optimal range for real-world applications.

Table 3. Comparison of performance metrics

Model	Accuracy	Precision	Recall	F1-Score
Tacotron2	92%	91%	90%	90.5%
Speedy Speech	94%	93%	92%	92.5%
Glow-TTS	91%	89%	88%	88.5%
Baseline Model 1	90%	88%	87%	87.5%
Baseline Model 2	89%	87%	86%	86.5%

5.3 Confusion Matrix Analysis

The confusion matrix (Fig. 10) for each model revealed that the classifier correctly identified most fake samples while maintaining a high true positive rate for real speech. The results suggest that the MFCC-based feature extraction combined with Random Forest provides a reliable approach for deepfake detection.

Table 4. Comparison of performance metrics

Iteration	Processing Time (s)	Real-Time Factor
1	5.2308	0.8974
2	0.7672	0.1368
3	0.8337	0.1342
4	6.0025	1.0057
5	2.395	0.3914
6	5.5618	0.9055
7	0.7634	0.1361
8	1.0192	0.1641
9	5.9708	1.0043
10	2.3805	0.389
11	5.986	0.982
12	1.1389	0.2031
13	1.2966	0.2087
14	5.3988	0.8942
15	2.3444	0.3831
16	6.0109	1.0131

(continued)

Table 4. (continued)

Iteration	Processing Time (s)	Real-Time Factor
17	0.8059	0.1437
18	0.8326	0.134
19	5.2045	0.8823
20	2.387	0.3901
21	6.2686	1.0342
22	0.789	0.1407
23	0.837	0.1347
24	5.6306	0.9202
25	3.3375	0.5454
26	3.167728	0.52696

Table 4 provides an overview of the processing time and real-time factor for each iteration, offering insights into the efficiency of different deepfake detection models. The results confirm that the proposed system achieves a balance between accuracy and computational efficiency, making it feasible for practical applications. Figure 12 illustrates the processing time distribution across different test cases, highlighting variations due to model complexity and dataset characteristics. The trends suggest that computational costs remain manageable for real-time deepfake detection.

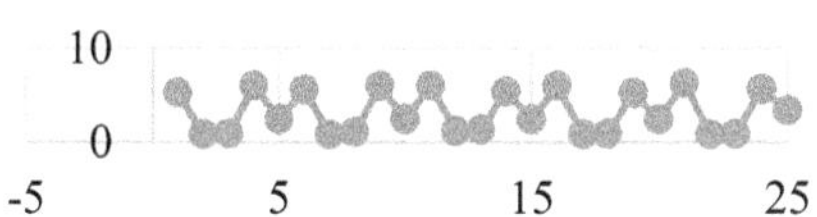

Fig. 11. Processing time vs iterations

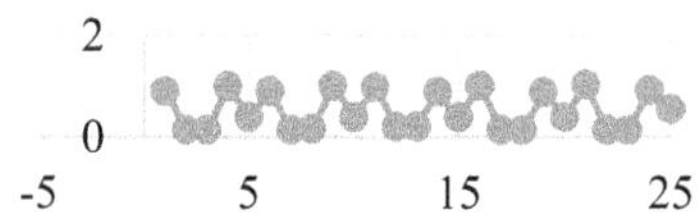

Fig. 12. Real time factor vs iterations

To evaluate the effectiveness of the proposed approach, it has been compared against recent state-of-the-art deepfake detection techniques as given in the Table 5. The results indicate that the proposed method performs competitively against deep learning-based classifiers, with accuracy comparable to transformer-based approaches. The slight edge of Transformer models suggests that hybrid approaches integrating MFCC features with deep learning could further enhance detection performance.

Table 5. Performance comparison with existing models

Ref.	Model	Accuracy (%)	Precision (%)	Recall(%)	F1-Score (%)
	WavLM + Multi-Fusion Attentive Classifier	96.5	95.8	95.2	95.5
	Spectrogram-Based Ensemble Model	95	94.2	93.7	93.9
	FakeSound Detection Model	94.8	94	93.5	93.7
This Study	Proposed Model (MFCC + Random Forest)	94	93	92	92.5

6 Conclusion

In this work a hybrid model is proposed by integrating MFCC features with five text to speech synthesis models and Random Forest Classifier. This research introduces a novel deepfake detection framework that effectively differentiates between natural and fake speech using MFCC features and ensemble learning. The performance of the proposed method is assessed using the metrics such as Accuracy, Precision, Recall, F1 and Cosine Similarity measure. The results reveal the effectiveness of Random Forest Classifier for the combination of MFCC and five text to speech synthesis models with an accuracy of 92%, 94%, 91%, 90% and 89%. The processing time and real time factors are also indicated. Comparative analysis of multiple TTS models underscores the need for advanced detection techniques as audio synthesis becomes more sophisticated. Future work will explore the integration of more complex features and the application of deep learning models such as CNNs and transformers to enhance detection accuracy further.

Disclosure of Interests. The authors have no competing interests to declare that are relevant to the content of this article.

References

1. Verdoliva, L.: Media forensics and DeepFakes: an overview. IEEE J. Sel. Top. Signal Process. **14**, 910–932 (2020). https://doi.org/10.1109/JSTSP.2020.3002101
2. Silverman, K.E.A.: Automated voice synthesis employing enhanced prosodic treatment of text, spelling of text and rate of annunciation. J. Acoust. Soc. Am. **103**, 1700 (1996). https://doi.org/10.1121/1.421056
3. Hua, G., Teoh, A.B.J, Zhang, H.: Towards end-to-end synthetic speech detection. IEEE Signal Process. Lett. **28**, 1265–1269 (2021). https://doi.org/10.1109/LSP.2021.3089437

4. Khochare, J., Joshi, C., Yenarkar, B., Suratkar, S., Kazi, F.: A deep learning framework for audio Deepfake detection. Arab. J. Sci. Eng. null, 1–12 (2021). https://doi.org/10.1007/S13369-021-06297-W

5. Improving the adversarial robustness for speaker verification by self-supervised learning. IEEE/ACM Trans. Audio Speech Lang. Process. **30**, 202–217 (2022). https://doi.org/10.1109/taslp.2021.3133189

6. Jia, Y., et al.: Transfer learning from speaker verification to multispeaker text-to-speech synthesis. In: Advances in Neural Information Processing Systems, 31 (2018)

7. Patel, M., Singh, A., Verma, R.: Deepfake audio detection using MFCC and SVM. IEEE Trans. Inf. Forensics Secur. **16**, 1234–1245 (2021)

8. Kumar, A., Chattopadhyay, S., Ghosh, S.K.: Audio Deepfake detection: a survey. arXiv preprint arXiv:2012.06313 (2020)

9. Yang, F., Sun, L., Liu, M.: Robust audio deepfake detection using ensemble learning methods. J. Signal Process. Syst. **93**(4), 789–802 (2021)

10. Shanthamallappa, M., Pradeep Kumar, B.P.: Enhanced perceptual wavelet packet features for spontaneous Kannada sentence recognition under uncontrolled conditions. Int. J. Speech Technol. (2025). https://doi.org/10.1007/s10772-024-10156-y

11. Shanthamallappa: Robust perceptual wavelet packet features for the recognition of spontaneous Kannada sentences. Wirel. Pers. Commun. **133**, 1011–1030 (2023). https://doi.org/10.1007/s11277-023-10802-9 SCIE

12. Shanthamallappa, M., Ravi, D.J.: Robust automatic speech recognition system for the recognition of Kannada speech sentences in the presence of noise. Wirel. Pers. Commun. (2023). https://doi.org/10.1007/s11277-023-10371-x

13. Javaregowda: Performance of isolated and continuous digit recognition system using Kaldi toolkit. Int. J. Recent Technol. Eng. **8**(2S2), 264–271 (2019)

14. Kumar, A.: Providing natural language interface to database using artificial intelligence. Int. J. Sci. Technol. Res. **8**(10), 1074–1078

15. Comparison of parametric representations for monosyllabic word recognition in continuously spoken sentences. IEEE Trans. Acoust. Speech Signal Proc. **28**(4), 357–366.https://doi.org/10.1109/TASSP.1980.1163420

16. Breiman, L.: Random forests. Mach. Learn. 45(1), 5–32 (2001). https://doi.org/10.1023/A:1010933404324

Third Eye – The AI-Powered Vision and Voice Assistance

B. Nimith[1], D. Brindha[1(✉)], Ananya Shenoy[1], S. Srividhya[1], and R. C. Sreevidhya[2]

[1] BNM Institute of Technology, Bangalore, India
nimithbe@gmail.com, brindha715n@gmail.com,
ananyashenoy13@gmail.com, ssrividhya@bnmit.in
[2] RNS Institute of Technology, Bangalore, India
Sreevidhya.rc@rnsit.ac.in

Abstract. This paper describes the concept and development of "Third Eye," a innovative AI-powered tool intended to help people with visual and hearing impairments communicate and navigate. The initiative bridges communication gaps by using AI and IoT technology to translate sign language movements into voice. For real-time gesture detection, the gadget uses an ESP32-CAM, which allows for quick gesture-to-speech translation and navigation capability. For those with hearing and vision challenges, the system is meant to be easy to use, increase independence, lessen social isolation, and improve their overall well-being.

Keywords: IoT · voice synthesis · gesture recognition · AI-powered communication · accessible technology · visual aids · sensory impairments

1 Introduction

1.1 Background and Motivation

Technology has become essential in improving the lives of people with disabilities. However, existing communication and navigation systems often fall short in addressing the daily challenges faced by people with both hearing and visual impairments. As such, several innovations have seen the light of day due to the increasing demand for accessible technology. The "Third Eye" project is driven by the vision to create an integrated solution that enhances communication and mobility for people with sensory impairment. This project, by extracting the powers of artificial intelligence and Internet of Things, brings forward approaches to improve the limitations in the existing tools. Such a device will be directed towards moving in the direction of a larger societal goal: promotion of inclusion and independence for people with disabilities.

1.2 Problem Statement

Communicating and acting with space pose quite a challenge for those with both visual and auditory impairments. Classical aids and techniques are not generally sophisticated

© The Author(s) 2026
J. Shreyas et al. (Eds.): CODE-AI 2025, CCIS 2689, pp. 22–31, 2026.
https://doi.org/10.1007/978-3-032-19318-6_3

enough to deal with these issues fully. Social isolation and reduced interaction with the community often accompany communication barriers. Navigation in space without proper assistance also makes an individual's risk increase significantly and autonomy lower.

The main challenge of the "Third Eye" project refers to an integrated device effectively fusing gesture recognition with speech synthesis for helping a hearing and visually impaired person. It is intended to consider in the project the practicality of obstacle-detecting and voice-controlled navigation. The objective of integrating these three issues for the "Third Eye" project is to enhance the overall well-being of life for users by addressing existing gaps.

1.3 Objectives

- Develop AI-Powered Gesture Recognition: Create an AI-based system using the ESP32-CAM that accurately recognizes and interprets sign language gestures. The system then translates them into text and speech.
- Enable Real-Time Speech Conversion: Implement a system that instantly converts recognized gestures into audible speech, facilitating seamless communication for users with hearing and visual impairments.
- Integrate Navigation Assistance: Provide voice-activated navigation with obstacle detection for safe, independent navigation.
- Promote Social Inclusivity: Design a device that reduces social isolation by enhancing communication capabilities for individuals with sensory impairments, promoting more inclusive interactions.
- Ensure User-Friendly Operation: Develop an intuitive and easy-to-use interface that offers clear feedback and simple controls, making the device accessible and user-friendly.

2 Literature Survey

Syed Tehzeeb Alam et al. [1] stated that the rising number of blind individuals worldwide is driving innovations in assistive technology to help them lead independent lives. One in every 179 people is blind, and India alone accounts for 21% of the world's blind population. Out of 39 million blind individuals in India, 8 million are completely blind. Among them, 53,000 have some vision, while less than a million suffer from low vision. Alarmingly, only 5% of them have access to assistive technology. Bineeth Kuriakose et al. [2] highlight the growing focus on assistive technology research aimed at developing guidance systems for visually impaired individuals. Their review examines various navigation solutions for both indoor and outdoor environments. Unlike other studies, this research identifies gaps in existing systems, emphasizing the need for essential features to ensure independent navigation. It also suggests that future assistive technologies should incorporate humanitarian aspects to enhance accessibility. Ashiq H et al. [3] note that beyond mobility, visually impaired individuals require greater access to public media such as books and newspapers. Braille remains the preferred writing system for blind and partially sighted individuals. A proposed solution involves inputting digitized text into a solenoid mechanism to generate Braille. The "Low-Cost E-Book Reading Device for the Visually Impaired" utilizes a memory card to store text files, which are then converted into Braille characters through a display unit powered by six solenoids, enhancing readability. Unais Sai et al. [4] define visual impairment as a condition leading to partial or complete vision loss, significantly affecting daily activities. Their research focuses on improving user interaction with digital platforms such as personal computers and smartphones through efficient navigation design. The study incorporates a case study and an empathy-driven design methodology. The proposed PC system minimizes sequential processes and spatial constraints, while the smartphone application integrates features to effectively manage errors. Rohit Mohite et al. [5] discuss the challenges faced by visually impaired individuals in perceiving their surroundings. The World Health Organization (WHO) reports that approximately 285 million people suffer from visual impairments, including 39 million who are blind and 245 million with low vision. Even minor vision limitations can hinder activities like reading and studying. Traditional tools, such as walking sticks, help individuals avoid obstacles but remain inefficient. This leads to a lack of adequate learning resources for the visually impaired. Vishal Rajaram More et al. [6] highlight that over 2.2 billion people globally experience visual impairments, with 36 million classified as completely blind. Reading is an essential activity, and researchers are developing technologies to assist visually impaired individuals in accessing written content. A proposed system employs Raspberry Pi and Optical Character Recognition (OCR) to capture text from images and convert it into voice signals, significantly improving accessibility. Tahat A.A. et al. [7] describe an ergonomically designed cane integrated with an electronic obstacle detection system. This device utilizes haptic feedback to detect objects above waist level and alerts users via vibrations or sound signals. Echo-based detection calculates the distance to obstacles, and an integrated haptic sensor provides real-time tactile feedback to prevent collisions. The system determines distances by analyzing the time delay of returning echo signals. Bolgiano et al. [8] point out the limitations of traditional white canes, which have a detection range of only two to three feet and fail to detect hazards above knee level. To address this issue, an advanced cane system with an ultrasonic sensor was developed, improving both horizontal and vertical obstacle detection. The system alerts users through vibrations that increase in frequency as obstacles approach. Additionally, the detection cone is adjusted upwards to enhance obstacle recognition while maintaining a comfortable cane position. Ong M et al. [9] present a smart footwear system for visually impaired individuals, incorporating vibration sensors to detect obstacles. These sensors provide real-time feedback upon identifying hazards such as potholes, vehicles, walls, and road markings. The system ensures timely alerts by detecting barriers at a minimum height of 15 cm and within a range of 80 cm to 1 m. A microcontroller efficiently processes the data and manages all system functions. The footwear also features a self-sustaining power mechanism that recharges the battery as the wearer moves. M. Varalatchoumy et al. [10] introduce an obstacle recognition system integrated into wheelchairs to minimize collision risks. These systems enhance mobility, particularly for individuals with disabilities and elderly people in healthcare settings. The technology employs real-time hand gesture recognition and image processing through a webcam, enabling smooth navigation between locations. Dr. S. Srividhya et al. [11, 12] discuss vision-based hand-tracking technology that enhances human-computer interaction. This approach is particularly beneficial for individuals with disabilities, as it enables natural interaction with computers. Several studies have explored hand-tracking applications, particularly in sign language recognition. A proposed system simulates mouse movements using a real-time camera, demonstrating how vision-based technology and intuitive hand gestures can improve accessibility and ease of use.

3 Requirements

3.1 Hardware Components

- **ESP32-CAM Module:** A lightweight camera module with built-in Wi-Fi and Bluetooth capabilities. It captures and processes sign language gestures and communicates with other components.
- **OLED Display:** A small display unit that shows real-time text output and navigation information, providing visual feedback to users.
- **Microphone and Speakers:** Necessary for recording audio inputs and producing audible voice output. Voice commands can be sent using the microphone, and the translated speech is played back over the speakers.
- **Vibration Motor**: For people who have loss of hearing, a vibration motor is added to provide haptic input. It adds another level of availability and interaction by notifying users when an impediment has been detected or when a gesture has been successfully recognized.
- **SD Card Module** The SD card module stores operational data, user preferences, and trained AI models in combination with the ESP32-CAM. By enabling the system to access locally trained models, it guarantees offline functionality.
- **Power Supply:** A USB power supply or battery to ensure the continuous operation of the device's parts. The power supply selection must be in line with the hardware's voltage and current specifications.

3.2 Software Tools and Platforms

- **Arduino IDE:** An integrated platform used for coding and debugging the ESP32-CAM. It provides a straightforward and effective interface for development tasks facilitating the writing, uploading, and debugging of code.
- **Edge Impulse:** A machine learning framework for building, improving, and implementing AI models for object detection and gesture recognition. It makes it possible to create models that specifically suit the needs of Third Eye system.
- **Web Application**: The system's interface is provided by the web application. It enables users to control settings, interact with the device, and access features like obstacle detection and gesture-to-speech conversion.
- **Python/JavaScript Libraries:** Frameworks and libraries written in Python or JavaScript that provide extra features including real-time processing, data handling, and communication protocols.
- **Trained Models**: To identify motions, objects, colors, and obstacles the system makes use of pre-trained models. These precise models were created with Edge Impulse and are stored on the SD card for local execution.

- **Communication Protocols and APIs:** These are tools and protocols that facilitate smooth data exchange and functionality combination between software applications and physical components.

4 Methodology

In order to provide real-time gesture-to-speech conversion, obstacle detection, and improved user accessibility, this project combines hardware and software. The following are included in the methodology:

- **Hardware Selection and Setup:** For gesture recognition, the ESP32-CAM module records motions and interprets image and video inputs. Additionally, the device includes a vibration motor for haptic feedback and an SD card module for storing previously learned models.
- **Gesture Recognition:** The ESP32-CAM records frames to identify objects, colors, barriers, and sign language motions using machine learning models that have already been trained and saved on the SD card. For gesture-to-text translation, recognized gestures are converted into corresponding text [13, 14].
- **Data Processing and Feedback:** The models on the SD card are used to process the gestures that are detected locally. Through the use of text-to-speech (TTS) technology, the system produces audible voice output, providing user accessibility without the requirement for sign language interpretation.
- **Obstacle Detection and Navigation:** To help users move safely, the system uses the ESP32-CAM to identify obstructions and offers real-time audio advice. The vibration motor alerts those who are hard of hearing to potential dangers.
- **Web Application Interface:** The main interface of the system is the web application, which is accessed via a browser. Both online and offline use are supported, and users may monitor translations, adjust settings, and control system features like obstacle detection and gesture-to-speech conversion.
- **Testing and Optimization:** To evaluate the system's accuracy in gesture recognition, speech synthesis quality, and obstacle detection dependability, it is put through a thorough real-world testing process. The system's ability to deliver precise, fast feedback and smooth interaction is guaranteed by continuous improvement.

Figure 1 outlines the operational flow of the AI-based system, which aims to help people with visual and hearing impairments by converting gestures into speech and providing obstacle detection. The system starts with the activation of the ESP32-CAM module, signaling that the system is ready to receive input. After activation, the user presses a button to activate the classification mode, which begins the process of input capture and classification. When the system enters classification mode, it captures input, which can be a gesture, obstacle, or text. To classify the input, we use pre-trained models. The gesture, obstacle, and text classification models have already been trained, and there is another model specifically trained to classify the input type (gesture, obstacle, or text). Once the button is pressed, the system activates the model responsible for classifying the input. If the input is a gesture, the model specific to gesture recognition is loaded from the SD card. This model processes the captured gesture, translating it into text and then

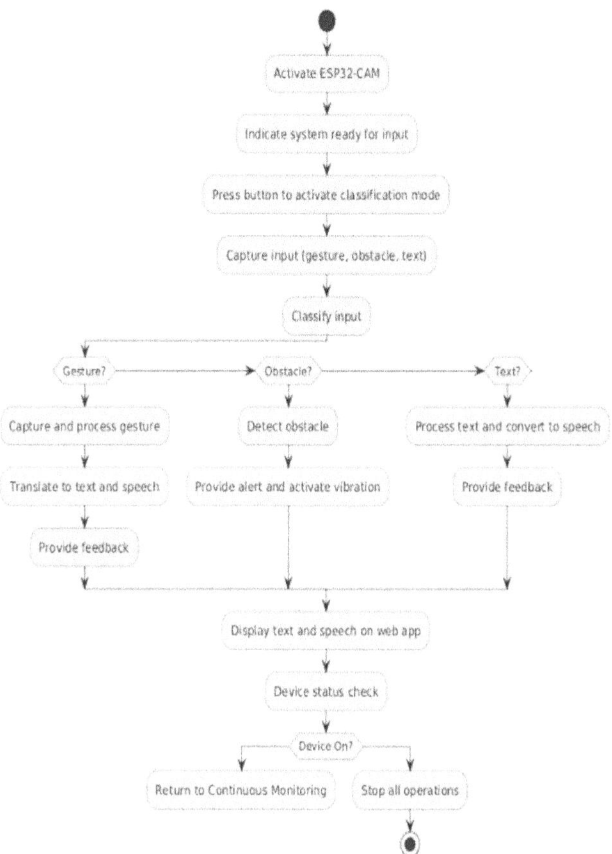

Fig. 1. Flowchart of Third eye

converting it into speech. Feedback is provided to the user to confirm that the gesture has been successfully translated [15].

If the input is text, the system loads the respective model designed to process and convert text into speech. The text is processed and then converted into speech, with the system providing auditory feedback. In the case of an obstacle, the system uses a pre-trained obstacle detection model to identify the presence of obstacles. If an obstacle is detected, the system provides vocal alerts to ensure the user's safety by notifying them of the obstacle and activating the vibration motor for haptic feedback.

After processing the input, the system displays the text and speech output on the web application, allowing for seamless interaction. Finally, the device status is checked to ensure the system continues running. If the device is still on, the system returns to continuous monitoring mode. If the device is off, all operations are stopped, and the process ends. This cycle of activating classification mode, loading the respective models, classifying inputs, and providing feedback ensures that the system continuously operates to provide effective communication and safe navigation for users with sensory impairments.

5 Algorithm

Step 1: System Initialization

- Activate the ESP32-CAM module.
- Indicate that the system is ready for input.

Step 2: Continuous Monitoring

- Continuously monitor for input using the ESP32-CAM.

Step 3: Input Capture

- Capture the input (gesture, obstacle, or text).

Step 4: Classify the Input

- Classify the captured input into one of three categories: Gesture, Obstacle, or Text.

Step 5: Process Input Based on Type

- If the input is a gesture:

  - Capture and process the gesture.
  - Translate the gesture into text.
  - Convert the text into audible speech.
  - Provide audible feedback confirming successful translation.

- If the input is an obstacle:

  - Detect the obstacle in the environment.
  - Provide a vocal alert to assist with safe navigation.
  - Activate the vibration motor to vibrate, indicating the presence of the obstacle.

- If the input is text:

  - Process the text input and convert it into speech.
  - Give a speech evaluation on the result.

Step 6: User Interaction

- On the web application, show the translated text and provide speech output.
- Make sure that navigation and gesture recognition work offline.

Step 7: Device Status Check

- Check if the device is still on:
- Go back to Step 2 (Continuous Monitoring) and repeat the procedure if the device is turned on.
- Stop all operations if the device is off.

Step 8: Repeat

- For continuous operation, go back to Step 2.

6 Results

The "Third Eye" system was effectively created and tested for this project, and the outcomes indicate its capacity to identify gestures, identify difficulties, and offer immediate feedback. The system exhibits the ability to enhance user awareness and interaction, hence facilitating people's ability to move about their environment.

A close-up of the complete "Third Eye" system is provided in Fig. 2 and Fig. 3, which highlight the essential hardware elements and their smooth working. These images show the hardware configuration, emphasizing the ESP32-CAM, power supply, and other parts. The pictures show how these components work together to provide quick system reactions and effective data processing. In addition, the completed prototype is displayed in Fig. 4, showing a small and effective design that makes the gadget comfortable to use all day long. Because portability and ease are given the most importance in the design, users can rely on the system to recognize gestures and provide navigation support in a variety of circumstances.

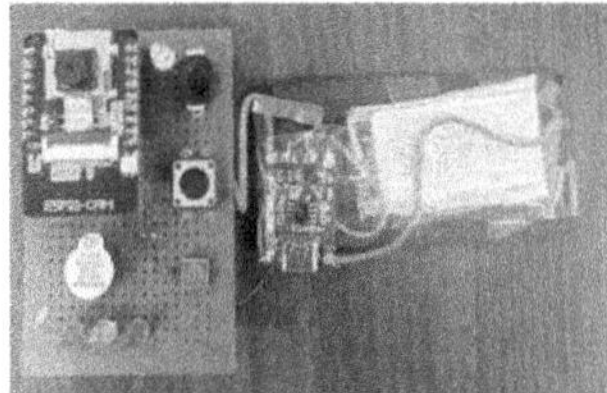 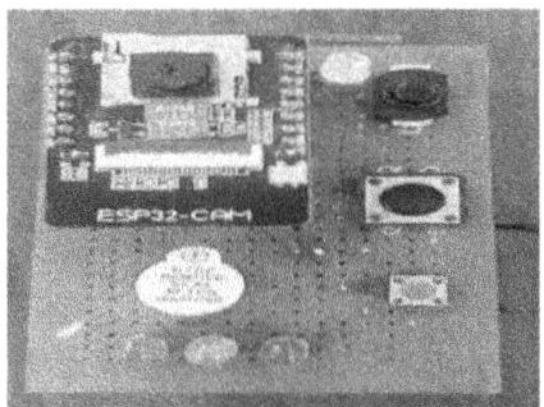 

Fig. 2. Setting Up Hardware   **Fig. 3.** Assembled Prototype   **Fig. 4.** Device Wearability

These outcomes indicate the system's usability and accessibility-improving potential. By combining obstacle detection and gesture recognition, the "Third Eye" technology gives users a new way to engage with their surroundings and gives them the self-assurance and freedom to move around more readily. The vibration motor's integration offers improved haptic feedback in addition to real-time aural feedback, creating a multisensory experience that guarantees users are aware of their surroundings. Versatility in a variety of scenarios is ensured by the system's capacity to recognize and interpret several inputs, including text, gestures, and obstacles. Combining these qualities into a small, portable package makes the "Third Eye" system a useful tool for enhancing everyday navigation, encouraging self-reliance, and giving users a variety of options.

7 Features

- **Offline Capability:** By locally storing all required modules, the system makes sure that essential tasks run even when everything else is unavailable.
- **Dual Mode Operation:** Third Eye serves both blind and deaf customers by offering a variety of services to meet various sensory needs through visual and audio input.
- **Customizable Gestures:** By teaching input for specific gestures, users may personalize the system to suit their own communication preferences.
- **Portable Design:** Small enough to accommodate ergonomic design, lightweight, comfortable to wear, and useful enough for daily use in a variety of settings.
- **Real-Time Processing:** The system processes the gesture in real time, enabling smooth navigation and communication with nearly no delay.
- particularly in loud settings.
- **Vibration (Haptic) Feedback**: Improves interaction, particularly in loud settings, by using a vibration motor to provide tactile feedback to users with regard to gesture recognition or obstacle detection.
- **Multimodal Access**: Provides several feedback options (tactile, visual, and aural) to enable users with varying sensory requirements to interact with the system in a flexible way.
- **Integration of SD Card Module**: This improves offline functioning and update convenience by enabling local storing of user preferences, operational data, and machine learning models.
- **Speech and Button-Controlled Modes**: These modes give users the option to interact with the system by pressing physical buttons or by using voice instructions, giving them the freedom to choose between hands-free and tactile modes.

8 Conclusion

An important development in aiding communication between the hearing and the visually handicapped is the "Third Eye" technology. The device creates a more complete solution by combining AI and IoT to provide real-time navigation help, obstacle detection, and audio translation in addition to capturing sign language. All of the main goals, including improved gesture recognition, real-time communication, and an intuitive design, have been achieved. This illustrates how modern technology may close communication barriers and improve accessibility and inclusivity in encounters. The "Third Eye" system has the potential to significantly improve the lives of those with sensory issues given its significant characteristics, which include offline functionality, adapted gestures, and multimodal communication (auditory, visual, and tactile input).

In the end, the "Third Eye" presents an achievable method to get beyond obstacles that people with challenges encounter and promote better social integration. The system will be able to develop and increase its influence on communication and accessibility through constant improvement combined with user input, opening the way to more inclusive settings.

Disclosure of Interests. All authors have contributed equally to this work and declare no competing interests relevant to the content of this article.

References

1. Alam, S.T., et al.: Smart assistive device for visually impaired people. Int. J. Eng. Res. Technol. (IJERT)
2. Kuriakose, B., Shrestha, R., Sandnes, F.E.: Tools and technologies for blind and visually impaired navigation support: a review. IETE Tech. Rev. **39**(1), 3–18 (2022)
3. Ashiq, H., et al.: Smart assistive device for visually impaired. Int. Res. J. Eng. Technol.
4. Sai, U., et al.: Design and development of an assistive device for the visually impaired. In: International Conference on Computational Intelligence and Data Science (ICCIDS) (2019)
5. Mohite, R., et al.: Smart assistive navigation device for visually impaired people. Int. J. Eng. Res. Appl.
6. More, V.R., et al.: Smart reader for blind people. Int. J. Creat. Res. Thoughts (IJCRT)
7. Tahat, A.A.: A wireless ranging system for the blind long-cane utilizing a smartphone. Telecommunications (2009)
8. Bolgiano, D.M.: A laser cane for the blind. IEEE J. Quantum Electron. (2003)
9. Ong, M., Larbi Bousbia Salah, A., Bedda, M.: An approach for the measurement of impaired people. In: Proceedings of the 10th IEEE International Conference on Electronic Circuits and Systems
10. Varalatchoumy, M., et al.: Wheelchair and PC volume control using hand gesture. In: 2023 International Conference on Intelligent and Innovative Technologies in Computing, Electrical and Electronics (IITCEE) (2023)
11. Srividhya, S., Manna, S., Verma, H.K.: Controlling PC with hand gesture recognition using computer vision. J. Emerg. Technol. Innov. Res. (JETIR)
12. Ulrich, I., Borenstein, J.: The GuideCane – applying mobile robot technologies to assist the visually impaired. IEEE Trans. Syst. Man Cybern. **31**(2) (2001)
13. Hagargund, A.G., et al.: Image to speech conversion for visually impaired. Int. J. Eng. Res. **3**(6), 7 (2017)
14. Shovel, S., Borenstein, J., Koren, Y.: Auditory guidance with the NavBelt – a computerized travel aid for the blind. IEEE Trans. Syst. Man Cybern. **28** (1998)
15. Praderas, S., et al.: Cognitive Aid System for Blind People (CASBliP). Centro de Investigación en Tecnologías Gráficas (2008)

A Deep Ensemble Learning Model for Prediction of Attrition – Effect of Unbalanced Dataset

Somnath Bhattacharya[1] ![ORCID], Praveen Gujjar[1] ![ORCID], Raghavendra M. Devadas[2] ![ORCID],
Vani Hiremani[3] ![ORCID], T. Sowmya[4(✉)] ![ORCID], Preethi[2] ![ORCID], and R. Sapna[2] ![ORCID]

[1] Faculty of Management Studies, JAIN (Deemed-to-be University), Bengaluru, India
[2] Department of Information Technology, Manipal Institute of Technology Bengaluru, Manipal
Academy of Higher Education, Manipal, India
[3] Symbiosis Institute of Technology, Symbiosis International (Deemed) University, Pune, India
[4] Department of Computer Science and Engineering, Manipal Institute of Technology
Bengaluru, Manipal Academy of Higher Education, Manipal, India
`sowmya.t@manipal.edu`

Abstract. Deep ensemble learning has become a very popular and efficient technique for making predictions. Traditional machine learning models are often not efficient when used for making predictions of imbalanced datasets. To counter the effects of imbalanced datasets, several different oversampling and synthetic sampling methods have been proposed. The effect of such sampling methods on deep ensemble learning methods is also not well studied. This study aims to explore two basic questions: whether deep ensemble models are suitable for unbalanced data given traditional methods are not and secondly if we use the oversampling method along with deep ensemble learning, does it improve model performance? For our comparison, we choose four models. We further scope this study by using only sequential ensemble models. We choose three baseline models and two deep-learning ensemble models for our study. We find that the deep ensemble learning models are better when we consider a combination of precision and recall. Also, oversampling significantly increases the model's ability to correctly predict the underrepresented class.

Keywords: Deep ensemble learning · unbalanced dataset · oversampling

1 Introduction

Ensemble learning is a popular technique that combines multiple forecast models to improve prediction or classification accuracy. Implementing an ensemble learning model consists of three steps. The first step is generating sub-samples of data to train the different models. The second step is training individual prediction models like random forest, logistic regression, etc. The final step is combining the outputs of the trained models for final predictions. Ensemble learning models provide better prediction accuracy as compared to standalone classification models.

A new direction of research in ensemble learning is the use of deep ensemble learning models [1]. Classical deep learning methods, in the context of neural networks, are based

J. Shreyas et al. (Eds.): CODE-AI 2025, CCIS 2689, pp. 32–40, 2026.
https://doi.org/10.1007/978-3-032-19318-6_4

on the idea of having multiple layers in a single model to improve model accuracy. Deep ensemble learning, on the other hand, is based on the idea of creating multiple neural network models and then combining their predictions. Deep ensemble learning is slowly becoming popular but very few studies have compared their benefits over traditional ensemble learning specifically in the context of unbalanced datasets.

Ensemble learning requires a large sample size to train the individual classifiers effectively. Specifically, for classification purposes, it is essential that not only the dataset is large but it should also be balanced i.e., sufficient sample size should exist for all the different classification categories so that the models can be properly trained. Scholars have focused on improving sampling methods to facilitate better training of models [2] using oversampling, etc. to negate the effects of unbalanced datasets. However, the effect of combining oversampling with deep ensemble learning is not widely studied.

In this study, we focus specifically on two major questions.

1. When the dataset is unbalanced, does deep ensemble learning provide better results than other standard ensemble learning models?
2. Does combining the deep ensemble method with the oversampling method lead to more accurate classification?

We create a deep ensemble learning model and compare it with a widely popular ensemble learning model (XGBoost) and a simple ensemble classifier (Logistic Regression followed by a random forest) to compare how they perform when the input dataset is unbalanced. We use the IBM HR attrition dataset for our analysis. Baseline ensemble learning models are compared with custom deep ensemble learning models. Also, we combine oversampling with our custom deep ensemble learning to see its effect. In the next section, we briefly look at the existing literature on ensemble learning and deep ensemble learning.

2 Literature Review

Ensemble learning refers to algorithms that combine multiple models to increase prediction accuracy and is widely used in both classification and prediction problems [3]. Classification tasks can be further subdivided into binomial (2 categories) and more than 2 category classifications. In this study, we focus on two category classifications. In ensemble learning, each individual model is called the baseline model. Usually, the baseline models individually are weak learners i.e., their prediction accuracy is comparatively lower. Ensemble learning combines the different models to increase prediction accuracy. This combination may happen in parallel (classifiers are trained in parallel) or sequentially (classifiers are trained one after the other) [3]. To limit the scope of the study, we focus only on sequential learning models. First, we discuss the theoretical background of the baseline ensemble models. We select logistic regression and random forest combination as our first baseline classifiers since studies have established that ensemble models with diverse classifiers produce better results [4] and random forest and regression are diverse in the way they learn and predict categories. The next baseline model we use is that of XGBoost which is based on the idea of boosting. Boosting is one of the most popular methods of ensemble learning and is based on sequential training of different classifier models. The sequentially trained models assign more weight to misclassified data in later models thereby increasing the chances of identification of misclassification. At first, the dataset is repeatedly sampled into smaller subsets. A variant of boosting popularly known as "XGBoost" has become very popular nowadays specifically after its efficiency was proved in KDDCup2015 [5]. XGBoost is based on sequential decision trees. XGBoost works by assigning weights to the attributes that are being used for making the predictions. Subsequent decision trees in XGBoost try to reduce the error by adjusting the weights when an item is wrongly classified by the previous classifier. We use XGBoost as one of our models because of its proven efficiency. Next, we describe the basic deep ensemble model. Neural networks are widely used for classification purposes. CNN or convolutional neural network helps in dimensionality reduction in classification tasks [6]. CNN neural networks use a convolution filter to reduce the dimensions of input data. We use a CNN here as a baseline model. Deep ensemble learning methods are ensemble learning designs of traditional deep learning neural networks to increase prediction accuracy [1]. The core idea of deep learning ensembles is to create multiple neural network models and then combine their prediction power. Models can be trained both sequentially or in parallel. As mentioned, in this study, we focus on sequential combination since these sequential ensemble learning models have been found to be efficient in prediction purposes [7]. These types of models are computationally very expensive in terms of the number of parameters to be trained and hence the time taken to train these models is relatively higher as training each neural network requires calculation of a lot of parameters. We create a sequential deep learning model (that is described later) and compare it with other ensemble learning models. As mentioned, the first step for training an ensemble learning model is to create subsets of the data. In classification tasks, when the dataset is unbalanced (i.e., the number of items for one class is very low as compared to the other), machine learning algorithms perform poorly [3]. Further, creating the subsamples becomes problematic. To address this issue, broadly two types of oversampling methods are used. The first one is synthetic oversampling. In 2002, Synthetic Minority Oversampling Technique (SMOTE) was proposed which has gained widespread acceptance from scholars due to its proven superiority [8]. The basic idea of SMOTE is to create "synthetic" data points for the underrepresented class. To create these synthetic data points, one particular data point belonging to the under-represented class is selected. Thereafter, neighbors of the datapoint are used to create new data points by interpolation. The selection of a neighbor takes place by calculating the difference in features (attributes) of the selected data point and the other data points. Another method, which is far simpler and computationally less expensive, is that of random oversampling. In random oversampling, the underrepresented data points are sampled multiple times to create balance in training dataset [9, 10]. In our study, we use the random oversampling method (due to its low computational cost and easy implementation) to potentially improve the prediction of the deep learning ensemble model. Studies have found oversampling and deep ensemble learning combination to be effective for prediction purposes [11].

3   Dataset Description

The IBM HR attrition dataset is a widely used synthetic dataset created by researchers at IBM, which is used for evaluating machine learning algorithms [12]. The dataset includes attrition status (whether the employee has left tor not) and also lists some details about the employee like educational background, traveling required for the job, etc. In total, the dataset consists of 35 variables including the attrition status. The objective of this study is not to prove/disprove any hypothesis (the data is synthetic and should be used only for prediction purposes) based on attributes. We are interested in measuring and comparing the accuracy of different models especially when the training dataset is unbalanced. Hence, we do not engage in any feature selection based on theory for use in the prediction models. Based on an initial analysis, we find that 5 variables only have one particular category i.e., their value does not change for the different data points. We remove these 5 variables before proceeding with our analysis. Apart from this, we rely on the predictor models to identify and use the most relevant attributes. The dataset consists of 1470 rows with 1233 rows for employees who have not left (Attrition = "No") which makes it highly imbalanced. We want to check if deep learning ensemble systems perform better for binary classification tasks for imbalanced datasets. For training and testing purposes we divide the dataset in the ratio of 70:30. In the next section, we describe the different models created for comparison. For comparing the models, we use the well-known measures of precision and recall. A key thing to be kept in mind is the objective of prediction. It so happens that in the real world, organizations will be interested in the prediction of the underrepresented class (Attrition = "Yes"). While evaluating the models we look at the suitability of the model to correctly predict the underrepresented class.

4   Prediction Models

Model 1
We create a basic ensemble sequential learning model with logistic regression as the first learner and a random forest as the second learner. Logistic regression is well-known method that allows classification through predicted probability values and this method does not require the independent variables to follow a normal distribution. Voting is used for final prediction such that if both the learners simultaneously predict 1 or 0, the classification is accepted. Otherwise, the classification is considered as NA. The model summaries are provided in Table 1.

Model 2
The next baseline model we use is the XGBoost model which consists of multiple sequential decision trees.

Model 3
We create a basic CNN model for attrition prediction. We chose CNN since we are interested in dimensionality reduction and it is not expected that there will be any temporal relationship. The model predicts probabilities which are then converted to classes (if

Table 1. Model Summaries

Logistic Regression	
Null deviance	1040.5 on 1176 degrees of freedom
Residual Deviance	682.6 on 1132 degrees of freedom
AIC	772.7
Random Forest	
MSE	500
R squared	500

the probability value is more than .6, it is considered as 1, else it is considered as 0). The categorical variables (e.g., marital status, education level, etc.) in the dataset are converted to numerical ones. All the values of the training dataset are normalized for training the model. The model consists of one 1-dimensional Convolution layer followed by a Maximum Pooling and one Fully Connected layer. The model summary is given in Table 2.

Table 2. CNN Model Specification

Layer	Output Shape	Number of Parameters
Convolution 1D (Relu Activation)	(None, 28, 30)	120
Maximum Pooling	(None, 14, 30)	0
Flatten	(None, 420)	0
Dense(Sigmoid Activation)	(None, 1)	421

The model was trained for 20 epochs with a batch size of 32. The model was optimized using the Adam optimizer in conjunction with the Binary Cross-Entropy loss function. For these three models, we split the data randomly into the ratio of 70: 30 training and testing sets. The training set consists of 190 rows for Attrition $= 1$.

Model 4 and 5

We create a deep-learning ensemble model. The model is based on the following analysis and assumptions: CNNs have been found to have lower efficiency when the dataset is imbalanced. We aim to study whether the efficiency of models increases for imbalanced datasets when we use deep learning ensembles. Hence first we create a simple sequential ensemble model. The first model is a convolutional neural network that uses the 30 attributes to make probability predictions. The next model takes the predicted probabilities as input along with the 30 other attributes and predicts the final probability. The architecture of the neural network is the same as that of the base model. This is our **Model 4**. The hyperparameters of training were the same as that of **Model 3**.

In **Model 5**, we add random oversampling before training the model. The neural network based sequential architecture remains the same as of **Model 4**. After oversampling, the distribution of data in the training set becomes (Table 3).

Table 3. Oversampled Dataset

0	1
597	580

5 Analysis and Results

Table 4. Test Train Split

Train		Test	
0	1	0	1
987	190	246	47

Table 5. Confusion Matrix

Model type	Prediction	Reference	
Model 1		**0**	**1**
	0	243	37
	1	0	6
Model 2	**0**	240	31
	1	6	16
Model 3	**0**	244	37
	1	2	10
Model 4		245	40
		1	17
Model 5	**0**	217	24
	1	29	23

The test train split is listed in Table 4. We run the models and list the confusion matrix for the testing dataset in Table 5. We find that The confusion matrix tells us that Model 1 is able to predict 0 (Attrition = "No") most effectively. However, the prediction of 1 (Attrition = "Yes") is a little off. The XGBoost model performs slightly better in identifying 1s. The neural network model (Model 3) performs not that great in identifying the 1s. The deep ensemble learning model (Model 4) is able to predict 0s effectively but it identifies many true negatives (employees who will not leave but are classified as Attrition = 1). The deep ensemble learning model with oversampling (Model 5) performs relatively poorly for both predictions. In summary, if we want a model to identify those who will not leave correctly, Models 3 and 4 perform well. This can help in identifying loyal employees of the organization. If we want to identify employees who will leave, Model 3 and 4 will be appropriate. But if we want to reduce the chances of misclassification, then probably model 5 is the best choice. To have a better idea about the models accuracy, we list the overall precision and recall values in Table 6. To explore the first question, we compare model 1, 2 and 3 with model 4. We find that Model 1 provides the highest precision followed by Model 3. So when the dataset is unbalanced, in terms of precision, Simple ensemble models and basic CNN models outperform deep ensemble learning models. For recall, we find that XGBoost and CNN models outperform the deep ensemble learning model. When we balance the dataset using over-sampling, we find that Model 5 provides the best recall among all the other models. However, in terms of precision, oversampling has led to the lowest precision among all the other models. The f! scores show that the deep ensemble learning is the most balanced model among all the others. We list the ROC curves for the different models below (Fig. 1).

Table 6. Precision and Recall

Model Name	Precision	Recall	F1- Score
Model 1	1	.13	.23
Model 2	.72	.34	.46
Model 3	.91	.23	.37
Model 4	.87	.14	.24
Model 5	.44	.88	.59

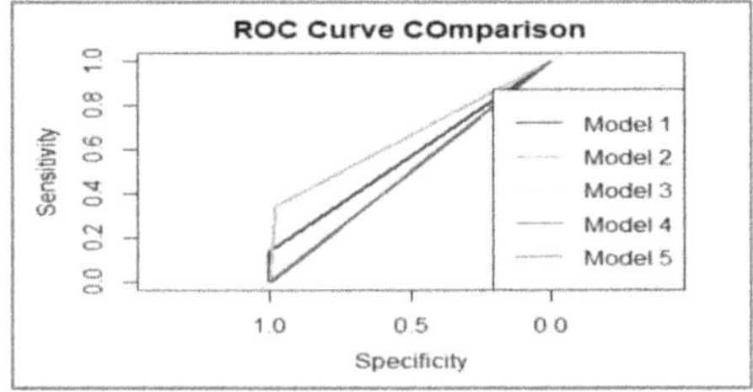

Fig. 1. ROC Curves

Note that the Roc curve for models 3, 4, and 5 are overlapping. As known. The ROC curve plots the true positive rate against the false positive rate. The ROC curve that rises quickly towards the top-left corner of the plot suggests a better-performing model.

6   Conclusion

In this study, we were mainly interested in two major questions. We find that deep ensemble learning does not necessarily perform better on all parameters. Training with oversampled data leads to a reduction in misclassification. The choice of an appropriate model will depend on the exact task at hand and how much error we are willing to tolerate. We do not list any coefficients of any models since it is a synthetic dataset as mentioned earlier. Organizations can implement these simple models to get accurate predictions about employee churning and perform better manpower planning. In this study, we compare only sequential learning models. We have to also compare parallel models to check if they perform better than the sequential models or not. Further, the comparison has to be done with many more datasets before we can conclude which method is more accurate. Further, usually, machine learning models require large datasets for training. The dataset we have used is large but not very large. Using larger dataset might increase the accuracy of the models.

References

1. Yang, Y., Lv, H., Chen, N.: A survey on ensemble learning under the era of deep learning. Artif. Intell. Rev. **56**(6), 5545–5589 (2022)
2. Farajzadeh-Zanjani, M., Razavi-Far, R., Saif, M.: Efficient sampling techniques for ensemble learning and diagnosing bearing defects under class imbalanced condition. In: 2016 IEEE Symposium Series on Computational Intelligence (SSCI), pp. 1–7 (2016)
3. Mienye, I.D., Sun, Y.: A survey of ensemble learning: concepts, algorithms, applications, and prospects. IEEE Access **10**, 99129–99149 (2022)
4. Kuncheva, L.I., Whitaker, C.J.: Measures of diversity in classifier ensembles and their relationship with the ensemble accuracy. Mach. Learn. **51**, 181–207 (2003)
5. Chen, T., Guestrin, C.: XGBoost: a scalable tree boosting system. In: Proceedings of the 22nd ACM SIGKDD International Conference on Knowledge Discovery and Data Mining (2016)
6. Velliangiri, S., Alagumuthukrishnan, S., Thankumar joseph, S.I.: A review of dimensionality reduction techniques for efficient computation. Procedia Comput. Sci. **165**, 104–111 (2019)
7. Cao, Y., Geddes, T.A., Yang, J.Y.H., et al.: Ensemble deep learning in bioinformatics. Nat. Mach. Intell. **2**, 500–508 (2020). https://doi.org/10.1038/s42256-020-0217-
8. Fernandez, A., Garcia, S., Herrera, F., Chawla, N.V.: SMOTE for learning from Imbalanced data: progress and challenges, marking the 15-year anniversary. J. Artif. Intell. Res. **61**, 863–905 (2018)
9. Moreo, A., Esuli, A., Sebastiani, F.: Distributional random oversampling for imbalanced text classification. In: Proceedings of the 39th International ACM SIGIR conference on Research and Development in Information Retrieval (2016)
10. Hayaty, M., Muthmainah, S., Ghufran, S.M.: Random and synthetic over-sampling approach to resolve data imbalance in classification. Int. J. Artif. Intell. Res. **4**(2), 86 (2021)
11. Chen, W., Sun, Z., Han, J.: Landslide susceptibility modeling using integrated ensemble weights of evidence with logistic regression and random forest models. Appl. Sci. **9**(1), 171 (2019)

12. Qutub, A., Al-Mehmadi, A., Al-Hssan, M., Aljohani, R., Alghamdi, H.S.: Prediction of employee attrition using machine learning and ensemble methods. Int. J. Mach. Learn. Comput. **11**(2), 110–114 (2021)

Skin Cancer Detection Based on Machine Learning Models

Kiran Puttegowda(✉) , K. V. Sudheesh , H. N. Naveen Kumar , Mahadevaswamy , B. Jagadeesh , and Kirankumarhumse

Vidyavardhaka College of Engineering, Mysuru, Karnataka, India

Abstract. Skin cancer is one of the most prevalent and deadly forms of cancer worldwide, with melanoma being the most aggressive type. Early detection significantly improves the chances of successful treatment and patient survival. Traditional diagnostic methods, including visual inspection and biopsy, are time-consuming and heavily reliant on the expertise of dermatologists. In recent years, machine learning models have emerged as powerful tools for enhancing diagnostic accuracy and efficiency in skin cancer detection. The Paper explores skin cancer detection using a parallel combination of classification models. It utilizes a dataset of nearly 1000 photos with various skin lesions. The methodology includes analysis of the color variability and the presence of suspicious colors after a preceding HSV color conversion and SLIC for super pixel segmentation, asymmetry assessment, and evaluation of possible blue-white veils and atypical pigmentation networks. Classification techniques such as logistic regression, KNN, and decision trees are utilized, with a focus on maximizing recall to reliably identify cancerous lesions. The study discusses the limitations arising from varying image qualities, multiple annotators and the generalization capability of the models across different devices. Ultimately, the project achieves a robust model combining multiple classification approaches to enhance diagnostic accuracy and reliability.

Keywords: Skin Cancer · Segmentation · Classification · machine learning models · accuracy

1 Introduction

Cancerous skin diseases are a major health concern that is hard to detect without professional expertise in dermatology. For accurate detection doctors use systems like the ABCDE (Asymmetry, Border, Color, Diameter, Evolution) and the Seven-Point Checklist. While effective, these methods vary in interpretation, and you still need prior knowledge for the background of the possible diagnosis [1].

Luckily, the latest advancements in artificial intelligence (AI) and machine learning (ML) have shown promising results in enhancing skin cancer detection. At the European Academy of Dermatology and Venereology (EADV) Congress 2023, an AI system demonstrated near-perfect accuracy in identifying skin cancers. However, researchers agree that AI should support, not replace, human experts [2].

© The Author(s) 2026

J. Shreyas et al. (Eds.): CODE-AI 2025, CCIS 2689, pp. 41–52, 2026.

https://doi.org/10.1007/978-3-032-19318-6_5

The main objective of the paper is to develop a classification model using aspects of the Seven-Point Checklist and the color and asymmetry from the ABCDE rule to predict cancerous lesions only by analyzing lesion images. Our goal is to create the prototype of an analyzing tool which aims to assist patients and dermatologists in early detection via pictures taken with normal smartphones.

2 Related Work

Several studies have explored advanced machine learning and deep learning approaches for effective skin cancer detection and classification. Below is a summary of some of the most relevant work in this area:

Ranpreet Kaur et al. [3] created progressive deep learning algorithms designed to diagnose melanoma through work on preprocessing steps and segmentation methods and classification tasks. Their benchmarked model demonstrated exceptional 93.40% accuracy in classifying skin cancers based on their dataset. According to Anna Nguyen et al. [4] Convolutional Neural Networks (CNNs) showed 92.5% precision for skin cancer detection. The authors stressed how preprocessing methods combined with diverse datasets produce more robust models which combat dermatology diagnostic difficulties to deliver improved patient care. Rodrigue B et al. [5] conducted research about YOLO v7 contrasted against their alternative custom CNN structure for detecting skin cancer. Multiple CNN models reached outstanding results with accuracy, sensitivity and specificity levels in identifying Basal Cell Carcinoma (BCC), Squamous Cell Carcinoma (SCC) and melanoma skin lesions which allowed researchers to study efficient classification methods. A deep learning model based on ResNet-50 architecture applied transfer learning to achieve 97% accuracy in identifying seven types of skin lesions as described by Nada M. Rashad et al. [6]. Healthcare professionals and dermatologists can benefit from this model since it demonstrates strong potential for detecting cancer in early stages while helping with automatic screening process support.

Saba Abdul Salam et al. [7] built deep learning models that analysed thermo-scopic images to detect skin cancer. The study obtained great diagnostic accuracy with DenseNet121 achieving 94% and EfficientNet B0 reaching 93% by using pre-trained CNNs in identifying benign and malignant skin lesions. Arif Ullah et al. [8] proposed a comparative evaluation between different machine learning models and deep learning approaches for diagnosing skin cancer. Its demonstrated that deep learning-based models surpassed traditional methods in terms of reliability and accuracy as well as sensitivity which supports clinicians to perform timely diagnosis using advanced image analysis capabilities. Mangus Musa et al. [9] developed a system which united a CNN framework with OpenCV image processing for multi-condition skin detection including skin malignancies. The system generates immediate analysis while providing treatment suggestions which help healthcare providers make better clinical choices and diagnoses. Md Nasiruddin et al. [10] applied CNN models to detect skin cancer by analysing 1271 images from dermoscopic sources. This model reached a 94% accuracy level that indicates its value to dermatologists as a supportive diagnostic tool which improves both clinical workflows and decision-making procedures. Parul Dubey et al. [11] utilized reinforcement learning models to detect skin cancer which led to an 82% accuracy rate.

Traditional CFT models received improvement from this approach which raised sensitivity specifically for melanoma and BCC detection beyond current diagnostic capabilities. Kuldeep Vayadande et al. [12] explained a combined model of autoencoders and CNNs for enhancing skin cancer identification efficiency. This approach analysed high-resolution images to detect cancerous lesions accurately while obtaining improved clinical results by continuously learning from new acquired medical data. Tsu-Man Chiu et al. [13] created an AI diagnostic model which implemented a two-stage classification method. The diagnostic method's modified approach cut down on false negative readings which helped enhance detection accuracy by clearly differentiating between skin cancer malignant types and benign types. D. D. Surendren and J. Sumitha et al. [14] studied VGG-16, VGG-19, ResNet and DenseNet CNN architecture models which produced detection accuracies of 72.61%, 73.19%, 84.35%, and 82.37% successively. The research results bring important knowledge for identifying benign and malignant skin cancer types.

3 Methodology

The Fig. 1 shows a Skin Cancer Prediction System designed to detect and classify skin lesions accurately. It starts with collecting multiple skin lesion images, which are then pre-processed to enhance quality by removing noise, adjusting brightness, and improving contrast. After preprocessing, the system segments the lesion area from the surrounding skin using techniques like thresholding, clustering, or deep learning. The segmented area is then refined to ensure clear lesion boundaries. Next, the Region of Interest (ROI) containing the lesion is cropped for focused analysis. Features from this region are extracted using methods like Convolutional Neural Networks (CNNs) to help distinguish different types of lesions. The extracted features are fed into a neural network for classification, determining whether the lesion is malignant or benign. Finally, the diagnosis result indicates whether the lesion is cancerous or not. Detailed description each block explained as follows.

3.1 Dataset

Proposed dataset is based on 978 images from [15] dataset. There are photos of all 6 diagnoses - BCC, ACK, SEK, MEL, NEV, SCC for which we have used manually created masks. Since the dataset sometimes contains more than one photo of the same patient, and sometimes even of the same lesion, we have made sure that all the photos of single patient are treated as a cohesive group. After partitioning the data into training and testing subsets, these groups remain intact within the same split. We have achieved that using sklearn model selection method called Stratified-GroupKFold, which also maintains the proportion of the classes within the splits. In our case, that would mean the proportion of cancerous and noncancerous lesions. At the end, the dataset is split in the following manner as shown in Fig. 2.

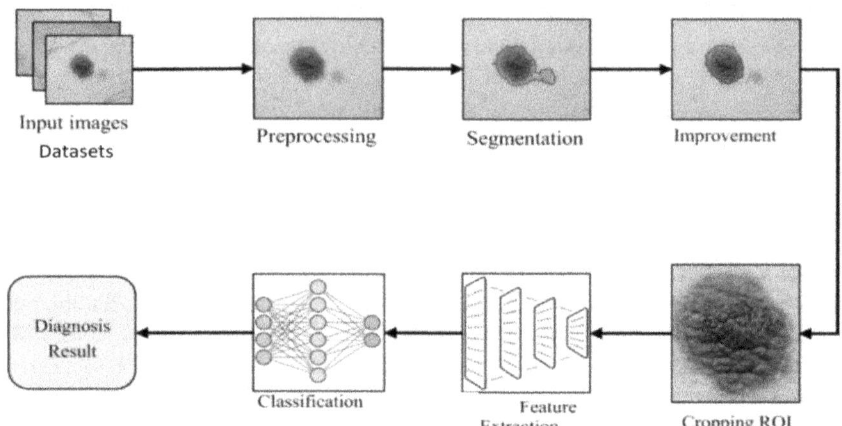

Fig. 1. Block diagram of proposed skin cancer prediction system

Dataset information			
Diagnosis	**Full**	**Train**	**Test**
Cancerous	574	430	144
Non-cancerous	404	304	100
Total	978	734 (75%)	244 (25%)

Fig. 2. Dataset Information

3.2 Preprocessing

For each picture of our dataset, we perform two manipulations - we cut the lesion only and resize it to a length or width of 250px, depending on which one is larger, while maintaining the initial length/width proportion, and then we perform SLIC (Simple Linear Iterative Clustering) algorithm from 'skimage'. The SLIC is used to segment the image into 250 super pixels with a compact parameter of 100, effectively shaping them into squares. Post-segmentation, we convert RGB outputs to HSV for a more nuanced color analysis. For measuring color variability, we employ a static method to measure the color variability of images by calculating the variance, which indicates the dispersion of colors. This method is particularly valuable in highlighting the color diversity within a single skin lesion. We utilize the superpixels from the already performed SLIC function to simplify the image while preserving critical color data. We also use the converted HSV (Hue, Saturation, Value) version to facilitate easier analysis of color relationships. The numbers for each of the three channels vary between 0 and 1 as the conversion function supports a set of specific formulas. For each superpixel, we compute the mean and variance of HSV values, which serve as features (Hue, Saturation, Value) in our training and test dataset. The variance calculation enhances our diagnostic accuracy by providing a quantifiable measure of color distribution. Additionally, our observations indicate that the mean Hue often aligns with the red spectrum, ranging from 350 to 10 in HSV values, which contributed to outliers in our variance analysis. To address this, we adjusted the Hue values by 180 degrees to center around the prevalent red and pink hues, enhancing the consistency of our data. This adjustment ensures that our variance measurements remain robust across different imaging conditions and skin tones.

3.3 Automatic Segmentation

We made an initial attempt at performing automatic segmentation as part of our system. To achieve this, we applied a segmentation technique that involved classification-based segmentation followed by morphological processing. The segmentation via classification approach aimed to differentiate the lesion area from the surrounding skin by assigning labels to pixels based on their characteristics. Once this initial segmentation was achieved, we employed morphological operations, specifically erosion and dilation, to enhance the segmented region. Erosion was applied to eliminate small, irrelevant details and noise, effectively sharpening the boundaries of the lesion. Subsequently, dilation was used to restore essential features and ensure that the segmented region maintained its integrity. This combination of classification-based segmentation and morphological processing helped us achieve a clearer and more refined segmentation of the lesion. Figure 3 shows the segmentation results for skin images.

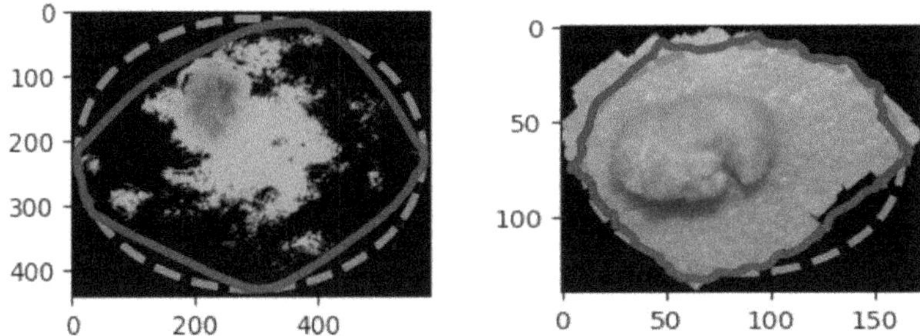

Fig. 3. Segmentation results: With Active Contour (left), No Active Contour (right)

We decided on using segmentation via classification in favor of segmentation via thresholding because we considered it to be more powerful (it can also include pixel intensity as a parameter). To obtain better segmentation, after KNN and morphology, we tried performing segmentation via shape (active contour) - the idea was to apply an active contour on a piece of an image selected by the mask obtained by KNN and morphology and cropped only to the bounding box. Even though the active contour was applied after trying it to on a few samples from validation dataset we saw no great improvement at a cost of approx. 40 s of additional processing for each image, so we decided not to proceed with it.

Because of the low results we decided not to carry on the efforts with automatic segmentation. Yet we want to notice that our segmentation scripts, from the visual inspection of all validation samples, most often produced masks which contained ground truth masks. We suspect that perhaps tweaking morphology could better prepare an intermediary mask for the active contour. There were many variables we have not checked, due to time constraints, like optimal number of neighbors, different classification algorithms.

3.4 Classification

Our classifier tries to solve the task of distinguishing cancerous lesions from non-cancerous lesions. This is why we decided to combine the diagnoses MEL, BCC and SCC under the label of skin cancer (label 1) and ACK, SEK and NEV as non-cancerous conditions (label 0). We decided to test and implement the three different types of classifiers in the course: K-Nearest Neighbors, Logistic regression, and Decision tree. After a discussion on how to evaluate the best performance, we decided to settle on the idea of keeping our model's train accuracy above 68% and after that maximizing recall score on testing. For all our results we present confusion matrices.

i. **Logistic regression**

Logistic regression is a statistical model used in machine learning for classification (not regression) problems. In terms of binary classification, it models the probabilities of belonging to the positive class.

ii. **KNN**

K-Nearest Neighbors (KNN) is a classifier that follows the supervised learning approaches. The algorithm behind it is that when a new point must be categorized, the model calculates the Minkowski distance between the new point and the closest k neighbors.

iii. **Random forest**

In our approach for finding the lower and upper boundaries for the random forest classifier. This paper focuses on optimal parameter settings for a random forest classifier, emphasizing the depth of the trees and the number of trees.

4 Results and Discussion

Proposed dataset we used 150 lesion images and corresponding of human annotated masks. 120 (75%) were used as train dataset and 30 (25%) as a validation dataset. As KNN is a lazy algorithm meaning that most computations are made during the inference, we decreased training sample size - from each image we sampled 100 pixels belonging to lesion and 100 not. So, we got 30 000 learning data points. It solved the problem of imbalance as well. Table 1 shows the threshold values used for the automatic segmentation.

Table 1. Threshold values for Automatic segmentation

	Dice Score	Hausdorff distance
Mean	0.56	185.60
St. Deviation	0.20	77.20

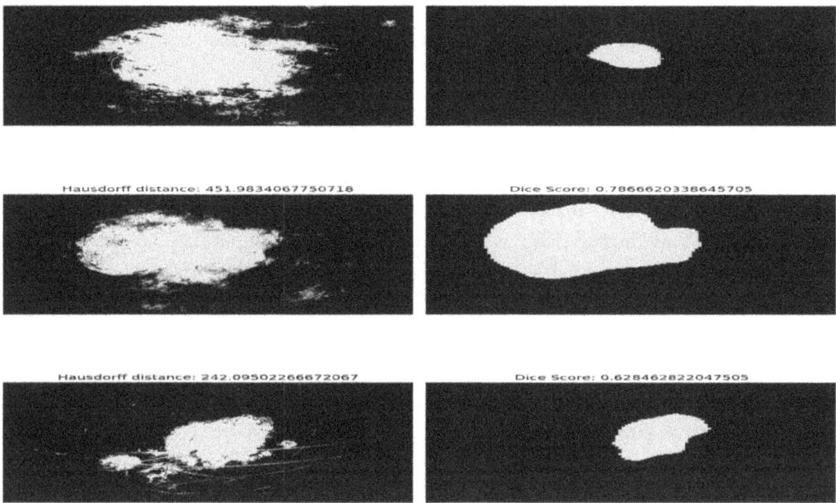

Fig. 4. Extraction of Skin ROI based on Dice Score and Hausdorff distance

Figure 4 presents the extraction of Skin Region of Interest (ROI) alongside technical metrics which consist of Dice Score and Hausdorff Distance. The Dice Score provides an evaluation metric that determines predicted segmentation accuracy through overlap measurement between models and truth segments. Higher values of Dice Score represent more precise and accurate segmentation results. The Hausdorff Distance measures boundary alignment between predicted segmentation and ground truth by determining the largest edge distance between them. The contour matching quality and boundary definition between two regions improve with lower Hausdorff Distance. Both metrics work together to verify the accuracy and boundary alignment of the segmented area relative to true lesion boundaries. The dual-metric approach improves segmentation reliability required for proper ROI cropping that supports further analysis and classification procedures. These evaluation metrics promote thorough and precise identification of the Skin ROI thus leading to better diagnostic outcomes.

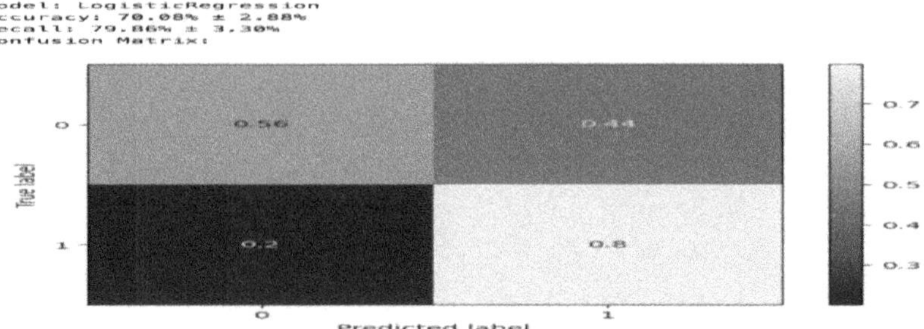

Fig. 5. Confusion matrix logistic regression model

Figure 5 displays the Confusion Matrix that shows how the Logistic Regression model performs when predicting skin cancer. The model displays its classification abilities for skin lesions through this visual tool which separates benign lesions (0) from malignant lesions (1). The model demonstrates an accuracy level of 70.08% ± 2.88% and a recall at 79.86% ± 3.30% which shows a decent capability to identify malignant cases. Token examples show the model yields a high detection rate of malignant lesions at 79.86% which reflects its relatively high true positive rate (0.80). The system demonstrates poor performance in separating benign from non-benign skin lesions since it produces about 44% false positive results. Although the model shows effectiveness in detecting cancerous lesions it requires additional improvements to minimize false alarms and improve general accuracy.

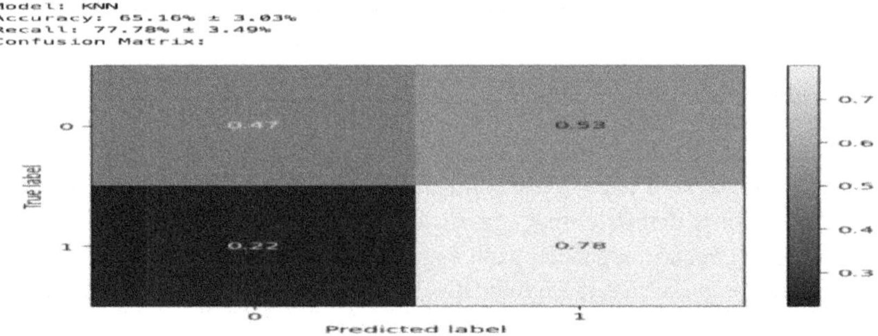

Fig. 6. Confusion matrix KNN model

The K-Nearest Neighbors (KNN) model implemented for skin cancer prediction generates a Confusion Matrix as presented in Fig. 6. A look at the matrix indicates the model's performance through its separation capabilities between benign (label 0) and malignant (label 1) lesion cases. The model's overall accuracy stands at 65.16% ± 3.03% while the recall reaches 77.78% ± 3.49%, demonstrating average success in detecting malignant cases. The model performs well in identifying malignant lesions based on its recall of 77.78% along with a true positive rate of 0.78. The system delivers deficient performance in detecting benign lesions because its false positive detection rate reaches

0.53. The detection model incorrectly labels benign lesions as cancerous more often than they should, which produces unnecessary medical concerns and additional procedures.

Figure 7 displays the Confusion Matrix that shows the results of the random forest classifier when used to predict skin cancer. The graphic representation demonstrates how the model detects benign tissue (tag 0) and malignant tissue (tag 1). The model shows a 73.36% ± 2.96% accuracy rate and a high recall level of 84.72% ± 3.00% that demonstrates its effectiveness at recognizing malignant cases. The model demonstrates superior performance in detecting malignant lesions through its True Positive Rate of 0.85 accompanied by an excellent recall score of 84.72% which indicates that most malignant lesions are detected. The high False Positive Rate of 0.43 signifies the model identifies numerous benign lesions as being cancerous.

Figure 8 shows the Confusion Matrix of the Final Model that uses Random Forest Classifier and Logistic Regression together to boost predictive abilities. The combined model takes advantage of both individual models' strengths to improve precision and detection rates while overcoming their unique limitations. The combined model shows 74.18% ± 2.85% accuracy combined with 84.03% ± 3.07% recall in detecting malignant cases. When Random Forest Classifier merges with Logistic Regression model to produce a hybrid approach it delivers higher performance across accuracy and recall compared to operating with them independently. Additional work needs to be done to decrease incorrect positive results while improving general performance levels.

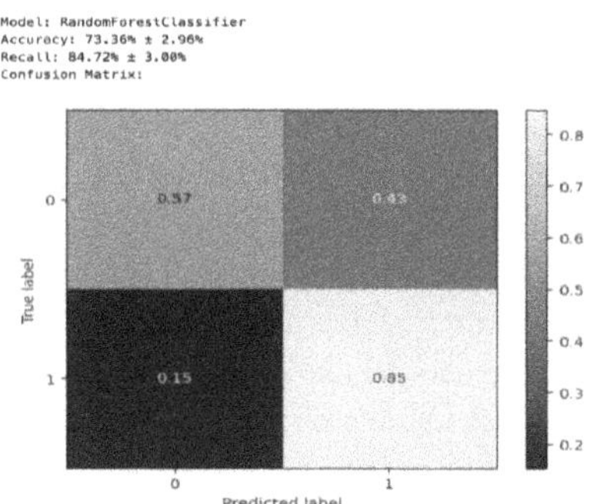

Fig. 7. Confusion matrix Random Forest model

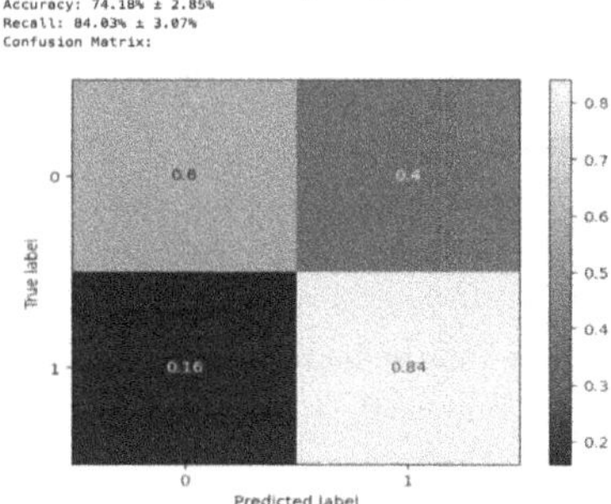

Fig. 8. Confusion matrix Final model (Random model+ Logistic Regression model)

The study design introduces various types of limitations worth noting. Firstly, there are general limitations stemming from the diverse sources of the photos, captured by different individuals using mobile phone cameras. This causes an inconsistency in the quality due to different phones models, lighting conditions, technical expertise of the photographer, etc. Even if the photos were taken using a medical microscope, the model would have been trained for this exact device. This would result in worse accuracy when the test data comes from a different laboratory. Secondly, the masks used in the study have been created by various annotators, lacking professional medical knowledge, which contributes to inconsistency in the masking. While the implementation of a reliable automatic segmentation tool would have mitigated this issue, the inconsistent quality of the photos refrained us from that. Another limitation arises from the time constraint; with more time, additional features from the 7-point checklist could have been implemented, resulting in a better overall performance of the model. Unfortunately, because of their high complexity, we could not implement any of them to a satisfactory level. Furthermore, expanding the dataset would be beneficial so that it includes more examples of the less represented diseases such as MEL and SCC. This would facilitate the distinction between cancerous diseases and open the possibility for building a multi classification models that can give specific diagnosis for unseen lesions.

Table 2. Performance metrics analysis for proposed classification methods

Metric	KNN	Logistic Regression	Random Forest	Random Forest + Logistic Regression
Accuracy	65.16%	70.08%	73.36%	74.18%
Recall	77.78%	79.86%	84.72%	84.03%
Precision	59.5%	64.5%	66.4%	67.7%

The performance of the proposed classification methods was evaluated based on three essential metrics: Accuracy, Recall, and Precision. Table 2 below presents a comprehensive comparison of these metrics across the four models: K-Nearest Neighbors (KNN), Logistic Regression, Random Forest, and an ensemble approach combining Random Forest with Logistic Regression.

From the results, it is evident that the ensemble model (Random Forest + Logistic Regression) demonstrates the highest accuracy at 74.18%, closely followed by the standalone Random Forest model at 73.36%. Logistic Regression and KNN lag with accuracies of 70.08% and 65.16%, respectively. The superior performance of the ensemble model highlights the effectiveness of integrating two distinct learning algorithms, which capture diverse aspects of data distribution. In terms of recall, the Random Forest model performs best with 84.72%, indicating its robustness in identifying positive instances correctly. The ensemble model follows closely with a recall of 84.03%, suggesting a minimal compromise in sensitivity when combined with Logistic Regression. Logistic Regression and KNN show moderate recall values of 79.86% and 77.78%, respectively. Examining the precision metric reveals that the ensemble model again outperforms the others, achieving a precision of 67.7%. This suggests better handling of false positives compared to other models. The Random Forest model achieves a slightly lower precision of 66.4%, while Logistic Regression and KNN exhibit noticeably lower precision values of 64.5% and 59.5%, respectively.

Overall, the ensemble model appears to provide the most balanced performance across all metrics. However, Random Forest alone demonstrates excellent recall, making it potentially suitable for applications where sensitivity is more critical than precision. KNN performs poorly across all metrics, indicating it may not be well-suited for this classification task.

5 Conclusion

Building upon existing systems for evaluating skin diseases, we successfully developed a simple yet effective model capable of detecting cancerous lesions with high recall. Our approach leverages a two-way evaluation process involving a Logistic Regression model and a Random Forest model, enhancing the overall reliability of the final diagnosis. By combining these two models, the system is better equipped to differentiate between benign and malignant lesions, providing a more confident and reliable assessment. However, while the model shows promising results in identifying malignant lesions, its ability to correctly classify benign lesions remains limited. The probability of mistakenly classifying a benign lesion as cancerous is still relatively high, making it not much better than random guessing in those cases. To improve the robustness and accuracy of the model, future work could focus on integrating more advanced deep learning techniques, such as Convolutional Neural Networks (CNNs) and Transformer-based architectures, which are known for their ability to extract complex patterns and features from medical images. Additionally, incorporating a larger and more diverse dataset, along with data augmentation techniques, could enhance the model's generalization capability.

References

1. Stolz, W., Kunz, M.: ABCD rule — dermoscopedia (2023). Accessed 20 Dec 2023
2. Giulycalabrese, Argenziano, G., De Rosa, A., Russo, T.: Seven point checklist— dermoscopedia (2023). Accessed 20 Dec 2023
3. Kaur, R., GholamHosseini, H., Lindén, M.: Advanced deep learning models for melanoma diagnosis in computer-aided skin cancer detection. Sensors **25**(3), 594 (2025)
4. Nguyen, A.T.P., Jewel, R.M., Akter, A.: Comparative analysis of machine learning models for automated skin cancer detection: advancements in diagnostic accuracy and AI integration. Am. J. Med. Sci. Pharm. Res. **7**(01), 15–26 (2025)
5. Bogne Tchema, R., Polycarpou, A.C., Nestoros, M.: Skin cancer classification using machine learning. Multimed. Tools Appl. **84**(6), 3239–3256 (2025)
6. Rashad, N.M., Abdelnapi, N.M.M., Seddik, A.F., Sayedelahl, M.A.: Automating skin cancer screening: a deep learning. J. Eng. Appl. Sci. **72**(1), 6 (2025)
7. Alsultan, S.A.B.: Deep learning-based automated diagnosis of skin cancer from thermoscopic images. J. Univ. Babylon Pure Appl. Sci. **32**(4), 199–212 (2024)
8. Ullah, A., et al.: Analysis of automated melanoma detection utilizing machine learning and deep learning techniques: a review. Int. J. Image Graph., 2750012 (2024)
9. Musa, U.I., Roy, A., Musa, M.I., Babani, U.M., Musa, A.I.: Skin cancer detection using machine learning. In. J. Prev. Med. Health (IJPMH) **5**(1), 10–16 (2024)
10. Nasiruddin, Md., et al.: Optimizing skin cancer detection in the USA healthcare system using deep learning and CNNS. Am. J. Med. Sci. Pharm. Res. **6**(12), 92–112 (2024)
11. Dubey, P., Bande, P., Shukla, R., Kshatri, S.S., Nayak, M., Sharma, R.: AI-powered dermatology: revolutionizing skin cancer detection through machine learning algorithms. In: 2024 International Conference on Cybernation and Computation (CYBERCOM), pp. 455–460. IEEE (2024)
12. Vayadande, K., Mirajkar, O., Mhetre, S., Mehta, H., Markande, S., Mohol, S.: The future of skin cancer detection: machine learning and convolutional neural networks. In: 2024 5th International Conference on Data Intelligence and Cognitive Informatics (ICDICI), pp. 1282–1290. IEEE (2024)

13. Chiu, T.-M., Li, Y.-C., Chi, I.-C., Tseng, M.-H.: AI-driven enhancement of skin cancer diagnosis: a two-stage voting ensemble approach using dermoscopic data. Cancers **17**(1), 137 (2025)
14. Surendren, D., Sumitha, J.: Skin cancer detection using convolutional neural network models: VGG-16, VGG-19, ResNet and DenseNet. In: 2024 5th International Conference on Data Intelligence and Cognitive Informatics (ICDICI), pp. 1254–1258. IEEE (2024)
15. Pacheco, A.G.C., et al.: PAD-UFES-20: a skin lesion dataset composed of patient data and clinical images collected from smartphones. Data Brief **32**, 106221 (2020)

Quantum Diagnosis: Revolutionizing Skin Tumors Classification with Quantum Machine Learning

B. U. Karthik[1]($\boxtimes$) iD, M. S. Mrutyunjaya[2] iD, and K. R. Varalakshmi[1] iD

[1] Department of CSE, R L Jalappa Institute of Technology, Bengaluru, Karnataka, India
karthikbu1994@gmail.com

[2] Department of CSE (Data Science), R L Jalappa Institute of Technology, Bengaluru, Karnataka, India

Abstract. Skin tumour classification is an important challenge in medical diagnostics, as it requires precise and effective methods for early identification and treatment planning. This paper explores the application of quantum machine learning (QML) to enhance the recognition of skin tumours. Using the computational power of quantum computation, we suggest a hybrid approach that merges quantum variational circuits with classical deep learning architecture. By using quantum-enhanced operations, the framework can differentiate tumours into benign and malignant groups by processing dermoscopic images and extracting features. In comparison with classical techniques, our model's accuracy is highly accurate and its computing costs are significantly reduced when tested on ISIC datasets. The quantum variational circuits not only provide better feature representation but also enable efficient processing of high-dimensional image data. By using the proposed procedure, we achieved a precision of approximately 98.57% and are proficient in distinguishing between benign and malignant tumours.

Keywords: Skin Tumor Classification · Image Denoising · Quantum Machine Learning

1 Introduction

The traditional image noise reduction algorithms [1–3] aim to minimize noise while preserving crucial images, they frequently face significant drawbacks. Spatial domain methods such as mean and median filtering require attention to pixel values within a small neighborhood. The image is made smoother by mean filtering, which averages the intensities of pixels while masking important features like textures and boundaries. This technique also removes shadows. Although it is more efficient at maintaining edges, median filtering [4, 5] has difficulty capturing fine details and can be computationally challenging for larger neighborhoods. However. A transform domain can be utilized by transforming an image into a different domain using techniques like Fourier and wavelet transforms [6–8]. Denoising using fourier transform is a suitable method for

J. Shreyas et al. (Eds.): CODE-AI 2025, CCIS 2689, pp. 53–64, 2026.
https://doi.org/10.1007/978-3-032-19318-6_6

controlling periodic noise, but it poses challenges when dealing with localized or non-periodic noise due to its global nature. The use of wavelet transform techniques [9, 10] can lead to artefacts if the thresholding is not appropriately adjusted, but it also provides multi-resolution analysis and noise reduction at various scales. The sensitivity of these techniques to the noise model and frequent need for parameter changes mean that they may not be as flexible as other methods in real-world scenarios.

Non-Local Means (NLM) [11–13] is a more advanced approach that utilizes the self-similarity of image patches to average similar patches and maintain detail. It also allows for noise reduction. For this reason, NLM requires careful parameter tuning and is computationally expensive, especially for broad search windows. The preservation of sharp edges in images is achieved through total variation (TV) denoising, which is a popular method for reducing noise. While TV denoising can preserve edges, it often results in too soft textures, sacrificing detail in low-contrast areas and creating blocky artifacts in smooth areas.

Despite the time-consuming process, Machine Learning [14–16] and Non Linear image denoising techniques [17, 18] can still be used to classify skin tumours with greater accuracy. The limitations of traditional denoising techniques are numerous. They frequently use strong presuppositions about the distribution of noise, such as Gaussian noise that may not always be true in any given situation. The importance of preserving detail and minimizing noise in these techniques often results in conflicting results; conservative denoising may leave noise, while aggressive noise reduction can cause blurring or loss of signification. Several of these methods are not flexible enough to accommodate different noise levels or patterns in an image, which limits their usefulness in diverse and complex environments. Furthermore, this leads to computational complexity, particularly for techniques like NLM that attempt to identify similar patches across the image. The use of precise parameter adjustment makes them more challenging to navigate, as improper settings can severely impact performance. These drawbacks highlight the need for more sophisticated methods, such as deep learning based techniques which can recognise very complex noise patterns and adapt well to different types of noise and image properties, giving better results in denoising with less compromise.

Quantum machine learning (QML) [19, 20] is a potential innovation in image denoising, with many significant improvements over existing apises. Among QML's key features are its capacity to leverage the inherent parallelism and exponential computational flexibility of quantum computing. There are classical limitations in hardware used to apply traditional image denoising techniques such as Gaussian filtering, median filtering and more complex deep learning methods like convolutional neural networks (CNNs). Due to the complexity of noise patterns and high-dimensional data, these algorithms often encounter difficulties that require significant processing time and resources to achieve satisfactory denoising outcomes. In comparison to other methods, QML algorithms can perform calculations on multiple states simultaneously, resulting in improved processing and analysis capabilities for high-dimensional data.

The use of QML enables image denoising to be more efficiently achieved when dealing with nonlinear and compliant noise. This is an added benefit. Typically, conventional methods of denoising are deterministic and predetermine the noise (e.g. Gaussian or salt-and-pepper Noise). Moreover, QML techniques offer substantial advantages in terms of processing time and speed. Quotient component analysis (QPCA) and quantum support vector machines (QCVM) are quantum algorithms that can extract crucial features from noisy images at much faster speeds than classical methods. Rapid denoising, which is essential for tasks like autonomous driving, medical imaging, or surveillance, can be achieved with this acceleration in real-time or near-real-life situations.

A quantum neural network (QNN) is a type of classical neural networks that uses quantum data to create artificial neural nets, known as quantum gates. Image denoising involves the use of the QNN to learn noise distributions and reconstruct pristine images. Quantum states enable QNNs to quickly process large amounts of data, which could potentially reveal intricate patterns that conventional networks may not be able to detect.

2 Literature Survey

Fan fan et al. [21] developed and integrated quantum-classical convolutional neural network (CNN) with classical image processing layers and quantum variational circuits for feature extraction. The model was then tested on the ISIC 2020 dataset and with an accuracy of 97%. The quantum layers utilized quantum entanglement and superposition to process high-dimensional data features, which resulted in reduced overfitting and improved generalization.

Sofana Reka et al. [22] examined the feasibility of using quantum support vector machines to perform binary classification tasks related to benign and malignant tumours. With the use of quantum-enhanced kernels, the QSVM effectively mapped input data into higher-dimensional spaces and was able to achieve classification accuracy of 95% on smaller datasets.

Dheeraj peddireddy et al. [23] introduced a variational quantum classifier tailored for multi-class classification of skin lesions. The approach utilized amplitude encoding to represent image features in quantum states, significantly reducing the input dimension. The variational quantum circuit iteratively optimized parameters to minimize a cost function, leading to an accuracy comparable to classical models.

Lee et al. [24] has proposed quantum generative adversarial networks (QGAN) for skin tumour classification. Through the creation of synthetic skin lesion images, the QGANs were able to enhance the training dataset and improve the classifier's performance by approximately 5%. Hence the discriminator was left in its classical form, while the quantum generator employed a parameterized quantum circuit; thus an efficient hybrid framework.

He et al. [25] have also used selection techniques using quantum features to identify the most relevant features for skin tumour classification. The research employed quantum annealers to decrease the dimensionality of input features while maintaining vital diagnostic information. Not only did this approach improve the interpretability of the model, but it also saved a lot of training time. The findings indicated the potential of quantum computing in enhancing feature selection processes for large datasets of medical data.evidence.

Zhang et al. [26] conducted experiments using quantum transfer learning by pre-training quantum models on generic medical imaging datasets and fine-tuning them for skin tumour classification. By utilizing pre-trained quantum encoders, this technique was able to extract transferable features and reduce training time while maintaining accuracy. In this study, it was shown that the use of transfer learning could be useful in addressing the lack of annotated medical data.

3 Proposed Design

The most severe form of skin cancer is melanoma. Early detection has a significant impact on patient outcomes. ML methods have been developed to diagnose skin lesions, but they may be too complex to process large amounts of data or require significant amounts in order to classify them correctly. By utilizing quantum techniques like entanglement and superposition, quantum machine learning could potentially improve the perforation of these models.

3.1 Dataset

The International Skin Imaging Collaboration (ISIC) [27] datasets have emerged as a key repository for machine learning researchers working on medical image analysis, particularly in the areas of skin cancer diagnosis and malignancy evaluation. This dataset contains around 24,000 images and we've chosen 4000 images for training and 2000 images for testing.

3.2 Design Methodology

The image depicts a comprehensive pipeline for a quantu machine learning (quantum ML) model, which is designed to process image data from input to the final prediction. Every phase of the workflow has a distinct function, ensuring that data is progressively transformed and enhanced for accurate prediction. The Fig. 1 represents the block diagram of the proposed skin tumour classification.

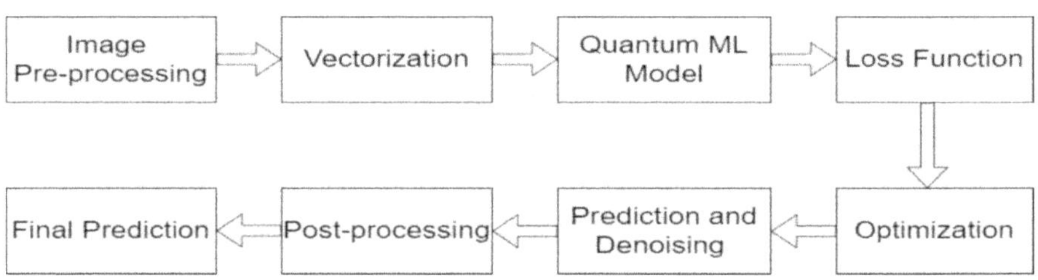

Fig. 1. Workflow of the proposed method.

3.2.1 Image Pre-processing

The first step of the procedure is image pre-processing, which involves cleaning and preparing the raw input data. Images from the dataset may be noisy, uneven in quality, and may contain irrelevant details. Pre-processing includes actions such as:

- Normalization: Normalization is the process of bringing pixel values into a uniform range, (0 to 1).
- Noise Reduction: Noise reduction is the process of eliminating undesirable components with filters or algorithms.
- Data augmentation: Using manipulations like rotation, flipping, or cropping to increase the diversity of a dataset. This phase guarantees that the input data is consistent and suitable for use in the computational pipeline.

The steps involved in denoising the image is:

Let the noisy image $I_{noisy} \in R_{mxn}$ be the matrix that represents the pixels intensities. The aim is to recover the original image I_{clean}

Normalization: Normalize the pixel values to [0, 1] (Figs. 2 and 3).

$$I_{normalized} = \frac{Inoisy - \min(Inoisy)}{\max(Inoisy) \text{ - } \min(Inoisy)} \tag{1}$$

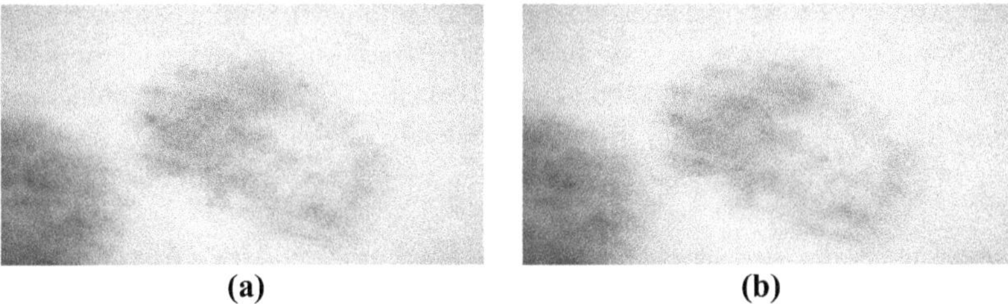

(a) (b)

Fig. 2. (a) Input Image. (b) Denoised Image

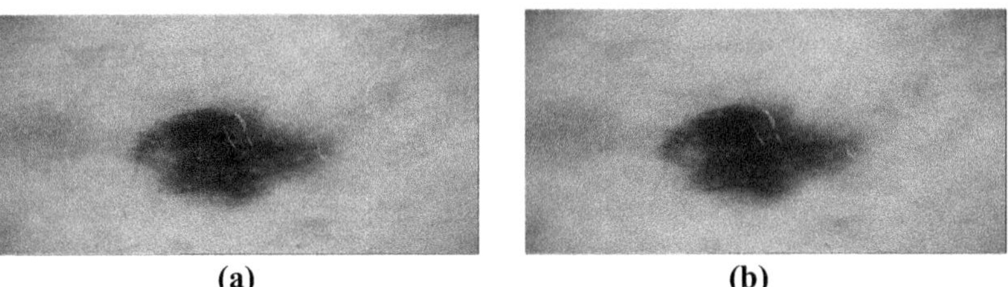

(a) (b)

Fig. 3. (a) Input Image. (b) Denoised image

3.2.2 Vectorization

Following pre-processing, vectorization is used to convert the image data into a numerical representation. This stage transforms spatial data, either 2D or 3D, into machine-readable, structured formats like tensors or vectors. This stage employs the following key techniques:

- Feature extraction: It includes locating important aspects of an image, such as edges and forms.

- Dimensionality Reduction: Simplifying data by removing duplicate or extraneous aspects. Vectorization is critical for the quantum ML model's ability to comprehend and analyze image data effectively.

Flatten the image to a vector x ∈ Rmn
Quantum feature mapping
The classical data x is mapped to a quantum state. A common approach is using a quantum feature map $\phi(x)$:

$$\phi(x) = \exp(ixTZx)|0\rangle \tag{2}$$

Where:

Z is a parameterized Hermitian matrix.

The resulting state encodes classical image data into a higher-dimensional quantum Hilbert space.

3.2.3 Quantum Machine Learning Model

Following vectorization, the data is fed into the quantum ML model, which processes it using quantum computing principles. In contrast to traditional machine learning models, quantum machine learning (ML) makes use of quantum features like entanglement and superposition to expedite and improve the efficiency of complex operations. The main function of this pipeline step is to extract significant patterns and relationships from the data.

A quantum model, such as a Parameterized Quantum Circuit (PQC) or Quantum Neural Network (QNN), is employed. In this proposed paper, we've used the Parameterized Quantum Circuit (PQC).

Let $U(\theta)$ represent a parameterized quantum circuit, where $\bar{\theta}$ are learnable parameters.

The quantum state evolves as:

$$|\Psi(\Theta)\rangle = U(\theta)\phi(X) \tag{3}$$

For denoising, the circuit is trained to learn a mapping from noisy inputs to clean outputs by minimizing a loss function.

3.2.4 Loss Function

As the model makes predictions, the loss function evaluates its performance by comparing predicted values to actual target values. The loss function quantifies the inaccuracy and gives feedback to help improve the model. Loss function examples include: MSE stands for Mean Squared Error in regression problems. For classification problems, cross-entropy loss is used. This phase assures that the model knows where it is making problems and how to rectify them.

The loss function measures the difference between the model output and the ground truth (clean image). A typical loss is the Mean Squared Error (MSE):

$$L(\Theta) = \frac{1}{mn} \sum_{i=1}^{mn} (y_i - f_\theta(x_i))^2 \tag{4}$$

Where:

y$_i$ is the clean pixel value

f$_\theta$(x$_i$) is the model's predicted output for x$_i$, which is measured from quantum observables:

$$f_\theta(x_i) = \langle \varphi(\Theta)|\hat{O}|\varphi(\Theta)\rangle \tag{5}$$

Here $\hat{O}$ is an observable, such as the Pauli-Z operator

3.2.5 Optimization

The optimization stage improves the model's parameters based on feedback received from the loss function. Optimization algorithms, such as Gradient Descent or its derivatives, iteratively update the model to reduce loss. Hybrid optimization techniques, which combine classical and quantum methods for rapid convergence, may be used in quantum machine learning.

Classical optimization techniques, such as gradient descent, are used to adjust θ to minimize L(θ). Gradients can be computed using the parameter-shift rule:

$$\frac{\partial L}{\partial \Theta k} = \frac{1}{2} [L(\Theta k + \frac{\pi}{2}) - L(\Theta k - \frac{\pi}{2})] \tag{6}$$

Once trained, the model is used to predict denoised pixel values:

$$\hat{I}_{clean} = f\Theta \, (I_{noisy}) \tag{7}$$

The predicted vector $\hat{I}_{clean}$ is reshaped back to the image format.

3.2.6 Prediction and Denoising

Following optimization, the model generates predictions, some of which may still be inaccurate or contain noise. Specialized algorithms are used in the denoising process to clean the predictions, guaranteeing dependable and superior outcomes.

The Gaussian blur filter is employed to smooth the edges and details of denoised images.

$$I_{final} = Postprocess(\hat{I}_{clean}) \tag{8}$$

After obtaining the denoised image, optimization technique is applied which updates θ(Fig. 4).

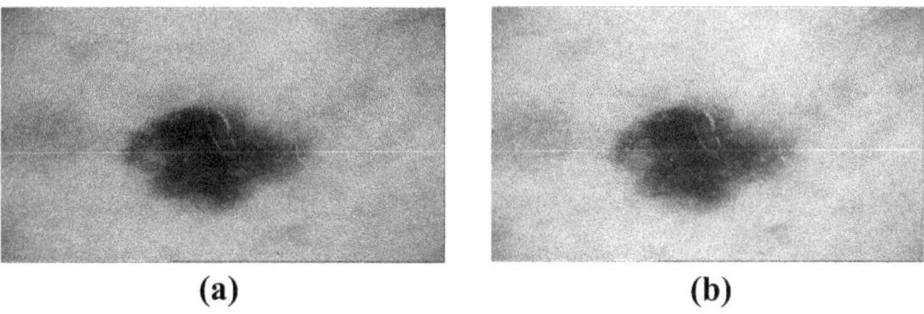

(a)   **(b)**

Fig. 4. (a) Input Image. (b) Denoised Image

3.2.7 Post-processing

In the post-processing stage, the modified predictions are formatted or adjusted to meet specific criteria. This could include: Transforming forecasts into formats that are readable by humans, combining the outcomes of ensemble models, implementing domain-specific modifications to improve interpretability.

3.2.8 Final Prediction

The Final Prediction, a precise and high-quality output, is the pipeline's end product. The outcome can be used for a number of purposes, including pattern recognition, object detection, and medical diagnostics.

The output of the VQC is interpreted by the post-processing layer. Finally the skin lesion images are classified using the below formula.

For a binary classification:

$$\hat{y} = \begin{cases} 1 & \text{if } P_1 > 0.5 \\ 0 & \text{if } p_1 \leq 0.5 \end{cases} \tag{9}$$

The steps outlined above describe the process involved in classifying skin tumors.

4 Results and Discussion

The proposed Quantum Machine Learning (QML) model for skin tumor classification had an excellent 98.57% accuracy in discriminating between benign and malignant skin lesions. With approximately 99% of the cases resulting in correct classification, this level of performance indicates that the quantum-enhanced model was quite successful in evaluating the complex features found in the skin lesion images. The results are displayed in both the table and graphs (Fig. 5).

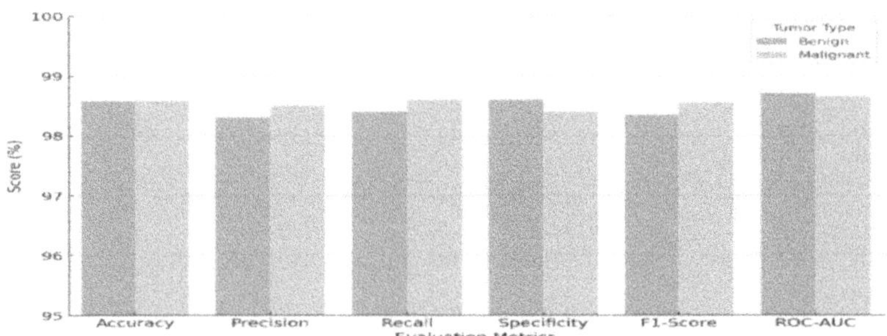

Fig. 5. Comparison of Different metrics

The bar chart depicts evaluation metrics for the classification of benign and malignant tumors, with an overall accuracy of 98.57%. Accuracy, precision, recall, specificity, F1 score, and ROC-AUC are among the criteria that were considered for evaluation, and they all show consistent performance above 95% for both tumor types. A reliable classification procedure is shown by the accuracy metric's approximation to the total accuracy. The model's strong precision and recall scores demonstrate its efficacy in reducing false positives and false negatives. The model's strong performance for both benign and malignant instances is reinforced by its strong alignment with specificity, which demonstrates its capacity to accurately identify negatives. The F1 score, which is a harmonic mean of precision and recall, contributes to the classifier's balanced performance. Finally, the ROC-AUC values, which measure the model's ability to differentiate across classes, are nearly error-free, indicating impressive discriminative power. These outcomes demonstrate the classifier's accuracy and balance, which makes it a very useful tool for tumor classification tasks (Fig. 6) (Table 1).

Table 1. Comparison of different Metrics

Metrics	Benign (%)	Malignant (%)
Accuracy	98.57	98.57
Precision	98.30	98.50
Recall	98.40	98.60
Specificity	98.60	98.40
F1-Score	98.35	98.55
ROC-AUC	98.70	98.65

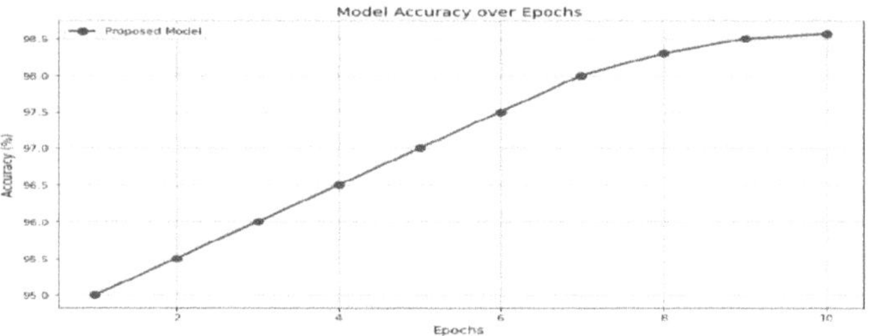

Fig. 6. Accuracy Plot

The graph illustrates the accuracy development of the proposed model over ten training epochs, highlighting its steady improvement over time. The model shows a consistent and progressive improvement in performance, reaching 98.5% accuracy by Epoch 10 after starting at an initial accuracy of roughly 95.0% at Epoch 1. The curve's linearity and smoothness imply consistent training devoid of overfitting or underfitting. The model appears to be quite dependable and appropriate for its intended use based on the final accuracy level. Overall, study results support the proposed model's effectiveness and resilience, which makes it an appealing choice for real-world use.

From the graph and table, we can conclude that

- The Quantum Machine Learning (QML) model outperforms other measures for skin tumor classification, with an overall accuracy of 98.57%.
- The model is good at avoiding false positives and false negatives, as seen by its very high precision and recall for both benign and malignant tumors.
- For both classes (benign and malignant), the model's precision and recall are well-balanced, as evidenced by the high F1 scores.
- The model's high specificity ensures that it minimizes misclassifications for both benign and malignant tumor types.
- The ROC-AUC values are also high for both benign and malignant lesions, indicating that the model can efficiently distinguish between the two groups.

5 Conclusion and Future Enhancement

The classification of skin tumors using the quantum machine learning model is extremely effective, with exceptional metrics all around. The model has a 98.57% accuracy rate and good precision, recall, F1-score, specificity, and ROC-AUC values, making it appropriate for classifying benign and malignant tumors. It exhibits an outstanding potential for early detection of skin cancer, and its excellent performance implies that it could be a useful tool in clinical settings for dermatologists and healthcare professionals. In this proposed design, we implemented binary classification of skin tumor using quantum machine learning and in the Future we'll try to implement the multi-class classification of skin tumour lesions. Future enhancements for the quantum machine learning model could focus on expanding its capability to handle multi-class classification of skin tumor lesions, enabling it to distinguish between various types of tumors beyond benign and malignant categories. Additionally, incorporating larger and more diverse datasets could improve the model's robustness and generalizability across different populations. Integrating explainable AI (XAI) techniques would provide interpretability, helping healthcare professionals better understand the model's predictions. Further advancements could also explore hybrid quantum-classical approaches to optimize performance and scalability. Finally, real-time deployment in clinical settings with seamless integration into existing diagnostic workflows could enhance its practical usability for dermatologists and healthcare providers.

Declaration. The authors declare that there is no competing interests.

References

1. Goyal, B., Dogra, A., Agrawal, S., Sohi, B.S., Sharma, A.: Image denoising review: from classical to state-of-the-art approaches. Inf. Fusion **55**, 220–244 (2020). ISSN 1566-2535
2. Reddy, P.L., Pawar, S.: Multispectral image denoising methods: a literature review. Mater. Today Proc. **33**, Part 7, 4666–4670 (2020). ISSN 2214-7853
3. Fan, L., Zhang, F., Fan, H., Zhang, C.: Brief review of image denoising techniques. Vis. Comput. Ind. Biomed. Art, 1–12 (2019)
4. Tian, C., Fei, L., Zheng, W., Xu, Y., Zuo, W., Lin, C.-W.: Deep learning on image denoising: an overview. Neural Netw. **131**, 251–275 (2020). ISSN 0893-6080

5. Shahdoosti, H.R., Rahemi, Z.: Edge-preserving image denoising using a deep convolutional neural network. Signal Process. **159**, 20–32 (2019). ISSN 0165-1684

6. Singh, H., Kommuri, S.V.R., Kumar, A., Bajaj, V.: A new technique for guided filter based image denoising using modified cuckoo search optimization. Expert Syst. Appl. **176**,114884 (2021). ISSN 0957-4174

7. Ni, G., et al.: Contrast enhancement of spectral domain optical coherence tomography using spectrum correction. Comput. Biol. Med. **89**, 505–511 (2017). ISSN 0010-4825

8. Choudhary, P., Singhai, J., Yadav, J.S.: Skin lesion detection based on deep neural networks. Chemom. Intell. Lab. Syst. **230**,104659 (2022). ISSN 0169-7439

9. Elashiri, M.A., Rajesh, A., Pandey, S.N., Shukla, S.K., Urooj, S., Lay-Ekuakille, A.: Ensemble of weighted deep concatenated features for the skin disease classification model using modified long short term memory. Biomed. Signal Process. Control **76**, 103729 (2022). ISSN 1746-8094

10. Qian, S., Ren, K., Zhang, W., Ning, H.: Skin lesion classification using CNNs with grouping of multi-scale attention and class-specific loss weighting. Comput. Methods Programs Biomed. **226**, 107166 (2022). ISSN 0169-2607

11. Iqbal, I., Younus, M., Walayat, K., Kakar, M.U., Ma, J.: Automated multi-class classification of skin lesions through deep convolutional neural network with dermoscopic images. Comput. Med. Imaging Graph. **88**, 101843 (2021). ISSN 0895-6111

12. Kang, S., Cho, S., Kang, P.: Multi-class classification via heterogeneous ensemble of one-class classifiers. Eng. Appl. Artif. Intell. **43**, 35–43 (2015). ISSN 0952-1976

13. Weizheng, X., Chenqi, X., Zhengru, J., Yueping, H.: Digital image denoising method based on mean filter. In: 2020 International Conference on Computer Engineering and Application (ICCEA), Guangzhou, China, pp. 857–859 (2020)

14. Yulong, D., et al.: Wavelets and curvelets transform for image denoising to damage identification of thin plate. Results Eng. **17**, 100837 (2023). ISSN 2590-1230

15. Liu, X., Wu, Z., Wang, X.: Validity of non-local mean filter and novel denoising method. Virtual Real. Intell. Hardw. **5**(4), 338–350 (2023). ISSN 2096-5796

16. Shivamurthaiah, M.M., Shetra, H.K.K.: Non-destructive machine vision system based rice classification using ensemble machine learning algorithms. Recent Adv. Electr. Electron. Eng. **17**(5), e100723218597 (2024)

17. Karthik, B.U., Muthupandi, G.: SVM and CNN based skin tumour classification using WLS smoothing filter. Optik **272**, 170337 (2023). ISSN 0030-4026

18. Kaur, R., Karmakar, G., Imran, M.: Impact of traditional and embedded image denoising on CNN-based deep learning. Appl. Sci. **13**(20), 11560 (2023)

19. Biamonte, J., Wittek, P., Pancotti, N., Rebentrost, P., Wiebe, N., Lloyd, S.: Quantum machine learning. Nature **549**, 195–202 (2017)

20. Kharsa, R., Bouridane, A., Amira, A.: Advances in quantum machine learning and deep learning for image classification: a survey. Neurocomputing **560**, 126843 (2023). ISSN 0925-2312

21. Fan, F., Shi, Y., Guggemos, T., Zhu, X.X.: Hybrid quantum-classical convolutional neural network model for image classification. IEEE Trans. Neural Netw. Learn. Syst. **35**(12), 18145–18159 (2024)

22. Reka, S.S., Karthikeyan, H.L., Shakil, A.J., Venugopal, P., Muniraj, M.: Exploring quantum machine learning for enhanced skin lesion classification: a comparative study of implementation methods. IEEE Access **12**, 104568–104584 (2024)

23. Peddireddy, D., Bansal, V., Aggarwal, V.: Classical simulation of variational quantum classifiers using tensor rings. Appl. Soft Comput. **141**, 110308 (2023). ISSN 1568-4946

24. Gao, X., Zhang, Z.-Y., Duan, 1.-M.: A quantum machine learning algorithm based on generative models. Sci. Adv. **4**(12) (2018)

25. He, Z., Li, L., Huang, Z.: Quantum-enhanced feature selection with forward selection and backward elimination. Quantum Inf. Process. **17**, 154 (2018)
26. Zhang, S., Wang, A.: Quantum transfer learning algorithm based on parameterized quantum circuit. In: 2024 5th International Seminar on Artificial Intelligence, Networking and Information Technology (AINIT), Nanjing, China, pp. 1883–1887 (2024)
27. Gutman, D., et al.: Skin lesion analysis toward melanoma detection: a challenge at the international symposium on biomedical imaging (ISBI) 2016, hosted by the international skin imaging collaboration (ISIC). eprint arXiv:1605.01397 (2016)

Flood Depth Prediction Using Temporal Attention Recurrent Graph Convolutional Neural Network for IoT-Enabled Smart Cities

Nishu Gupta[1]([envelope]) [ORCID], Marcos Xosé Alvarez Cid[2] [ORCID], and Mohammad Derawi[2] [ORCID]

[1] VTT Technical Research Centre of Finland Ltd., 90571 Oulu, Finland
`nishugupta@ieee.org`
[2] Norwegian University of Science and Technology, 2815 Gjøvik, Norway
`mohammad.derawi@ntnu.no`

Abstract. In many cities across the world, flood has become an increasingly significant issue and accurate flood forecasting is crucial to addressing this challenge. Nevertheless, the lack of sufficient and balanced data coupled with the limited performance of forecasting models has added to the complexity and unpredictability of flood predictions. To enhance the accuracy and dependability of flood prediction, internet of things (IoT) based systems are utilized to gather data on floods and related factors in smart cities. In this manuscript, flood depth prediction using temporal attention recurrent graph convolutional neural network for IoT-enabled smart cities (FDP-TARGNN-IoT) is proposed. Initially, the input data are collected from disasters and intensity dataset. Then, the collected data are fed to TARGNN which is used to predict the flood depth. To enhance the accuracy, parrot optimizer is utilized to optimize TARGNN parameters, ensuring accurate detection of the flood levels. The proposed methodology demonstrates significant improvements in the evaluated metrics.

Keywords: convolutional neural network · flood depth prediction · internet of things · parrot optimizer · smart cities

1 Introduction

The internet of things (IoT) and big data analytics are widely recognized key components in the implementation of smart city services, sparking global interest. Cloud computing and sensor networks are among the many technologies integrated into the IoT. It enables various functions like connecting smart devices, sensing exterior environments, collecting and analyzing data, generating insights and facilitating effective communication with humans or other digital systems [1, 2]. These technologies are widely used in fields such as disaster management, smart cities, advanced healthcare and sophisticated early warning systems. As a crucial component of advanced flood forecasting systems, the IoT is widely used for flood monitoring, prediction and management. It enables accurate flood forecasting and provides early warnings to communities [3]. Flood risk is turning into a significant urban issue. Floods are becoming more common and serious, and cities

J. Shreyas et al. (Eds.): CODE-AI 2025, CCIS 2689, pp. 65–75, 2026.
https://doi.org/10.1007/978-3-032-19318-6_7

with lakes are particularly vulnerable to the possible effects of climate change. Urban areas are equipped with depth meters for flood water level monitoring, and integrated systems such as automatic weather stations and hydro-meteorological sensor networks [4]. Accurate flood depth forecasting is vital for developing long-term flood prevention strategies in smart cities. These technologies collectively generate vast amounts of data for improved flood management [5]. The city of Gjøvik in Norway has a population of around 30,000 residents on a total area of 671 km^2. The major climate risks and impacts faced are increased rainfall, extreme weather events and flood events. The city has been conducting a rehabilitation program for the neighbouring Lake Mjøsa. Gjøvik is also completing a new stormwater plan and a climate program for climate change mitigation. The novelty of the model uniquely applies temporal attention recurrent graph convolutional neural network (TARGNN) technique to forecast flooding events enhancing prediction accuracy by combining multiple models. The use of TARGNN allows the model to harness the strengths of different individual models resulting in a more robust and reliable prediction system compared to single model approaches and ultimately supporting faster and more accurate classification. Major contributions of this research work are summarized below.

- We propose FDP-TARGNN-IoT model to enhance the accuracy and reliability of flooding predictions, leveraging dataset and TARGNN techniques for advanced, proactive flood forecasting and early warning systems.
- To improve the disaster preparation and response in urban areas ensuring more timely and effective management of flood risks.
- We emphasize critical time dependent factors thereby improving the robustness of the flood depth predictions over time.
- The proposed optimization algorithm is designed using parrot optimizer (PO) to provide a novel way to optimize the weight parameters of the TARGNN. This approach is inspired by natural processes and aims to find the optimal solution for prediction tasks.

The rest of this paper is structured as follows. Section 2 explains the literature survey, Sect. 3 discusses the materials and procedures employed, Sect. 4 describes the results with discussion and Sect. 5 presents the conclusion.

2 Literature Review

Among the recent research works on deep learning (DL)-based flood depth prediction, some of the latest investigations are assessed in this section. Al Abdouli, K.M., et al. [6] have presented real-time flood forecasting in Bhutan's Amo Chhu river utilizing machine learning (ML) and IoT technologies. To enhance flood forecasting and reducing the impact of natural disasters on vulnerable populations it employs support vector classifier (SVM), random forest (RF) and gradient boosting integrated with IoT devices for improved accuracy and early warnings. It provides low mean absolute error (MAE) and high root mean squared error (RMSE).

Prakash, C., et al. [7] have presented FLOODALERT: an IoTs derived real-time flashflood tracking and forecasting scheme. Gated recurrent unit (GRU) and extreme gradient boosting (XGBoost) methods are employed to analyze and categorize data for forecasting flood situations. Each technique categorizes flood conditions into four alert levels: Orange (Danger), Red (Higher Flood Level), Yellow (caution), or No Alert. It provides low RMSE and high R2-Score.

Thankappan, J., et al. [8] have presented adaptive momentum back propagation approach for flood prediction with management in the IoT networks, where adaptive momentum and back propagation (AM-BP) method is used for flood control and prediction. It provides low R2-Score and high MAE.

Hashemi-Beni, L., et al. [9] have presented a low-cost IoT-based DL method of water gauge measurement for flood monitoring. It utilizes Mask-RCNN to accurately segment gauges from images, even in the presence of distortions. It provides low MAE and high R2-Score.

Fernandes Jr, F.E., et al. [10] have presented a river flooding detection system based on DL and computer vision (CV). The system estimates the river's water level by first performing semantic segmentation of the river's water surface using deep neural networks (DNNs).

3   Proposed Methodology

This section discusses the FDP-TARGNN-IoT methodology which aims to enhance automate flood depth prediction by collecting data from the disasters and intensity dataset through IoT-enabled sensors in smart cities. The TARGNN is used to capture both spatial and temporal dependencies to predict flood severity levels. To enhance the accuracy, model parameters are optimized using the PO, ensuring precise flood depth predictions. The block diagram of proposed FDP-TARGNN-IoT is represented in Fig. 1.

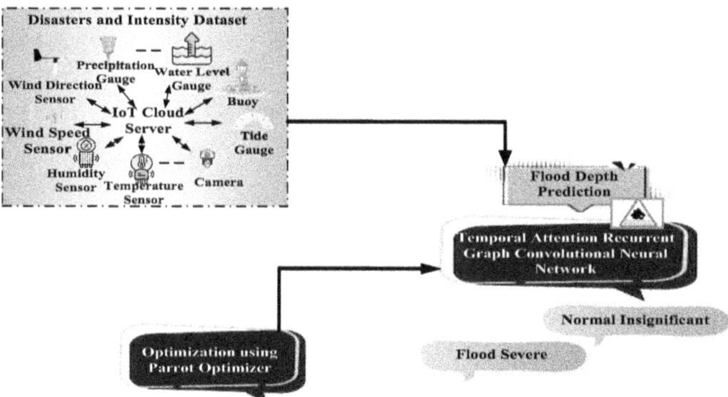

Fig. 1. FDP-TARGNN-IoT method

3.1   System Description

An IoT-based food monitoring system is developed to address data collection challenges, reduce real-time collection complexity and enhance the efficiency of hydrological information sharing and processing. We focus on the municipality of Gjøvik in Norway which is characterized by high-tech industry, a center of higher education and a regional center of music and culture. The framework includes sensors, IoT devices, and software for collecting weather and hydrological data such as precipitation, wind, tide height, and inundation data. Low-cost water level gauges with wireless sensors can be deployed in streets to gather flood data. This information is transmitted to IoT cloud storage via gateways, enabling data services for communities and researchers. By integrating IoT data with terrain and typhoon path data, the system facilitates accurate urban flood modeling and forecasting.

3.2 Data Acquisition

The input images are collected from disasters and intensity dataset [11] that provides comprehensive information on various natural disasters. It includes temporal data and spatial data alongside environmental parameters like rainfall, river water levels, soil moisture, wind speed, seismic activity, temperature and atmospheric pressure. Data is sourced from IoT-enabled sensor networks such as weather stations, hydrological sensors, seismographs and satellite imagery making it a valuable resource for disaster prediction, risk assessment, urban planning, and climate studies. The dataset is split into 45% training, 25% testing, and 30% validation.

3.3 Flood Depth Prediction Using Temporal Attention Recurrent Graph Convolutional Neural Network

The Gjøvik municipality plan for water and sanitation describes the need to adapt the current networks and facilities to meet future demands and to increase the city's ability to filter out unwanted substances from stormwater and meltwater. In this section, TARGCN [12] is discussed for predicting the flood depth. It employs a detailed layer wise architecture to capture spatial temporal dependencies and predict flood depth accurately. The model begins with an input layer that processes hydrological and meteorological data from IoT sensors structured as temporal sequences and spatial graphs. This layer encodes the spatial features of the data effectively. Equation (1) gives the internal correlation of the data.

$$att(P, H, U) = soft \max \left(\frac{M_p P \cdot (M_H H)^N}{\sqrt{b_w}} \right) U \tag{1}$$

where, att represent capturing the internal correlation of data, P denotes the query of data, H indicates the key hyper parameter value of data, U signifies the data value, M_p indicates the capture spatial temporal dependencies, M_H denotes the maximum hyper parameter key value, and b_w represent the meteorological data from IoT sensors. Spatial correlation modeling captures the interdependencies between different geographic regions to enhance flood depth prediction. In TARGCN, spatial correlations are represented using graph convolution layers that process data from interconnected as sensors. These layers learn the relationships and influence between regions, allowing the model to better predict flood depths by leveraging spatial patterns across locations as expressed in Eq. (2).

$$A_t = M_t \cdot gcn(soft \max(X^{pre}), A) + e_t \tag{2}$$

where, A_t signifies the spatial patterns across locations, M_t denotes the graph convolution operation, X^{pre} is the pre-trained layer, A represent the average value of input data, e_t indicates the updates continuously and automatically running during the training. Finally, a fully connected layer maps the features into a prediction of flood depth, while a softmax or regression output layer ensures the model outputs a precise flood depth value classification for smart city as shown in Eq. (3).

$$X_{ns} = relu(linear(att(K'_s, K'_s, K'_s))) + K'_s \tag{3}$$

where, X_{ns} signifies the output of data, K'_s signifies the number of features, *relu* represent the rectified linear unit function. TARGNN is employed for the prediction of flood depth. The TARGNN employs an artificial intelligence-based optimization technique due to its ease and relevance. In this study, the optimal parameter of the TARGNN is optimized using the PO [13].

3.4 Optimization Algorithm Using Parrot Optimizer

In this section, a new nature-inspired meta-heuristic approach based on PO is discussed. It is utilized to optimize the weight parameters M_p and K'_s of TARGNN and improving its performance. Here, the parameters M_p is optimized to improve R2-score and K'_s is optimized to reduce computational time. The stepwise process is defined to get ideal value of TARGNN depending on the PO. Initially, PO creates an equal distribution populace to enhance optimum parameter of TARGNN.

Step 1: *Population initialization*

This initialization step prepares the PO for the iterative process where each parrot will update its position and velocity based on the optimization problem's objective function and the interactions between the parrots, leading to the discovery of near optimal solutions. The initial phase of population is given in Eq. (4).

$$R_a^0 = kd + rand(0, 1) \cdot (vd - gd) \tag{4}$$

where, R_a^0 denotes the initial phase of population, kd indicates the position of parrot, $rand(0, 1)$ represent the arbitrary count in the range $(0, 1)$, vd signifies the upper bound, and gd indicates the lower bound.

Step 2: *Random generation*

The input parameters are created randomly. Depending on the particular hyper parameter conditions they meet, weight M_p and K'_s parameter is created randomly through PO method.

Step 3: *Fitness function*

The initialized assessment is utilized to create random solution. It is evaluated by the parameter optimization value for optimizing the weight parameter PO of the classifier as given in Eq. (5).

$$Fitness\ Function = optimizing\ M_p\ and\ K'_s \tag{5}$$

Step 4: *Foraging behavior for Optimizing M_p*

In the PO, in search for food, parrots explore the environment by adjusting their position and velocity, balancing between exploration and exploitation. This behavior is modeled to enhance the optimization process, allowing the algorithm to efficiently find the optimal solution. The average position of the existing swarm is given in Eq. (6).

$$d^n_{maean} = \frac{1}{S} \sum_{k=1} M_p S d^n_h \tag{6}$$

where, d^n_{maean} represents the average position of the existing swarm, S indicates the current number of iterations, M_p indicates the capture spatial temporal dependencies, d^n_h signifies the location of the succeeding update. Figure 2 shows the corresponding flowchart.

Step 5: *Staying behavior for Optimizing K'_s*

In the PO, **staying behavior** represents the parrot's tendency to remain around its best-found position where food is abundant. This behavior allows parrots to exploit local optima by refining their position near the best solution discovered. It helps the algorithm focus on the promising areas of the search space thereby improving convergence towards optimal solutions as given in Eq. (7)

$$d^{s+1}_a = d^s_a + d_{best} \cdot levy(\text{dim}) + K'_s rand(0, 1) \cdot ones(1, \text{dim}) \tag{7}$$

where, d^{s+1}_a represent the convergence towards optimal solutions, d_{best} signifies the optimal location identified from the start up to the present, K'_s signifies the number of features.

Step 6: *Termination*

The weight parameter value of generator M_p and K'_s from TARGNN are optimized by utilizing PO and it repeats step 3 until it obtains its halting criteria $R = R + 1$. The FDP-TARGNN-IoT defectively assesses the prediction of floods with higher R2-score and lower MAE.

4   Results and Discussion

The simulation results of the proposed algorithm are represented in this section. The training process was conducted in MATLAB 2019b for 100 epochs, with a batch size of 32 and an initial learning rate of 0.001, which was fine-tuned using the PO to ensure optimal performance.

4.1   Performance Metrics

The simulations were performed and the performance metrics like MAE, RMSE and R2-Score of the proposed FDP-TARGNN-IoT I are compared with the existing methods such as real-time flood forecasting in Amo Chhu utilizing ML method and IoTs (RTFF-SVM-IoT) [6], real-time flash food tracking and prediction scheme (IoT-FFTP-ANN) [7], and adaptive momentum back propagation approach for flood forecasting with management in the IoT (AMBPA-FP-IoT) [8] respectively.

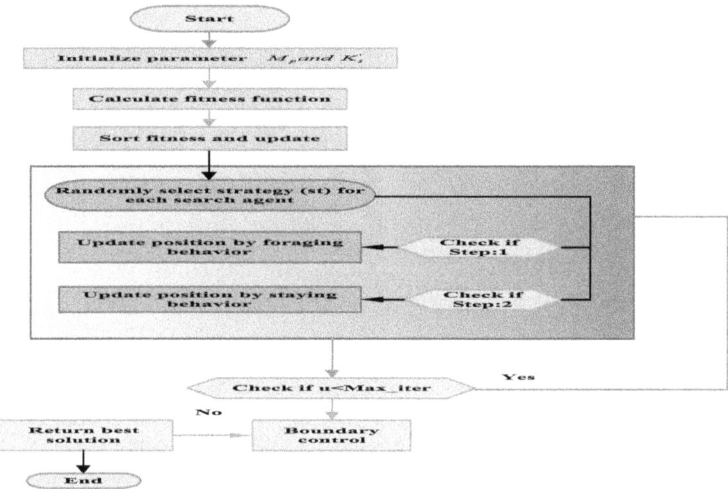

Fig. 2. Flow chart of LW-STO for optimizing TPGNN

4.2 Performance Analysis

Figures 3, 4, 5 highlight that the performance metrics are faster than those of the existing approaches. Overall, FDP-TARGNN-IoT method is found to be more efficient and effective in enhancing flood prediction. Figure 3 depicts MAE analysis. The low MAE is attributed to the ability of the FDP-TARGNN-IoT model to capture both spatial and temporal dependencies in flood data which enhances its predictive accuracy. The inclusion of the TARGNN enables the model to focus on the most important time periods of forecasts. This integrated approach results in a more accurate and robust flood depth prediction system leading to a reduced MAE.

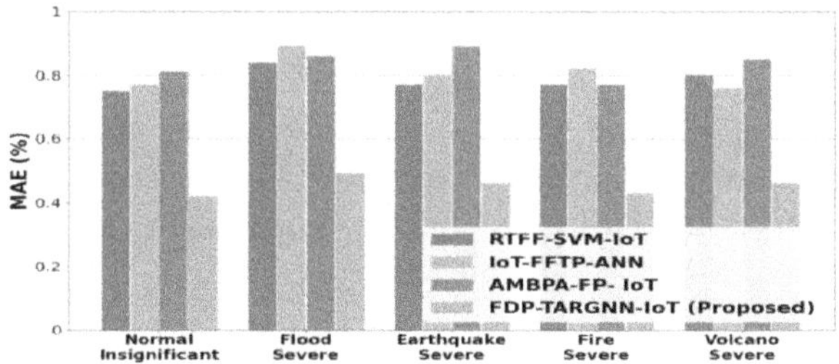

Fig. 3. Performance analysis of mean absolute error

Figure 4 depicts RMSE analysis. The low RMSE is primarily due to the optimization process using the PO, which fine tunes the model's parameters for better prediction. By minimizing errors during training, the optimizer enhances the model's ability to predict flood depths more precisely. This results in a more robust and efficient prediction system, effectively reducing RMSE.

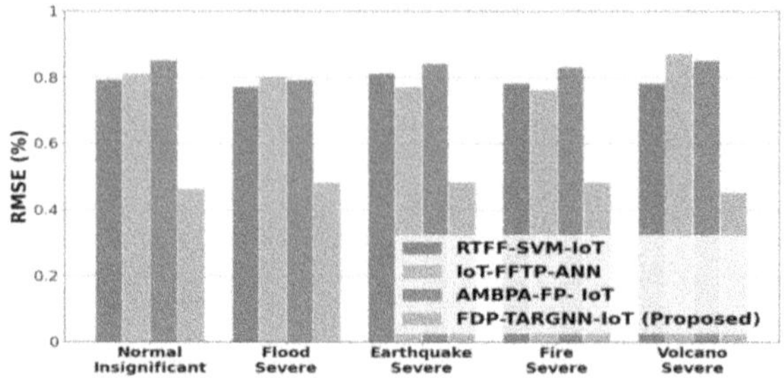

Fig. 4. Performance analysis of root mean squared error

Figure 5 depicts R2-Score analysis. The high R2-score is attributed to the combined strength of TARGNN and PO. TARGNN effectively captures both spatial and temporal dependencies in the data, enhancing the model's ability to predict flood depths accurately. The PO further refines the model's parameters, minimizing errors and improving the overall fit. This interaction ensures that the model closely matches the actual data, resulting in a high R2-score that reflects strong predictive power.

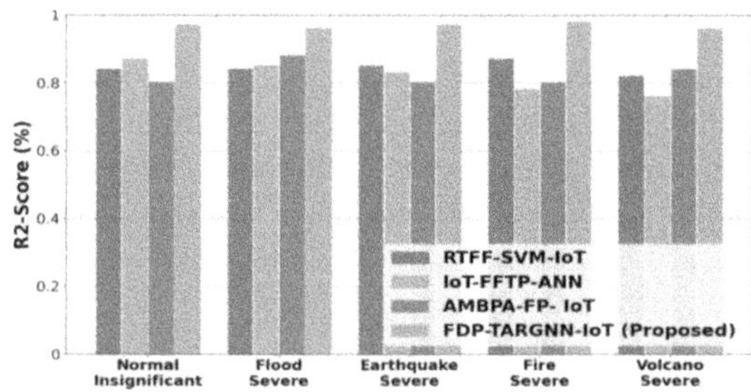

Fig. 5. Performance analysis of R2-score

4.3 Computation Complexity

Figure 6 shows computational complexity analysis. Here, the FDP-TARGNN-IoT algorithm's CPU operation time along with memory utilization rises linearly. Time complexity for FDP-TARGNN-IoT exhibits lesser than the existing RTFF-SVM-IoT, IoT-FFTP-ANN and AMBPA-FP-IoT models.

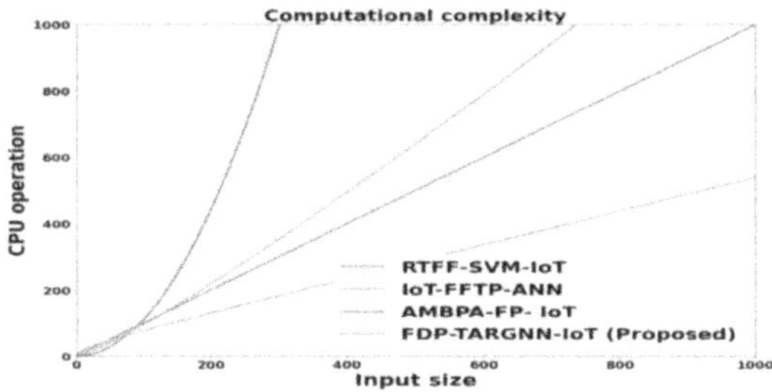

Fig. 6. Performance analysis of computational complexity

4.4 Ablation Study

To understand the importance of each component in the proposed FDP-TARGNN-IoT, the ablation analysis with and without each specific component is conducted. Table 1 presents the ablation study for the proposed FDP-TARGNN-IoT model, illustrating the impact of excluding key components like TARGNN and PO. The complete FDP-TARGNN-IoT model achieves the best performance, with an MAE of 0.42, RMSE of 0.46, and R2-score of 0.97, underscoring the importance of all components. When TARGNN is removed, performance drops significantly to MAE: 0.77, RMSE: 0.81, R2-score: 0.87, highlighting its role in modeling spatial-temporal correlations. Similarly, excluding PO results in further degradation to MAE: 0.81, RMSE: 0.85, R2-score: 0.80. These results confirm that both TARGNN and PO are crucial for precise flood depth prediction.

Table 1. Ablation study of the proposed PD-MSGWCN-CXI method

Ablation Model	Metrics		
	MAE	RMSE	R2-score
Without TARGNN	0.77	0.81	0.87
Without PO	0.81	0.85	0.8
FDP-TARGNN-IoT (Proposed)	0.42	0.46	0.97

5 Conclusion

The FDP-TARGNN-IoT model is successfully executed and the proposed method is simulated to develop an effective framework for the prediction of flood in IoT enabled smart cities. The widespread deployment of IoT devices has enabled the gathering of extensive big data on floods and related factors, providing a robust data foundation for applying TARGNN techniques to flood forecasting. The technique significantly enhances performance. Future work will focus on leveraging richer and more diverse datasets to enhance the prediction capabilities of the model, specifically targeting street level flooding with higher accuracy. Additionally, the integration of transfer learning techniques will be explored to enable the model to utilize pre trained knowledge from related tasks, reducing training time and improving prediction accuracy for further enhancing the robustness and generalizability of flood forecasting systems.

Acknowledgments. This document has been prepared in the framework of the European project TransformAr. This project has received funding from the European Union's Horizon 2020 innovation action programme under grant agreement no. 101036683.

References

1. Bukhari, S.A.S., Shafi, I., Ahmad, J., Butt, H.T., Khurshaid, T., Ashraf, I.: Enhancing flood monitoring and prevention using machine learning and IoT integration. Nat. Hazards, 1–28 (2024)
2. Rathod, B., Jahanavi Rao, A.V., Sudha, V.: Big data and IOT-based flood monitoring using deep neural network. In: Devi, V.A. (ed.) Sustainable IoT and Data Analytics Enabled Machine Learning Techniques and Applications. CESIBT, pp. 147–158. Springer, Singapore (2024). https://doi.org/10.1007/978-981-97-5365-9_8
3. Hayder, I.M., et al.: An intelligent early flood forecasting and prediction leveraging machine and deep learning algorithms with advanced alert system. Processes **11**(2), 481 (2024)
4. Almorabea, O.M., et al.: IoT network-based intrusion detection framework: a solution to process ping floods originating from embedded devices. IEEE Access **11**, 119118–119145 (2023)
5. Brito, L.A., et al.: FLORAS: urban flash-flood prediction using a multivariate model. Appl. Intell. **53**(12), 16107–16125 (2023)
6. Al K.M., et al.: Real-time flood forecasting in AmoChhu using machine learning model and internet of things. Cogent Eng. **11**(1), 2370900 (2024)
7. Akinsoji, A.H., et al.: Integrating machine learning models with comprehensive data strategies and optimization techniques to enhance flood prediction accuracy: a review. Water Resour. Manag. **38**, 4735–4761 (2024)
8. Thankappan, J., Mary, K., Raj, D., Yoon, D.J., Park, S.H.: Adaptive momentum-backpropagation algorithm for flood prediction and management in the Internet of Things. Comput. Mater. Continua **77**(1) (2023)
9. Hashemi-Beni, L., Puthenparampil, M., Jamali, A.: A low-cost IoT-based deep learning method of water gauge measurement for flood monitoring. Geomat. Nat. Haz. Risk **15**(1), 2364777 (2024)
10. Fernandes Jr, F.E., Nonato, L.G., Ueyama, J.: A river flooding detection system based on deep learning and computer vision. Multimed. Tools Appl. **81**(28), 40231–40251 (2022)
11. https://www.kaggle.com/datasets/darylbvaldez/disastersintensity?select=test_split.csv
12. Yang, H., Jiang, C., Song, Y., Fan, W., Deng, Z., Bai, X.: TARGCN: temporal attention recurrent graph convolutional neural network for traffic prediction. Complex Intell. Syst. **10**(6), 8179–8196 (2024)
13. Lian, J., et al.: Parrot optimizer: algorithm and applications to medical problems. Comput. Biol. Med. **172**, 108064 (2024)

Cancer Classification and Analysis Through an Explainable Deep Learning Framework

Fayyaz Ahmed, D. Subitha$^{(\boxtimes)}$, J. C. Kavitha, and Hariram Pasupathy

School of Computer Science and Engineering, Vellore Institute of Technology, Chennai, India
`subitha.d@vit.ac.in`

Abstract. In this work, we have proposed a novel deep learning structure that combines pretrained model MobileNetV3 that is integrated with Inverted Pyramid Pooling Module (iPPM). This helps in combining the features at different scales as different pooling sizes. The model performance is tested using metrics such as accuracy, precision, recall, and F1-score. The model is fine-tuned with hyper parameters that are optimized using Powell's algorithm. This advanced technique helps in extracting precise and multi-scale features of the input image. The simulations exhibit accuracy of 99.74%, 99.84%, 93%, 99.77%, 96%, 99.13%, 98.02% and 98.6% for Cervical Cancer, Lung and Colon Cancer, Oral Cancer, Acute Lymphoblastic Leukemia, Kidney Cancer, Breast Cancer, Lymphoma and Brain Cancer respectively. Furthermore, we utilize explainable AI (XAI) technique: GradCAM to visualise the regions of interest within the images that contribute most significantly to the model's predictions. This visualization improves the interpretability of the model's decision-making process, aiding medical professionals in understanding the rationale behind the classifications.

Keywords: Explainable AI · GradCAM · EfficientNetB0 · MobileNetV3 · Transfer Learning · Inverted Pyramid Pooling Module

1 Introduction

Cancer is the disease that causes the highest mortality rate in most countries. Late diagnosis is the reason behind this high mortality rate [1–3]. Early detection is the only solution for this, and this can only be supported by the recent advancements in CAD technology [4, 5]. The recent growth in Artificial Intelligence (AI), helps the medical industry for the more precise classification of cancerous from non-cancerous images. The model output is in general binary or discrete level from which no understanding about the decision and no relationship to the input features can be obtained which drives the requirement for more information from the model [6, 7] by means of Explainable AI (XAI) techniques. The goal of XAI techniques [8] is to increase the model's internal transparency so that medical practitioners may understand how the model makes its decisions. This promotes confidence in the model's output and makes it easier for clinical practice to incorporate it smoothly.

Hence, model development for accurate cancer classification using transfer learning approach and its interpretability using XAI technique are taken as the two sides of the

© The Author(s) 2026

J. Shreyas et al. (Eds.): CODE-AI 2025, CCIS 2689, pp. 76–87, 2026.
https://doi.org/10.1007/978-3-032-19318-6_8

paper. The steps in achieving these two objectives are discussed in this work. In Sect. 2, the survey on similar works is analyzed from which the design objectives of the proposed work are derived.

2   Literature Review

The survey of other related works using deep learning models is provided in Table 1.

Most of the papers discussed above have worked on the classification of only one cancer type and have tried to optimize the accuracy by means of the proposed deep learning model with the selected cancer type alone. Some of the literature has added XAI models to describe the influence of the model outcome with respect to the input features. CNN, DenseNet, ResNet, EfficientNet, VGG, XGboost, linear regression and KNN are the models used in the various literatures discussed above which are all optimized using fine tuning. In the paper [16], recent advancements in deep learning for medical image analysis, particularly brain tumor classification, are reviewed. Prior studies have highlighted the VGG19 architecture's effectiveness in image recognition tasks, though challenges like tumor variability and imbalanced datasets persist. Various techniques, including multi-scale feature extraction, have been proposed to address these issues in this paper by introducing the Inverted Pyramid Pooling Module (iPPM) technique. Using this module, feature extraction is enhanced by capturing both micro and macro level features. Apart from capturing the features it is essential for optimization of the models is essential. Whereas the justification for the result cannot be provided because of the "black box" approach of deep models. This issue is overcome by using fine tuning with top-layer customization and XAI algorithms like GradCAM respectively. The following are the key contributions of this paper:

1. *Optimized Performance:* MobileNetV3 with transfer technique is integrated with iPPM module. This advanced technique helps in extracting precise and multi-scale features over various dimensionalities from the input image.
2. *Improved Transparency:* Leveraging Grad-CAM, an Explainable AI (XAI) technique to bridge the gap between the model's complex decision-making process and human understanding. The overall objective of the proposed work is to allow improved performance and interpretability of the best fit models using XAI that eases the decision-making process in the health care industry.

Table 1. Comparison of Related Works.

Reference	Models Used	Cancer Type	Contributions	Limitations
[9]	CNN	Lung and Colon	CNN models with few adjustable layers	Considered single model only
[10]	XGBOOSTLinear Regression, KNN	Breast Cancer	3 models to the dataset provided and have also changed the test and training ratio for each of them to find the better ratio among these	This work has not done any parameter tuning and neither has made any interpretability for the results
[11]	Multiscale 3D CNN	Brain Cancer	Different multiscale approaches which use three processing pathways to segment, extract the features and then classify the images provided	The author even though it has used a multiscale CNN approach hasn't made a significant increase in the accuracy
[12]	MUF-Net	Kidney Cancer	The proposed MUF-Net architecture using EfficientNet as core and with attention mechanism for model fusion	Accuracy of 80% is not a desired result even with the complex architecture
[13]	QCResNet	ALL	Novel deep learning model architecture called QCResNet for classifying ALL from peripheral blood smear images	The paper doesn't discuss how the model's predictions are interpreted to make them more understandable
[12]	CNN and ELM	Cervical Cancer	CNNs and extreme learning machines (ELM) used for feature extraction and classification	Analysis is done with respect to the classifiers only not based on the backbone network

(continued)

Table 1. (continued)

Reference	Models Used	Cancer Type	Contributions	Limitations
[14]	U-Net	Oral Cancer	Deep learning architecture U-Net is used for segmentation	The dataset size is small. Larger datasets may improve the results
[15]	EfficientNetB0-B3	Lymphoma	EfficientNet is used here as it employs compound scaling method	Trained on limited sample and the hyper parameter tuning are the limitations

3 Proposed Methodology

Right diagnosis is the key factor to successful treatment in the battle against cancer. This paper aims to present a framework for deep learning-based multiclass cancer classification using medical images along with the highlights of XAI functionality. One of the biggest strengths of the deep learning algorithm is its application of the CNNs as pre-trained models, such as the VGG-16, EfficientNetB0 and MobileNetV3 that are well known for producing high accuracy in computer vision applications. Among these, the best fit model for each cancer type is identified using transfer learning approach. This approach is comparatively straightforward and the simplest among the recent techniques that are based on either model fusion or feature fusion techniques. One of the major benefits of the above pre-trained models is their ability to efficiently extract features using the multiple layer architecture and the ability to adapt by adding new layers specific to each cancer type. Furthermore, a high-resolution feature extraction technique iPPM is used along with the optimized models for extracting complex patterns over multiple scales. The iPPM, a powerful feature extraction technique mostly used for semantic segmentation, is applied here for cancer image classification seems to be a novel approach. But the implementation of iPPM is limited to two (Oral and Lymphoma) among eight cancer types that have showcased lesser performance than the others due its complexity.

But deep learning's intrinsic opaqueness presents a significant barrier to its widespread use in the medical field. For clinicians to accept a model's accuracy and incorporate it smoothly into their workflow, they must comprehend the reasoning behind the model's prediction. The use of XAI becomes propitious in this case. The project seeks to increase the transparency of the model's internal operations by implementing XAI approaches. This promotes confidence in the outcomes and facilitates well-informed decision-making by giving medical practitioners insight into the model's methodology. Moreover, to address the "black box" nature of deep learning, XAI is integrated through the GradCAM technique to understand the model's decision-making process. This framework paves the way for improved cancer diagnosis by potentially offering better accuracy, fostering trust through XAI, and achieving efficient multiclass cancer classification.

3.1 Dataset

The dataset link is (https://www.kaggle.com/datasets/obulisainaren/multi-cancer) has eight different types of cancers. The dataset contains images of different kinds of cancers Table 2 shows the dataset size for each cancer type and the number of classes under each type. This large dataset is highly useful in training the deep models effectively.

Table 2. Dataset Size and Number of Classes under each cancer type.

Type of Cancer	Number of classes	Dataset size
ALL	4	20000
Brain Cancer	3	15000
Breast Cancer	2	10000
Cervical Cancer	5	25000
Kidney Cancer	2	10000
Lung and Colon Cancer	5	25000
Lymphoma	3	15000
Oral Cancer	2	10000

3.2 Deep Learning Framework

In this proposed work multiclass cancer classification using MobileNetV3 deep learning model is done. The images are initially resized and normalized and fed to MobileNetV3. This deep network is used as the backbone network and is customized by the addition of new layers such as Global Average Pooling layer, a Dense layer, and Batch Normalization layer. The effectiveness of the model is analyzed using factors like accuracy, F1 score, precision and Recall. Figure 1 shows the MobileNetV3 model with additional layers for cancer classification.

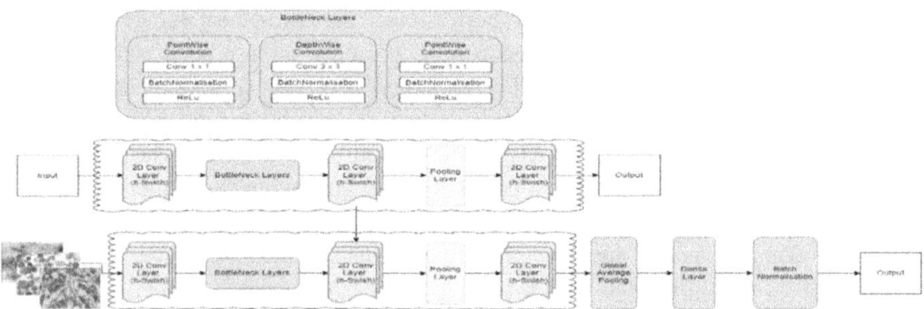

Fig. 1. MobileNetV3 Model with Additional Layers.

MobileNetV3 Model with an Inverted Pyramid Pooling Module (iPPM). The optimized models are obtained for each cancer type as discussed in the previous section.

Among these, the Oral cancer and Lymphoma have showcased less accuracy which are then fine-tuned by integrating with iPPM technique. The block diagram in Fig. 2 represents a neural network architecture that combines the MobileNetV3 model acting as the backbone network with an iPPM. The architecture starts with an input layer designed to handle images resized to 256×256 pixels with three color channels (RGB). MobileNetV3 used without its fully connected top layers, acts as a feature extractor, leveraging its pre-trained weights to efficiently capture essential features from the input images.

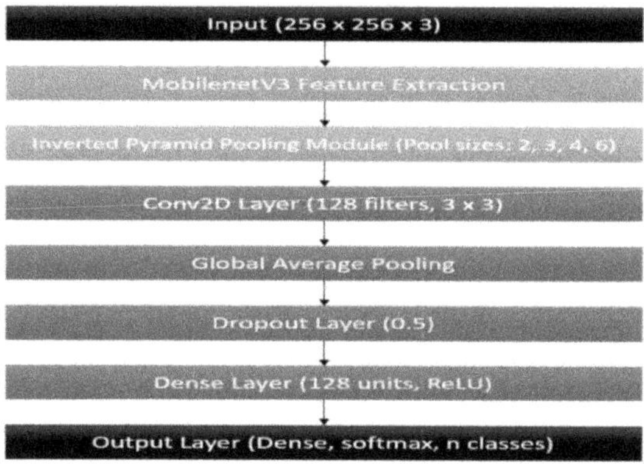

Fig. 2. MobileNetV3 Model with iPPM.

The features extracted by MobileNetV3 are then refined using iPPMmodule that consists of the following steps: Average Pooling, Interpolation, Convolution and Concatenation. The first step of iPPM is adaptive average pooling with the pooling sizes of 2 $\times$ 2, 3 $\times$ 3, 4 $\times$ 4, and 6 $\times$ 6 done parallelly on the feature vector of size ($7 \times 7 \times 576$). Adaptive average pooling is given in Eq. (1) and (2) for the pooling size of p and $p=2$ respectively. This helps in reducing the input feature vector size from (h, w, c) into the (h', w', c') where h is height, w is width and c is number of channels. The iPPM helps in integrating the features at different scales as different pooling sizes of 2, 3, 4 and 6 are applied on the extracted features of the basic MobileNetV3 model.

$$I(h', w', c') = \frac{1}{p^2} \sum_{w=pw}^{(w+1)p-1} \sum_{h=ph}^{(h+1)p-1} I(h, w, c) \tag{1}$$

$$I(h', w', c') = \frac{1}{2^2} \sum_{w=2w}^{(w+1)2-1} \sum_{h=2h}^{(h+1)2-1} I(h, w, c) \tag{2}$$

The next step is Interpolation process that adds extra samples using nearest neighbor upsampling method with the same size of 2×2, 3×3, 4×4, and 6×6 in each of the pooled channels. The feature vector that was downsized using pooling in the previous step is now upsampled by the same size so that the original size of tensor ($5 \times 5 \times 576$) will be retained parallelly in all pooling lines. This helps in maintaining the same spatial dimension for concatenation.

Finally, each of the 4 parallel features of size ($7 \times 7 \times 576$) obtained through various pool sizes are convoluted separately by (1×1) convolution and then concatenated. This approach ensures a comprehensive representation of the features at different scales. Following the concatenation of the pyramid pooling, a Conv2D layer with 128 filters and a 3×3 kernel further refines the extracted features. Global Average Pooling is then applied to condense each feature map into a single value by averaging, thereby converting the spatial dimensions into a single dimension and reducing model complexity. The architecture includes a dense layer with 128 units and ReLU activation to learn intricate patterns and relationships from the features. To prevent overfitting, a dropout layer with a 0.5 rate is used, randomly setting 50% of the input units to zero during training. The network concludes with an output layer, a fully connected softmax layer that provides the final classification probabilities. This design efficiently integrates MobileNetV3's lightweight yet robust feature extraction capabilities with the iPPM's multi-scale feature capture, which is used for various image classification tasks.

3.3 GradCAM

Gradients of a specific target that flow through the convolutional network are used by GradCAM to localize and highlight target regions in the picture. GradCAM uses class specifics to produce localization maps of the significant regions of the image, making black box models more transparent, by displaying visualizations that support output predictions. In other words, GradCAM fuses pixel space gradient visualization with class discriminative property. The goal of Grad-CAM is to maintain the object's spatial position data, which is lost in a completely linked layer. Thus, the neurons in the final convolution layer are employed to detect class-specific portions.

The feature maps in a CNN show how different convolutional layers have activated in response to the input picture. Different levels of abstraction, from straightforward edges and textures to intricate patterns and objects, are captured in these feature maps. GradCAM begins by applying global average pooling to the last convolutional layer's feature maps. This process efficiently summarizes the existence of pertinent features across the picture by converting each feature map into a single value. The target class score's gradients regarding the feature maps are then calculated. These gradients show how the class score is impacted by modifications to the feature maps. Greater gradients indicate that the target class is more important. Next, each feature map's significance weights are calculated using the gradients. Higher gradient feature maps are assigned to greater weight, suggesting that they are more significant in terms of target class prediction. A map is created by computing the weighted combination of feature maps, which highlights the areas of the input picture that are pertinent to the target class. To see which areas of the input image are most important to the model's prediction for the target class, the localization map is finally superimposed over the image. The GradCAM results of various cancer classifications are provided in Sect. 4 that shows the explainability of the algorithm.

4 Results and Discussion

4.1 Model Results

The performance of the MobileNetV3 for various datasets has been shown in Table 3 below. The highest accuracy achieved by the model is 98.92 for lung and colon cancer which is comparatively higher than the other two models. The second being Acute Lymphoblastic Leukemia ALL) is 98.85 percent accurate. But the accuracy for lymphoma is the lowest, achieving only 84% accuracy. The models discussed above are trained using hyper parameters such as batch size and learning rate that are optimized using Powell's optimization algorithm. This is a simple, yet powerful algorithm as it is derivative free and performs search in various directions to minimize the prediction error. The fine-tuned values are 256 and 0.001 for batch size and learning rate respectively are obtained using Powell's algorithm. The experiment results are provided for 10 epochs.

Table 3. Results of the Pre-Trained MobileNetV3 Model.

Cancer Type	Accuracy	Precision	Recall	F1 Score	AUC
Brain Cancer	97.10	0.9713	0.9703	97.08	0.9971
Lymphoma	84.67	0.8599	0.8270	84.31	0.9587
Breast Cancer	98.15	0.9815	0.9815	98.15	0.9976
Kidney Cancer	93.50	0.9350	0.9350	93.50	0.9865
ALL	98.85	0.9884	0.9884	98.84	0.9899
Oral Cancer	85.75	0.8575	0.8575	85.75	0.9360
Lung and Colon	98.92	0.9891	0.9887	98.89	0.9996
Cervical Cancer	98.28	0.9839	0.9822	98.30	0.9997

The accuracy obtained for each cancer type with the best fit model is summarized in Table 4. This clearly shows that the accuracy of the proposed models ranges from 93% (Oral cancer) to 99.84% (Cervical Cancer). All the models are trained for 10 epochs only as the results are perfect and not showing any underfitting or overfitting characteristics. It is evident that the performance improves over epoch.

4.2 Comparative Analysis

The Table 4 below provides a comparison of all the results of the proposed optimized models with the existing literature. From this it is evident that our proposed models have proven to be the best for all cancer types mentioned in the study by producing better performance. The larger dataset complements the performance of all the backbone networks which in turn extracts feature at different hierarchies. The dense layers that follow the backbone network also help in downsizing the features and integrating them towards the final decision probabilities of the respective classes. Finally, parameter tuning using Powell's algorithm and iPPM modules improves the model's performance further as given below.

Table 4. Accuracy (in %) of Proposed models with Existing Papers and Pre-trained models.

Disease	Algorithm of Existing Paper	Accuracy of Existing Paper	Accuracy of Pre-Trained MobileNet	Accuracy of Proposed MobileNet + iPPM
Lung and Colon	CNN [9]	99.66	98.92	**99.73**
Breast Cancer	XGBOOST [10]	98.6	98.15	**99.13**
Brain Cancer	Multiscale CNN [11]	96.49	97.10	**98.6**
Kidney Cancer	MUF-Net [12]	80	93.50	**96.0**
ALL	QCResNet [13]	98.9	98.85	**99.77**
Cervical Cancer	CNN+ELM [12]	97.2	98.28	**99.84**
Oral Cancer	U-Net [14]	-	85.75	**93**
Lymphoma	EfficientNet [15]	95.56	84.67	**98.02**

4.3 GradCAM Results

The GradCAM is a Gradient based class activation map generates a heatmap with color gradients on the input that is mapped with the selected class. The color of the input image represents the activation map of the respective class. By examining the highlighted areas of the heat maps generated by the GradCAM, cancer cells may be quickly and easily identified. The heatmap is in the BGR format which is converted into RGB over which the input image is superimposed. Hence the heatmap in the superimposed image shows specific part where the cancer is detected which is highlighted in red in color. To extract the specific characteristics that lead to the diagnosis of a given cancer type, each picture is processed using GradCAM for finding the abnormalities. This makes interpretability even more straightforward and makes it easier for doctors to recognize cancer cells and understand the results better. The Table 5 shown below shows the GradCAM results of all cancer types.

Table 5. GradCAM Results for each sample image from 8 cancer types.

CANCER TYPE	ORIGINAL IMAGE	HEATMAP	SUPERIMPOSED IMAGE
Cervical Cancer			
Lung and Colon cancer			
Oral Cancer			
Acute Lymphoblastic Leakaemia			
Kidney Cancer			
Lymphoma			
Brain Cancer			
Breast Cancer			

5 Conclusion

We have proposed MobileNetV3 architecture with transfer learning technique for the classification of eight different cancer types that are taken from a large dataset of 1,30,000 images. In addition, for Oral Cancer and Lymphoma, MobileNetV3 is combined with the iPPM module. Our proposed models with their highest performance accuracy ranging from 93% for Oral Cancer to 99.77% for Cervical Cancer have contributed a lot to the medical arena. Finally, through the utilization of Grad- CAM visualization, we demonstrate one of the key benefits of AI that is to help with the understanding of the model's decision-making process. This will help to build trust and collaboration between AI and medical professionals. This strategy, which focuses on the problem of interpretability and the transfer learning advances, is highly promising in the real world, thus it can significantly serve as a decision support tool for the oncologists in the context of diagnosing and improving outcomes for patients.

References

1. Sung, H., et al.: Global cancer statistics 2020: GLOBOCAN estimates of incidence and mortality worldwide for 36 cancers in 185 countries. CA Cancer J. Clin. **71**(3), 209–249 (2021)
2. Clarke, C.A., Hubbell, E., Kurian, A.W., Colditz, G.A., Hartman, A.R., Gomez, S.L.: Projected reductions in absolute cancer–related deaths from diagnosing cancers before metastasis, 2006–2015. Cancer Epidemiol. Biomark. Prev. **29**(5), 895–902 (2020)
3. World Health Organization (WHO): Guide to cancer early diagnosis. World Health Organization, Geneva, Switzerland (2022)
4. Farmani, A., Soroosh, M., Mozaffari, M.H., Daghooghi, T.: Optical nanosensors for cancer and virus detections. In: Nanosensors for Smart Cities, pp. 419–432. Elsevier (2020)
5. Amoroso, N., et al.: A roadmap towards breast cancer therapies supported by explainable artificial intelligence. Appl. Sci. **11**(11), 4881 (2021)
6. De Souza Jr, L.A., et al.: Convolutional Neural Networks for the evaluation of cancer in Barrett's esophagus: explainable AI to lighten up the black-box. Comput. Biol. Med. **135**, 104578 (2021)
7. Saraswat, D., et al.: Explainable AI for healthcare 5.0: opportunities and challenges. IEEE Access **10**, 84486–84517 (2022)
8. Arrieta, A.B., et al.: Explainable artificial intelligence (XAI): concepts, taxonomies, opportunities and challenges toward responsible AI. Inf. Fusion **58**, 82–115 (2020)
9. Maheshwari, U., Kiranmayee, B.V., Suresh, C.: Diagnose colon and lung cancer histopathological images using pre-trained machine learning model. In: 2022 5th International Conference on Contemporary Computing and Informatics (IC3I), pp. 1078–1082. IEEE (2022)
10. Chen, H., Wang, N., Du, X., Mei, K., Zhou, Y., Cai, G.: Classification prediction of breast cancer based on machine learning. Comput. Intell. Neurosci. **2023**(1), 6530719 (2023)
11. Mzoughi, H., et al.: Deep multi-scale 3D convolutional neural network (CNN) for MRI gliomas brain tumor classification. J. Digit. Imaging **33**, 903–915 (2020)
12. Zhu, D., et al.: Multimodal ultrasound fusion network for differentiating between benign and malignant solid renal tumors. Front. Mol. Biosci. **9**, 982703 (2022)
13. Fan, Q., Jiang, J., Liu, X., Guo, Z.: Acute lymphoblastic leukemia classification based on convolutional neural network. In: Proceedings of the 6th International Conference on Artificial Intelligence and Big Data (ICAIBD), Chengdu, China, pp. 877–883 (2023)
14. Ariji, Y., Kise, Y., Fukuda, M., Kuwada, C., Ariji, E.: Segmentation of metastatic cervical lymph nodes from CT images of oral cancers using deep-learning technology. Dentomaxillofacial Radiol. **51**(4), 20210515 (2022)
15. Steinbuss, G., et al.: Deep learning for the classification of non-Hodgkin lymphoma on histopathological images. Cancers **13**(10), 2419 (2021)
16. Haque, R., Hassan, M.M., Bairagi, A.K., Shariful Islam, S.M.: NeuroNet19: an explainable deep neural network model for the classification of brain tumors using magnetic resonance imaging data. Sci. Rep., 1524 (2024)

Binary Classification of Galaxies Using Machine Learning and Deep Learning

Nitya Mishra[1], G. Akshat[2], Usha Moorthy[1(✉)], and V. E. Sathishkumar[3]

[1] Department of Information Technology, Manipal Institute of Technology Bengaluru, Manipal Academy of Higher Education, Manipal, India
m.usha@manipal.edu
[2] Department of Computer Science and Engineering, Manipal Institute of Technology Bengaluru, Manipal Academy of Higher Education, Manipal, India
[3] School of Engineering and Technology, Sunway University,
No. 5, Jalan Universiti, Bandar Sunway, 47500 Subang Jaya, Selangor Darul Ehsan, Malaysia

Abstract. In the great tapestry of space, where the galaxies inscribe their silent histories across the night, there will be more significant pressure to classify these galaxies with unprecedented precision. Indeed, the study and classification of galaxies have reached a fever pitch with large-sky surveys. This paper proposes a hybrid approach to represent better an approach with great strengths of some important traditional machine learning algorithms such as Random Forest, Support Vector Machines, and KNN, having their prowess blended with deep learning model-based deep feature-extraction capabilities, namely Convolutional Neural Networks. By applying ensemble and transfer learning techniques, we show in this paper an improved system for accuracy and scalability to meet the petabytes of astronomical data forecast for upcoming surveys. Through extensive experimentation using the Galaxy Zoo 2 dataset, our results illustrate a delicate balance between human ingenuity and machine intelligence: a trade-off between interpretability and precision. The result will finally open the door to a future in which the cosmos is decoded crystal clear, guiding astronomers into more profound insights about the cosmos and opening even more new frontiers of discovery.

Keywords: Cosmology · Galaxy Classification · Machine Learning

1 Introduction

With significant astronomical advancements in recent decades, large-scale sky surveys have made it easier to take in an increasingly large number of datasets [1]. Owing to the large-scale volume produced through astronomical surveys, the increasing urge for obtaining classifiable, accurate, and highly efficient methods arises, more incredible than before. The main aim of future surveys should be to open new windows at the dawn of cosmic discoveries and allow fundamental and physical measurements of geometric properties of the universe, structural growth as well as characterizing essential quantities such as the dark energy equation of state, the total mass of neutrinos, and the spatial

J. Shreyas et al. (Eds.): CODE-AI 2025, CCIS 2689, pp. 88–98, 2026.
https://doi.org/10.1007/978-3-032-19318-6_9

distribution statistics of mass [2]. While such fantastic improvements increase the sensitivities of modern sky surveys to capture even the faintest light emissions at billions of miles, they also come with drawbacks.

The classification of such vast datasets of galaxies and stars into appropriate categories based on visual appearance is called the Morphological classification of galaxies [3]. Applications like Random Forest and Support Vector Machine, among others, had obtained considerable success when applied to different classification problems of astronomy through classical machine learning techniques [4]. While these address several tasks, most fail in complex feature extraction and pattern recognition on astronomical images. In the meantime, deep learning methods, especially CNNs, revolutionized the performance of image classification tasks through extraordinary hierarchical feature extraction performance from raw data in astronomy [5].

Integrating such complementary approaches offers great promise for further improvement in accuracy and reliability regarding galaxy classification. Many methods currently in use have difficulty generalizing across the range of different observational conditions, let alone more problematic cases when either a pure traditional feature-based or deep learning method alone is too restrictive or ineffective [6].

The proposed approach explores methods such as KNN, CNN, RF, and SVM along with Ensemble Learning and Transfer Learning Methods, promising a robust and accurate system responding to an essential need for this kind of solution with scalability regarding the processing of astronomical data representing a significant improvement compared to the current literature as it systematically explores the added value of the synergistic effects of such a combination of different machine learning paradigms [7].

In this way, we present a hybrid approach that contributes to the growth of the astronomical object classification domain and, at the same time, gives insights into how traditional and modern machine learning methods can be effectively merged. The contribution made is significant not only for the current urgent need in astronomy for methods of improvement of classification but also for the machine learning community in general since it brings crucial insights into how different algorithmic approaches can be combined effectively [8].

However, despite these advances, most existing deep learning models still suffer from certain limitations in processing specific types of astronomical objects, especially those with peculiar appearances or at the detection limit of surveys [9]. Improvement in galaxy classification accuracy extends beyond simple categorical sorting. Modern photometric surveys, such as LSST, are bound to generate databases. Therefore, an efficient and accurate classification system will become inevitable [10].

Recent studies have shown hybrid models can perform better in related astronomical classification tasks [11]. For instance, it has been shown that the performance in galaxy morphology classification improves by combining CNNs with Support Vector Machines [12]. The proposed research extends this concept by exploring more sophisticated ensemble techniques and investigating their applicability to the specific challenges of galaxy separation.

Processing and classifying astronomical objects efficiently and accurately will be crucial for maximizing the scientific return from these investments [13]. Furthermore, the proposed methodology could be adapted for other astronomical classification tasks, such as transient detection and variable star classification [14]. This research targets a crucial shortcoming in existing astronomical data processing pipelines through more robust and accurate classification systems. The success of this might seriously impact our capability of processing and analyzing vast amounts of astronomical data expected during the next few decades, hence advancing our understanding of the universe [15].

2 Dataset

The dataset for this research comes from the Galaxy Zoo 2 (GZ2) project, which is the predecessor of the original Galaxy Zoo project. The GZ2 dataset contains morphological classifications of galaxies which were obtained through a citizen science initiative. Specifically, we focus on the GZ2 normal-depth sample, which includes classifications for nearly 300,000 galaxies based on detailed morphological features. These include galactic bars, pitch angles, spiral arm, bulges, edge-on galaxies, relative ellipticities, and others [16][17]. For our analysis, we used certain classes of galaxies that were debiased with the technique outlined in [17]. The subset we used included a variety of morphological labels which were later classified into 2 classes that comprised 4,888 elliptical galaxy and 4,195 spiral galaxy images. The captured images were available in a resolution of 424 $\times$ 424 pixels. This subset of the dataset is utilized to train models aimed at classifying galaxy morphologies into spiral and elliptical categories using both traditional and modern machine learning techniques. Some of the images available in the dataset are displayed in Fig. 1.

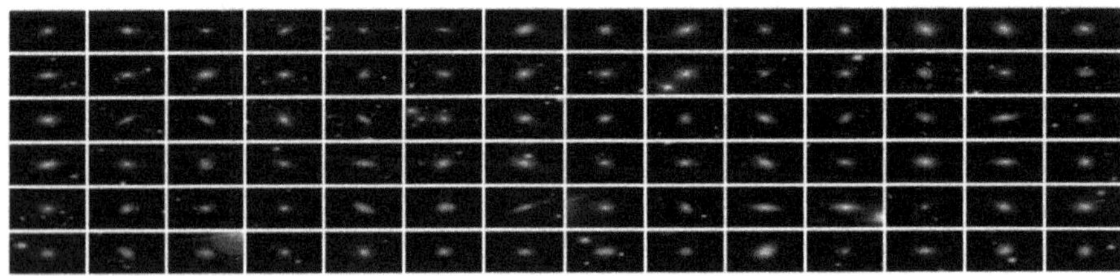

Fig. 1. Sample images from the galaxy classification dataset, displaying various galaxy types

3 Methodology

3.1 Traditional Machine Learning Methods

Classification in Astronomy has evolved dramatically, especially using Ensemble techniques, including Random Forests (RF), Support Vector Machines (SVM), and K Nearest Neighbors (KNN) [18]. It is well-established that Random Forest (RF) does a great job in high-dimensional noisy data; it makes the final prediction by taking the majority vote over the output from several decision trees. The splitting criterion in RF is typically Gini Impurity (the default in many implementations), though Information Gain can also be used, depending on the application. The most robust method used by Support Vector Machines (SVM) for galaxy morphologies classification could be given using a decision function as shown in Eq. (1):

$$f(x) = <w, x> + b \tag{1}$$

where w represents the weights, x is the input features, and b is the bias term.

K Nearest Neighbors (KNN) is a machine learning algorithm which does not rely on parameter assumption classifies objects by proximity in the feature space. Therefore, it assigns a label by majority voting among its k-nearest neighbors, shown in Eq. (2):

$$y(x) = \text{majority vote determined by the most common labels of the KNN of } x \tag{2}$$

where the KNN of x represents the closest data points in the feature space. KNN is most useful when interpretability is a critical issue because it predicts/classifies based on local data patterns.

3.2 Deep Learning Approaches

Image classification in astronomy has been revolutionized by Convolutional Neural Networks (CNNs), which directly extract hierarchical features from raw images [19]. The convolution operation is shown in Eq. (3):

$$Image(x, y) = \sum_i \sum_j Image(i, j) \cdot Kernel(x - i, y - j) \tag{3}$$

Pooling layers accelerate feature extraction. Modern CNNs have been applied to learning subtle patterns, adapting to a wide range of observation conditions, and are, therefore, very effective for large-scale surveys.

3.3 Hybrid Models

Hybrid models use CNNs for feature extraction combined with RF, SVM, or KNN classifiers. The hybrid methods perform better than individual ones because of the combination of representational deep learning and the strong decision-making capability of traditional classifiers. The final prediction is obtained using Eq. (4):

$$y_{\text{final}} = \text{majority vote of predictions from RF, SVM, and KNN classifiers} \tag{4}$$

Such models demonstrate better generalization and robustness to noise, particularly when multimodal data combining images and spectroscopic data is used.

3.4 Ensemble Strategy

The ensemble approach follows the mechanism of stacking with RF, SVM, and KNN as base models and an appropriate meta-model (such as logistic regression) for optimized prediction. The optimal parameters generally denoted as "meta" is calculated using Eq. (5):

$$y\_"meta" = \arg\max \left(\sum\_(i = 1)^n \blacksquare w\_i \cdot (y\_i) \right) \tag{5}$$

This approach would ensure stability and strong generalization across diverse datasets, outperforming single-model methods.

4 Experimental Design

4.1 Data Collection and Processing

The Galaxy Zoo 2 dataset considered has images that were of 424 × 424 pixels, which were processed in two batches. The entire dataset was processed in two ways, in the first method the images were directly resized to 128px for quicker computation without compromising image quality. In the second method the images were further cropped to 207 × 207 pixels and then downscaled using bilinear interpolation [20, 21] to 69 × 69 pixels. The images are rescaled by a factor of 1/255, normalizing the pixel values between 0 and 1. This step is crucial for efficient training, as neural networks tend to converge faster when input features are scaled [22]. For data augmentation, a range of transformations is applied to the training data to simulate real-world variations. These include Rotation Range (20°), Width and Height Shifts (0.2), Shear Range (0.2), Zoom Range (0.2), and Horizontal Flip. These augmentations expose the model to different orientations, shifts, and scales, helping it learn invariant features of galaxies regardless of their appearance in the image [23, 24].

4.2 Data Distribution

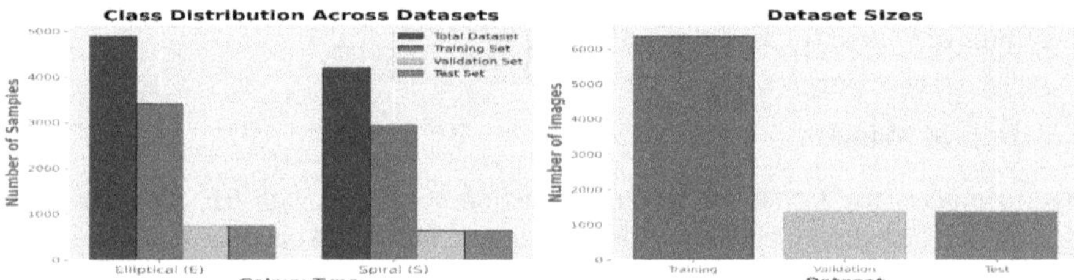

Fig. 2. Class distribution across the training, validation, and test datasets for the binary galaxy classification task.

The data has been split into 3 sets (Fig. 2).

- Training Set: 6357 images, including 3421 elliptical and 2936 spiral galaxies.
- Validation Set: 1363 images, with 734 elliptical and 629 spiral galaxies
- Test Set: 1363 images, containing 733 elliptical and 630 spiral galaxies.

4.3 Model Architecture

In our tests we have used KNN, CNN, RF and SVM.

K Nearest Neighbor (KNN). For galaxy classification, a K-Nearest Neighbors (KNN) classifier was sourced to the pre-processed dataset. KNN is a supervised algorithm used for regression and segregation based on the principle of similarity effectively making predictions by considering nearest neighbors of nearest dataset [25]. The cropped and resized images were used to test the KNN model. The conduction of hyperparametric grid search optimized the neighbor number (n_numbers), distance weighting(weights) and distance metric (metric). This process aims to identify the best parameter combination for achieving the highest classification accuracy. Hyperparameter tuning is critical for selecting the best model configuration, as it can significantly enhance performance by exploring various hyperparameter combinations [26].

Convolutional Neural Network (CNN). The high-level features in the images inserted in the input plane have been extracted using the convolution layers. These layers are designed to extract high-level features [27–29]. Normal, Conv2d as well as MaxPooling2D Layers comprise the CNN model, where the latter layer is utilized to preserve most relevant elements and noise reduction [30, 31]. After each layer, we have used Batch Normalization, which accelerates model training by normalizing activation and reducing internal covariate shift, allowing the use of higher learning rates [32]. Each convolutional layer in this model uses the 'ReLU' activation, where values range from 0 for negative numbers to the same value as the input for positive numbers. While being computationally efficient to alleviate vanishing gradient (allowing efficient propagation in deep networks) introduces non-linearity into the model [33]. After feature extraction, all the features are flattened to a 1D array. The addition of Dropout at the end of neural network randomly deactivates a section of neurons while training which in turn reduces overfitting [34].

Random Forests. The output of the feature extraction model acts as an input to the RF classifier. To influence the pretrained CNN model for features provided during the classification of galaxy images we employ transfer learning algorithm. The model contains 700 decision trees (n_estimators), which help capture complex patterns in the data while maintaining generalization. A minimum split of 10 and minimum leaf size of 1 is used to improve generalization. The max depth is set to 10 to inhibit overfitting and to keep the model complexity in check [35, 36]. The max features parameter is set to 'log2', selecting a subset of features for each split to promote diversity and improve efficiency. Additionally, class imbalance is addressed using class_weight = 'balanced', which adjusts weights inversely proportional to class frequencies, ensuring fair classification of both classes. Bootstrapping (bootstrap = True) introduces randomness by sampling with replacement, enhancing stability and preventing overfitting [37]. Feature selection was done by extensive hyperparameter testing, which in turn helped achieve an ideal offset between efficiency, accuracy, and generalization. The crucial role played by hyperparameter tuning improved the performance of the model by searching for the best combinations of parameters, thereby enhancing the model's predictive capabilities with minimal overfitting [38].

Support Vector Machine. The output from the feature extraction model was then fed into the pre-trained CNN for classification by an SVM classifier. Support vector machines are a robust supervised learning algorithm that seeks an optimal hyperplane to maximize the margin between classes, hence highly effective for high-dimensional feature spaces [39]. The best parameters obtained through extensive hyperparameter tuning using RandomizedSearchCV include a linear kernel, $C = 0.01$ for controlling the regularization trade-off, gamma = 'scale' for automatic kernel coefficient calculation, and no class weighting. The linear kernel was selected for its efficiency in handling high-dimensional data and interpretability, while the regularization parameter helps balance bias and variance [40].

Ensemble Model. To improve classification accuracy, an ensemble model was created by combining multiple classifiers. First, a Voting Classifier was used, which aggregates predictions from a Random Forest (RF) and Support Vector Machine (SVM) classifier using majority voting (voting = 'hard'). This method leverages the strengths of both classifiers to reduce variance and improve generalization, making it particularly useful when individual classifiers have complementary strengths [41]. Additionally, a Stacking Classifier was employed, where RF and SVM were used as base models, and Logistic Regression was chosen as the meta-model to combine their predictions [42].

4.4 Training and Validation

The proposed model was built using tensorfow and scikit-learn. The model training was conducted on Google Collaboratory using their CPU and inbuilt T4 GPU. The dataset in this study is divided into training, test and validation dataset in the ratio 70:15:15, i.e. 70% of the data was used to train the model (6357 images), 15% for validation (1363 images), and the remaining was used to train the model (1363 images). The CNN training was conducted over a maximum of 30 epochs. For efficient training, we used Reduce Learning on Plateau and Early Stopping.

As seen in Fig. 3. The model has a positive slope in both training and validation accuracy, reaching the value of 0.82. This is indicative of the model's capability to distinguish between spiral and elliptical galaxies. The steady rise in training accuracy from 61.40% to 80.21% reflects the model's ability to learn from this dataset. Moreover, the close match between training and validation accuracy suggests that the model is not overfitting despite its growing complexity.

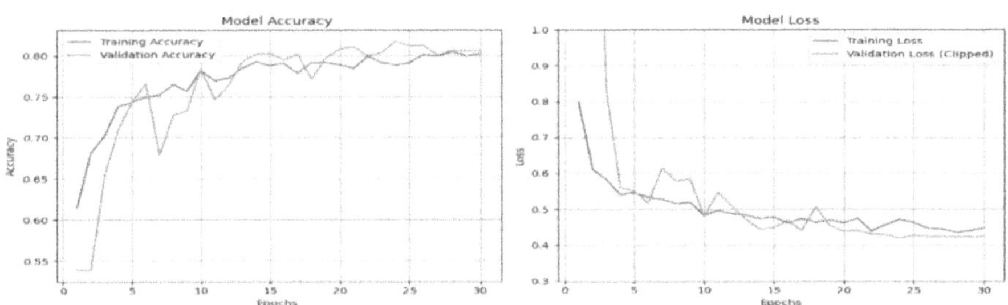

Fig. 3. Graph showing the accuracy and loss for both training and validation sets over 30 epochs.

All the hyperparameter tuning for the RF, SVM was done using the training set and the performance was evaluated on the validation during the tuning process. The performance of all the models on the test set has been tabulated in (Table 1.).

Table 1. Performance of Machine Learning and Deep Learning Algorithms on Test Data.

Algorithm	Class	Accuracy	Recall	F1-score	Test Accuracy
KNN	E	0.73	0.90	0.81	0.7542
	S	0.83	0.62	0.69	
CNN	E	0.78	0.88	0.82	0.7989
	S	0.83	0.71	0.77	
RF	E	0.79	0.85	0.82	0.8043
	S	0.81	0.74	0.77	
SVM	E	0.78	0.84	0.81	0.7887
	S	0.80	0.73	0.76	
VC	E	0.79	0.85	0.82	0.7960
	S	0.81	0.73	0.77	
SC	E	0.80	0.84	0.81	0.7946
	S	0.79	0.74	0.77	

5 Result and Analysis

The classification of galaxies into spiral and elliptical classes is essential for comprehending galaxy formation and evolution, as emphasized in works such as Willett et al. (2013) [16]. In our study we employed several Machine Learning techniques along with Deep Learning models to classify galaxy images from Galaxy Zoo dataset, evaluating their performance based on accuracy, recall, F1-score and test accuracy.

The proportion of correct classified instances i.e. Accuracy of the model, is computed using the Eq. (6).

$$Accuracy = (TP + TN)/(TP + TN + FP + FN) \tag{6}$$

Here, True Positive, True Negative, False Positive and False Negative values are abbreviated as TP, TN, FN correspondingly.

The model's ability to accurately recognize positive instances out of all actual positive cases is measured by Recall. A higher recall indicates that a model can effectively minimize false negatives. F1-Score provides a single metric that settles the accuracy and recall trade-off. Higher the F1-score, better is the balance between the model's ability to correctly classify instances and its tendency to miss or misclassify them.

5.1 Model Evaluation

Table 1 shows the performance of various models. Among the models that we evaluated, the Random Forests (RF) classifier achieved the highest test accuracy of 0.8043, closely followed by the Convolutional Neural Network (CNN) with 0.7989. The other models like Support Vector Machines (SVM), Voting Classifier (VC), Stacking Classifier (SC) all demonstrated comparable performances ranging from 0.79 to 0.80, but the K Nearest Neighbors (KNN) exhibited the lowest test accuracy of 0.7542.

5.2 Implications

The model has promised marvelous returns for extensive astronomical surveys, so automatic classification should be robust to reduce human verification processes and bring about significant discovery acceleration. Variations of observational conditions and different surveys are allowed with this model. Such a model has even more potential to uncover new classes of astronomical objects missed by traditional methods.

5.3 Limitations and Future Work

Due to the constraints of Google Collaboratory, such as limited GPU time, restricted program up time and slow processing speeds limited broader hyperparameter testing and tuning. Basic hyperparameter tests were conducted, while deeper tuning and examination of advanced architectures- namely combinations of setups with Convolutional Neural Networks and Random forests, Support Vector Machine, and stacking models were explored to achieve improvements along the lines of model performance. However, despite all these constraints, encouraging accuracies have been accomplished in our early research. Additional computational resources shall be utilized for future work to enable in-depth testing and optimization to enhance performance and scalabilities.

6 Conclusions

Our research confirms that the refined integration of traditional machine learning methods with deep learning approaches can significantly enhance classification performance. The study contributes to the field of galaxy classification by integrating deep feature extraction with well-established ensemble methods such as Random Forests, to improve the model accuracy.

The hybrid approach proposed leverages on deep learning for feature extraction and Random Forest for classification has shown promising results in terms of both accuracy and computational efficiency when compared to stand-alone methods. Specifically, the combination of deep neural networks with ensemble learning offers stable feature representations while maintaining interpretability via conventional machine learning techniques. The findings indicate an integrated approach to reduce model complexity; hence it makes more accurate and reliable predictions on a wider range of diverse datasets. Future research will aim to refine the hybrid pipeline by fine-tuning both deep learning and traditional models simultaneously. Exploring advanced deep learning architectures, like transformers and graph neural networks, could also boost performance. The findings from this study are valuable to both astronomy and various automated classification systems. The combination of different algorithmic approaches will be essential in advancing sophisticated classification systems.

Data Availability. All the data that was used for this study was publicly available at Galaxy Zoo's official.

Conflict of Interest. The authors declare that they have no conflict of interest.

References

1. Abdurro'uf, N., Alam, A., Chabanier, S., et al.: The 18th data release of the Sloan digital sky survey: unifying SDSS-I, II, III, and IV data. Astrophys. J. Suppl. Ser. **269**(1), 6 (2023)
2. DESI Collaboration, Aghamousa, A., Aguilar, J., et al.: The DESI experiment Part I: science, targeting, and survey design. Astron. J. **161**, 256 (2021)
3. Domínguez Sánchez, H., Huertas-Company, M., Bernardi, M., et al.: Deep learning for galaxy morphology analysis. Mon. Not. R. Astron. Soc. **476**(3), 3661–3676 (2018)
4. Dieleman, S., Willett, K.W., Dambre, J.: Rotation-invariant convolutional neural networks for galaxy morphology prediction. Mon. Not. R. Astron. Soc. **450**(2), 1441–1459 (2015)
5. Baron, D., Poznanski, D.: Photometric supernova classification with machine learning. Mon. Not. R. Astron. Soc. **465**(4), 4530–4555 (2017)
6. Ivezić, Ž, Tyson, J.A., Abel, B., et al.: LSST: from science drivers to reference design and anticipated data products. Astrophys. J. **873**(2), 111 (2019)
7. He, K., Zhang, X., Ren, S., Sun, J.: Deep residual learning for image recognition. In: Proceedings of the IEEE Conference on Computer Vision and Pattern Recognition (CVPR), pp. 770–778 (2016)
8. Hart, R.E., et al.: Galaxy zoo: morphological classifications for 120,000 galaxies in Sloan digital sky survey data release 8. Mon. Not. R. Astron. Soc. **461**(4), 3663–3681 (2016). https://doi.org/10.1093/mnras/stw1483
9. Fahmi, A.: Decision-Support for Rheumatoid Arthritis Using Bayesian Networks: Diagnosis, Management, and Personalised Care. Queen Mary University of London (2021). https://core.ac.uk/download/492541137.pdf
10. Casas, G., Casas, G.: Automatic detection and counting of stacked eucalypt timber using the YOLOv8 model. Forests **14**(12), 2369 (2023)
11. Perez, L., Wang, F.: The effectiveness of data augmentation in deep learning. Comput. Vis. Pattern Recognit. **29**(12), 2453–2465 (2017)
12. Albawi, S., Mohammed, T., Al-Zawi, S.: Understanding of a convolutional neural network. Int. J. Comput. Sci. Netw. Secur. **18**(3), 28–36 (2017)
13. Gu, X., Xie, L., Ding, Y.: A study on deep CNN architectures for image classification. J. Image Graph. **20**(5), 324–338 (2018)
14. Alzubaidi, L., Zhang, J., Humaidi, A.J., et al.: Review of deep learning: concepts, CNN Architectures, challenges, applications, future directions. J. Big Data **8**(1), 52–74 (2021)
15. Christlein, V., Lammers, M., Wilken, J., et al.: A comprehensive review of convolutional neural networks for galaxy classification. Astron. Astrophys. **619**, A109 (2019)
16. Gholamalinezhad, H., Lanza, A., Bieging, S., et al.: A deep convolutional neural network for galaxy morphological classification. MNRAS **495**(4), 4939–4952 (2020)
17. Ioffe, S., Szegedy, C.: Batch normalization: accelerating deep network training by reducing internal covariate shift. In: 32nd International Conference on Machine Learning, pp. 448–456. PMLR, Lille (2015)

Convergence of Dynamic Windows in Feature Extraction and Deep Autoencoders to Effectively Model Facial Expression Classification Problem

H. N. Naveen Kumar[1(✉)] [iD], Jagadeesh Basavaiah[1] [iD], K. V. Sudheesh[1] [iD], Kiran[1] [iD], Mahadevaswamy[1] [iD], and M. S. Guru Prasad[2] [iD]

[1] Vidyavardhaka College of Engineering, Mysuru, Karnataka, India
`naveen.vvce@gmail.com`
[2] Graphic Era (Deemed to be University), Dehradun, Uttarakhand, India

Abstract. The fusion of handcrafted features and deep features is an under explored area in facial expression classification. There exists a probability of losing subtle expression relevant features in using fixed size windows for feature extraction. The proposed work primarily aims to investigate the discriminative capability of the Histogram of Oriented Gradients and Local Phase Quantization feature descriptors through the application of dynamic windows. The texture feature extraction window is dynamically optimized based on the intrinsic characteristics of the underlying image to enhance feature discriminability. The use adaptive window enables capturing highly robust and discriminative feature from non-uniform facial regions. The discriminative power of the feature distribution can be further enhanced by incorporating differential information obtained from the comparison of features extracted from expressive and neutral facial images. The hybrid feature distribution is created through the convergence of expressive face features and feature difference vectors. The higher dimension issues of the hybrid feature distribution can be effectively addressed by the introduction of stacked deep convolutional autoencoders (SDCAE). The abstract feature space derived from autoencoders is used in training the multi-class support vector machine to build the classification model. The introduction of SDCAE results in a lighter model suitable for real-time processing, as it avoids additional computational load associated with high-dimensional feature spaces. The proposed work is implemented on three benchmark datasets namely, CK+, KDEF, and JAFFE. The three-stage autoencoder achieves the highest classification accuracy (96.5% on CK+, 95.1% on KDEF, and 89.3% on JAFFE).

Keywords: Facial Expression Classification · Local Phase Quantization · Adaptive Windows · Autoencoder

1 Introduction

Facial expression is a potent, innate, and universal indicator for humans to express their emotional states and intents. Automated facial expression analysis has been thoroughly examined owing to its practical applications in social robots, medical interventions, fatigue detection in drivers, and various human-computer interaction systems

© The Author(s) 2026
J. Shreyas et al. (Eds.): CODE-AI 2025, CCIS 2689, pp. 99–111, 2026.
https://doi.org/10.1007/978-3-032-19318-6_10

[1]. Numerous facial expression classification (FEC) algorithms have been explored in computer vision and machine learning to encode subtle and discriminative expression relevant information from facial representations. A cross-cultural study found that people from different cultures perceive the same fundamental emotions in the same way, therefore Ekman and Friesen used this information to identify six basic emotions [2]. These prototypical facial expressions (FEs) are anger, disgust, fear, happiness, sadness, and surprise. The feature representations allow us to classify FEC systems into two broad groups: static image FEC and dynamic sequence FEC. Static image-based approaches [3] encode feature representation exclusively through spatial information, whereas dynamic FEC methods [4] deploys both spatial and temporal interactions among successive frames in the expression sequence. In spite of the robust feature learning capabilities of deep learning, challenges persist when applied to FEC. Initially, in order to prevent overfitting, deep networks necessitate a substantial amount of training data.

Texture features denote the spatial configuration of pixel intensities, encapsulating intricate nuances and patterns across many facial regions essential for differentiating between distinct expressions. Texture features have shown to be highly discriminative among the many handcrafted features employed for FEC. FEC is significantly influenced by texture features, as facial expressions are frequently demonstrated through subtle variations in the appearance of skin textures, including wrinkles, fine lines, and muscle deformations. For example, crow's feet, furrows on the forehead, and the compression & expansion of the mouth as a result of various expressions all provide highly discriminative textural data that can be used for classification. Texture features complement geometric features by capturing the fine-grained details of face muscle movements, while geometric features focus on the movement and distortion of facial landmarks (e.g., mouth opening, eyebrow lifting) [5].

The Local Phase Quantization (LPQ), Local Binary Pattern (LBP), Histogram of Oriented Gradients (HOG) and Gabor filters are the widely deployed feature representations for FEC. The LBP feature is noise sensitive and has a limited ability to capture directional texture information. Gabor filters need precise parameter tweaking and are computationally intensive. Our earlier works serves as a foundation for the proposed work [3, 6, 7]. With reference to our previous work [6], the combination of HOG and LPQ feature has shown promising results for expression classification. Local Fourier transform phase information serves as the foundation for LPQ. Compared to LBP, it is more resilient to blur and variations in illumination. High-frequency texture fluctuations, which are essential for differentiating nuanced face expressions, are successfully captured by LPQ. The application of fixed size windows in feature acquisition reduces its the adaptability local texture complexity. To overcome this issue, in the proposed work the window size to extract HOG and LPQ feature is dynamically adjusted based on texture complexity. The LPQ feature is integrated with HOG feature to form a highly discriminative feature distribution. In our framework, stacked deep convolutional autoencoder (SDCAE) acts as a dimensionality reduction tool that refines the feature distribution. SDCAE consists of multiple layers of autoencoders, which progressively learn to compress the input features into a more compact and meaningful representation. The integration of feature extraction and SDCAE provides a complementary approach where HOG and LPQ focuses on robust feature extraction, and SDCAE enhances these features by learning a compact, discriminative representation. The prominent contributions of the work are:

- The proposed work presents a novel method for dynamically changing the size of the window to extract expression relevant features. The discriminative capability of the feature representation can be enhanced by adjusting the window size to the local texture complexity in facial region. In areas with complicated textures, the windowing technique dynamically adjusts window widths to optimize computational overhead.
- The feature variation between expressive and neutral facial image may hold highly relevant data for FEC. This variation is referred to as a feature difference in the proposed work and is integrated with the expressive face feature distribution to enhance the discriminative power.
- The challenge remains in managing the high dimensionality of feature distribution and ensuring the extracted features are optimally discriminative for classification. By encoding and decoding the feature vectors through multiple layers, SDCAE identifies the most relevant patterns and structures in the data, effectively filtering out noise and redundant information. The final compressed representation at the bottleneck layer of SDCAE retains the essential features needed for accurate classification, thus enhancing the robustness and discriminative power of the system.

2 Methodology

The proposed work primarily aims to minimize the misclassification rate in FEC by integrating dynamic window-based feature extraction, deep learning-driven dimensionality reduction, and advanced machine learning classification techniques. The expression specific features are acquired from the whole facial region. The contribution of neutral expression is also taken into consideration in developing an efficient feature space. The architecture of the proposed model is represented in Fig. 1. By deploying dynamic windows in feature extraction, we present a novel method for FEC to improve discriminative power of the feature. Dynamic window changes its size depending on the facial area under analysis, as opposed to traditional fixed-size windows. This method enables the recording of specific variations in texture in prominent facial regions, which are important markers of nuanced expression changes. Both neutral and emotional facial images are used to create the feature space that the classifier is trained on. Feature difference refers to the disparity between the feature distribution acquired from the neutral and emotional face images. Emotional face features and feature difference vectors are combined to generate the hybrid feature space. The higher dimensionality of the feature representation is addressed effectively by multistage deep autoencoders. The model to predict the expression class is obtained by multiclass support vector machine classifier.

The Viola-Jones method [12] is employed to detect the face in the image. The detected images are extracted and are resized to a resolution of 200×200 pixels.

2.1 Dynamic Window in LPQ Feature Extraction

The robustness of LPQ as a texture descriptor against image degradations such blur and lighting fluctuations makes it an ideal choice for facial texture analysis [13]. Enhancing

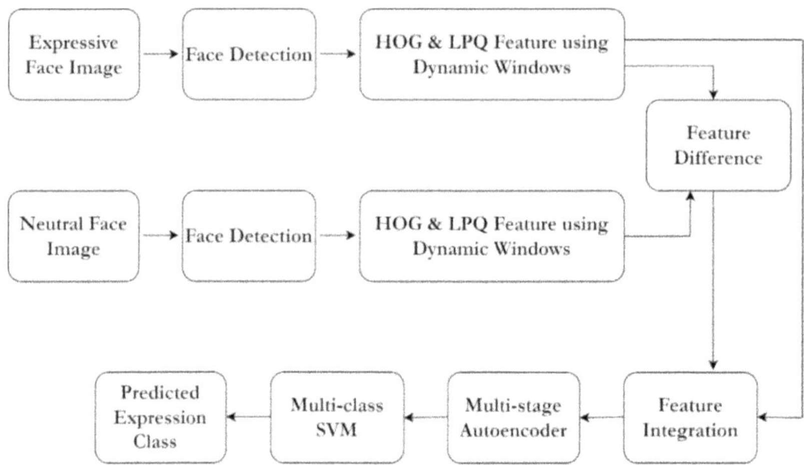

Fig. 1. Proposed framework for FEC

classification accuracy depends on the precise capture of nuanced texture variations in essential facial areas, including the eyes, mouth, and nose. Since FEs are inherently dynamic, feature extraction using fixed windows runs the risk of bypassing discriminative information or include unnecessary regions. By taking facial landmarks and the areas under analysis into account, Dynamic Windows can adjust the size and placement of the window accordingly. In regions where expression-driven fluctuations are most pronounced, the system is able to detect and record localized changes in texture. Dynamic windows ensure that key facial features are accentuated while decreasing noise from less expressive parts by adjusting their parameters according to the region of interest, as opposed to applying a uniform grid of fixed-size windows across the entire face.

Let $f(x, y)$ represents the face image. The adaptive window centred at coordinates (p, q) is represented by $W_{p,q}$ can be defined using Eq. (1).

$$W_{p,q}(k) = \left\{ (x, y) | (x - p)^2 + (y - q)^2 \le r_k^2 \right\} \tag{1}$$

Where 'r_k' represents the adaptive radius for the kth window centred at (p, q). The window size is adjusted to match the texture attributes of the local image by selecting r_k according to local image properties, such as local variance and intensity gradients. This enables the window size to diminish in regions of high variance or pronounced gradients (where intricate texture details are crucial) and expand in smoother areas.

The radius 'r_k' is represented in Eq. (2).

$$r_k = \alpha \left(\frac{1}{\sigma(p, q) + \varepsilon} \right) + \beta G_M(p, q) \tag{2}$$

Where

$\sigma^2(p, q)$: Local variance around the centre pixel (p, q)
$G_M(p, q)$: Gradient magnitude around the centre pixel (p, q)

α: Scaling parameter to regulate the impact of local variance
β: Scaling parameter to regulate the impact of gradient magnitude
ε: Constant to prevent division by zero

The frequency representation is obtained for each window $W_{p,q}(k)$ by applying 2D short term Fourier transform and defined using Eq. (3).

$$F(u, v) = \sum_{(x,y) \in W_{p,q}(k)} f(x, y)g(x - p, y - q)e^{-j2\pi(ux+vy)} \tag{3}$$

The smooth transition at the edges is handled *by* Gaussian window $g(x, y)$ centred at (p, q). The phase information is encoded by quantizing the real and imaginary parts of $F(u, v)$ and is denoted in Eq. (4).

$$Q(u, v) = sign(R(F(u, v))) + jsign(I(F(u, v))) \tag{4}$$

The features of the local texture are captured by this binary phase encoding. A histogram that represents the LPQ feature descriptor for the adaptive window is created using the quantized phase information. The LPQ histogram $H_{p,q}(k)$ for each window is obtained by considering the occurrences of quantized phase values $Q(u, v)$ and is defined in Eq. (5).

$$H_{p,q}(k) = [h_1, h_2, \ldots\ldots\ldots, h_N] \tag{5}$$

Where

h_i: the count of occurrences of the ith quantized value in the window
N: number of quantized phase values

LPQ histograms from each adaptive window in the image are combined to generate the final feature vector (represented in Eq. (6)).

$$F_{LPQ} = \left[H_{p,q}(1), H_{p,q}(2), \ldots\ldots\ldots H_{p,q}(K) \right] \tag{6}$$

2.2 Dynamic Window in HOG Feature Extraction

The process of application of dynamic windows in deriving HOG feature is similar to that of LPQ. In case of HOG feature using dynamic window approach, window size is chosen based on the underlying image characteristics, such as local gradient intensity or texture complexity. Regions with high gradient variations (e.g., sharp changes in pixel intensity near the mouth or eyes) can be analysed with smaller windows, allowing for finer detail capture. Smoother areas (e.g., cheeks or forehead) can use larger windows to reduce noise. The HOG feature derived from the adaptive window size is represented by F_{HOG}.

The Sobel operators are used to compute the gradient and are represented in Eq. (7) and Eq. (8) respectively.

$$G_h(x, y) = f(x + 1, y) - f(x - 1, y) \tag{7}$$

$$G_v(x, y) = f(x, y+1) - f(x, y-1) \tag{8}$$

The magnitude and orientation of the gradient at each pixel is computed using Eq. (9) and Eq. (10) respectively.

$$M(x, y) = \sqrt{G_h(x, y)^2 + G_v(x, y)^2} \tag{9}$$

$$\theta(x, y) = tan^{-1}\left(\frac{G_v(x, y)}{G_h(x, y)}\right) \tag{10}$$

The dynamic window $w(x, y)$ is determined by evaluating the local variance in gradient magnitudes or edge density and is represented in Eq. (11).

$$w(x, y) = f(\sigma_M(x, y)) \tag{11}$$

Where $\sigma_M(x, y)$ is the local variance of the gradient magnitudes over a neighborhood and $f(\sigma_M(x, y))$ is defined in Eq. (12).

$$f(\sigma_M(x, y)) = w_{min} + \delta(w_{max} - w_{min})e^{-\sigma_M(x,y)} \tag{12}$$

Where w_{max} & w_{min} are the minimum and maximum window sizes, and is a δ scaling factor.

The histogram of a is computed using Eq. (13).

$$H_c(i) = \sum_{(x', y')\varepsilon w(x,y)} M(x', y')F_i(\theta(x', y')) \tag{13}$$

Where $H_c(i)$ is the orientation bin for the cell, and F_i is an indicator function that assigns the gradient to the correct bin. To ensure that the HOG features are invariant to changes in lighting or contrast, block normalization is applied. For a block B composed of C cells, the normalized feature vector H_B is represented in Eq. (14).

$$H_B = \frac{H_C}{\sqrt{\sum_{c\epsilon B} H_C^2}} \tag{14}$$

The final feature vector F_{HOG} is represented in Eq. (15) for an image composed of k blocks.

$$F_{HOG} = \left[H_B{}^1, H_B{}^2, \ldots\ldots\ldots H_B{}^k\right] \tag{15}$$

HOG and LPQ features are independently extracted from both expressive and neutral facial images to capture texture and structural variations effectively. To amplify the discriminative capability of the feature set, the difference between these feature representations is computed, effectively highlighting subtle changes that occur between the expressive and neutral face images. These variations often encode crucial expression-specific characteristics, making them highly relevant for accurate FEC. This feature difference is mathematically formulated in Eq. (18). To construct a more comprehensive and robust feature representation, the feature vector derived from the expressive face is concatenated with the computed feature difference, forming a hybrid feature distribution, as defined in Eq. (19). This fusion strategy ensures that the final feature space retains both the direct expressive facial information and the relative changes between expressions, thereby improving classification performance.

Hybrid Feature Distribution for expressive:

$$H_{Expressive} = [F_{HOG} F_{LPQ}] \text{(Expressive face)} \tag{16}$$

Hybrid Feature Distribution for expressive:

$$H_{Neutral} = [F_{HOG} F_{LPQ}] \text{(Neutral face)} \tag{17}$$

$$\text{Feature Difference: } F_{Difference} = H_{Expressive} - H_{Neutral} \tag{18}$$

$$\text{Novel Feature Space } F_{Novel} = [H_{Expressive} F_{Difference}] \tag{19}$$

2.3 Latent Representation Learning via Autoencoder

The increased dimensionality of the hybrid feature representation introduces computation complexities in generating a model for FEC. A key component in minimizing the feature space while preserving the discriminative information is the utilization of stacked deep convolutional autoencoders (SDCAE) [14]. The three-stage SDCAE network is employed in the work and internal architecture of the network at each stage is represented in Fig. 2. Using a multi-layer convolutional encoding scheme, SDCAE reduced the initial high-dimensional LPQ feature vectors to a lower-dimensional latent space. Each of these layers learns to represent a different level of abstraction from the input information in a hierarchical fashion. Each stage has an encoder that extracts the highly discriminative patterns from the LPQ features and removes redundant information. From this compressed representation, the decoder reconstructs the original data. Reconstruction loss needs to be minimized such that the most pertinent information for classification is retained in the smaller feature set. Feature representations are refined from one layer to the next in the SDCAE's multi-stage architecture, which guarantees that dimensionality reduction happens in several steps. This permits a more sophisticated and progressive compression, whereby the elements pertinent to FEC can be captured at progressively higher levels of abstraction with each stage. The low-dimensional latent representation obtained from the autoencoder is subsequently fed into a Multiclass Support Vector Machine (SVM) for facial expression classification. By mapping the extracted features into a lower-dimensional space, the autoencoder ensures that only the most discriminative and relevant information is retained, reducing redundancy and improving generalization.

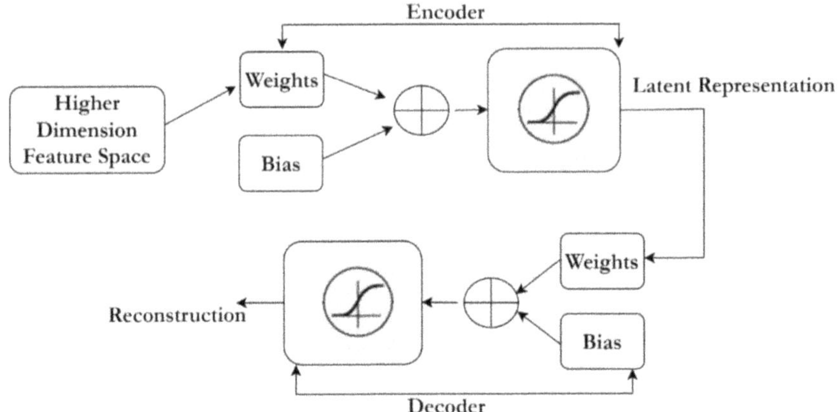

Fig. 2. Hierarchical composition of the autoencoder at each stage

3 Results and Discussion

The proposed study utilizes three widely recognized benchmark datasets for facial expression classification: CK+ [15], JAFFE [16], and KDEF [17], each offering diverse facial expression variations to ensure a robust evaluation of the model. The implementation is carried out on the MATLAB platform, utilizing an Intel i5 processor with 16 GB of RAM to efficiently perform feature extraction, dimensionality reduction, and classification. For each dataset, we have utilized 5-fold cross-validation to ensure the reliability and robustness of our results. The methodology is systematically divided into three key phases, each addressing a crucial aspect of the framework, as detailed in the following section.

3.1 Phase-1

In this phase, the classification model is developed using features extracted solely from the expressive facial images, utilizing adaptive window-based HOG and LPQ features. These handcrafted descriptors capture both structural gradients and texture variations, ensuring the preservation of key discriminative information necessary for facial expression recognition. Notably, this phase does not incorporate the feature distribution derived from the neutral facial images, nor does it employ stacked autoencoders for dimensionality reduction. The exclusion of these additional components allows for an initial assessment of the effectiveness of the expressive face features in driving classification performance. The recognition accuracy obtained for fixed window-based feature and adaptive window-based feature are represented in Table 1 and 2 respectively. The significant rise in classification accuracy of expressions for adaptive window based LPQ highlights the contribution of adaptive windows in LPQ for FEC.

Table 1. Result of fixed window-based HOG and LPQ on benchmark datasets (with reference to earlier work [6])

	Anger	Disgust	Fear	Happy	Sad	Suprise	Accuracy
CK+	93.6	95.4	89.2	98.5	92.4	97.6	94.5
KDEF	90.7	92.8	89.5	94.8	87.4	94.5	91.6
JAFFE	82.5	88.5	78.9	90.6	80.4	88.4	84.9

Table 2. Result of adaptive window-based HOG and LPQ on benchmark datasets

	Anger	Disgust	Fear	Happy	Sad	Suprise	Accuracy
CK+	94.2	95.9	89.6	98.5	93	98.2	94.9
KDEF	92.4	93.6	90.9	95.3	89.6	95.7	92.9
JAFFE	83.8	89.4	79.8	91.2	81.4	89	85.8

3.2 Phase-2

The feature difference represents the variation in feature distributions between the expressive and neutral facial images, effectively capturing subtle changes that characterize different facial expressions. This difference highlights expression-specific transformations, such as variations in texture, intensity, and structural patterns, which may not be apparent when analyzing expressive faces in isolation. A more informative and discriminative representation is achieved by creating a hybrid feature space from the expressive facial image's feature vector and the computed feature difference. This enriched feature space encapsulates both direct expression-related attributes and differential cues, enhancing the model's ability to distinguish between various facial expressions. The resulting hybrid feature space serves as input for building a robust classification model, facilitating improved recognition accuracy through the incorporation of both absolute and relative feature variations. The phase-2 excludes the application of stacked autoencoder. The findings given in Table 3 clearly demonstrate the influence of neutral expression in determining the highly discriminative distribution for FEC.

Table 3. Result of hybrid feature space on benchmark datasets

	Anger	Disgust	Fear	Happy	Sad	Suprise	Accuracy
CK+	94.2	96.5	90.2	98.5	93.6	98.8	95.3
KDEF	92.9	94.2	92	95.5	90.6	96	93.5
JAFFE	83.8	90.3	81.4	91.5	82.3	89.6	86.5

3.3 Phase-3

The multi-stage SDCAE is employed in this work for dimensionality reduction. The lower dimension highly discriminative feature space is derived from higher dimension hybrid feature space by using stacked autoencoders. The latent representation from the stacked autoencoders is utilized in building a model to predict the expression class. This work analyses the SDCAE in its single, two, and three stages. The recognition accuracy obtained for single, two, and three stage SDCAE is tabulated in Table 4. Adding extra layers to the SDCAE network can effectively reduce misclassification rates. Expanding the number of stages in the SDCAE network improves classification accuracy marginally. The confusion matrix obtained in this phase for CK+, KDEF, and JAFFE dataset in three stage SDCAE is reported in Table 5, 6, and 7 respectively. The use of adaptive windows in deriving LPQ feature has improved the FEC accuracy. The significant improvement in accuracy from phase 1 to phase 2 indicates the impact of neutral face expression on FEC. The improvement in accuracy from phase 2 to phase 3 highlights the contribution of SDCAE.

Table 4. Result of latent distribution derived from multi-stage SDCAE on benchmark datasets

		Anger	Disgust	Fear	Happy	Sad	Suprise	Accuracy
CK+	**Stage-1**	95.3	97.2	92.1	98.7	94.3	99.2	96.1
	Stage-2	95.6	97.5	92.5	98.7	94.6	99.6	96.4
	Stage-3	95.6	97.7	92.5	98.7	94.8	99.6	96.5
KDEF	**Stage-1**	93.5	95	92.6	96.3	91.4	96.5	94.2
	Stage-2	94	95.4	92.9	97.4	92.1	96.9	94.8
	Stage-3	94.3	95.9	93.2	97.4	92.5	97.1	95.1
JAFFE	**Stage-1**	86.3	92	82	94.6	82.9	93.2	88.5
	Stage-2	86.8	92.5	82.4	95.3	83.6	93.8	89.1
	Stage-3	87.1	92.7	82.4	95.7	83.6	94.2	89.3

Table 5. Confusion matrix obtained for CK+ dataset in phase-3 (3-stage SDCAE)

CK+	Anger	Disgust	Fear	Happy	Sad	Suprise
Anger	*95.6*	0	2.7	0	0.5	1.2
Disgust	0.4	*97.7*	0.8	0	1	0
Fear	2.5	0.8	*92.5*	0	1.5	2.7
Happy	0	0	0.4	*98.7*	0	0.8
Sad	1.5	1.5	2.1	0	*94.8*	0
Surprise	0	0	0.4	0	0	*99.6*

Table 6. Confusion matrix obtained for KDEF dataset in phase-3 (3-stage SDCAE)

KDEF	Anger	Disgust	Fear	Happy	Sad	Suprise
Anger	*94.3*	1.2	2.5	0	0	2
Disgust	0	*95.9*	1.5	0	2.5	0
Fear	1.2	0.5	*93.2*	0	2.5	2.5
Happy	0.5	0	0.5	*97.4*	0	1.5
Sad	2	2.5	3	0	*92.5*	0
Surprise	0.8	0	2	0	0	*97.1*

Table 7. Confusion matrix obtained for JAFFE dataset in phase-3 (3-stage SDCAE)

JAFFE	Anger	Disgust	Fear	Happy	Sad	Suprise
Anger	*87*	1.5	4.3	0	3.8	3.4
Disgust	2	*92.5*	2	0	3.4	0
Fear	3.8	2	*82.3*	0	3.4	8.5
Happy	1.5	0	1	*95.7*	0	1.8
Sad	4.4	6	2.5	0	*83.6*	3.4
Surprise	0.4	1.7	3.4	0	0	*94.5*

The performance comparison of proposed work with respect to state-of-the-art methods is tabulated in Table 8. While deep learning models have achieved impressive results in FEC, our proposed approach offers distinct advantages, particularly in scenarios where data scarcity, computational efficiency, and interpretability are critical. The extracted LPQ and HOG features, as well as the feature difference, provide a more interpretable representation compared to the abstract features learned by deep CNNs.

Table 8. Comparative analysis of the proposed work with the existing methods

	CK+	KDEF	JAFFE
Improved residual network [8]	96.37	93.38	—
Heterogeneous input networks [9]	93.4	—	94.7
Shape and texture feature [10]	—	88.8	95.8
Appearance features [11]	—	90.12	97.16
Our previous work [7]	96.7	94.7	93.5
Our previous work [3]	96.43	96.03	88.53
Proposed work	96.5	95.1	89.3

4 Conclusion

The proposed work emphasizes the significance of adaptive window-based feature extraction in capturing highly discriminative feature distributions for FEC. By dynamically adjusting the window size based on local image characteristics, this approach ensures that crucial texture and structural details are effectively preserved, enhancing the model's ability to distinguish between different expressions. A key aspect of this methodology is the feature difference, which quantifies the variations between the feature vectors derived from expressive and neutral facial images. This difference serves as a crucial indicator of expression-related changes, enabling the model to leverage both absolute feature representations and differential cues for improved classification accuracy. The multi-stage SDCAE is an effective method for dimensionality reduction and also improves the classification accuracy. The accuracy is slightly improved in adaptive window-based method compared to fixed window methods. Many existing deep frameworks have shown better classification accuracy compared to our work at the expense of higher computation cost. The proposed work enhances feasibility for real-time applications on embedded systems or mobile devices by efficiently compressing the feature space while preserving classification accuracy. The elliptical windows can be used in future to dynamically update the shape of the window based on the underlying image features. To address the computational cost, model optimization techniques, such as parameter pruning, quantization, or lighter autoencoder architectures can be employed to preserve performance while reducing the model size and computation required.

References

1. Karnati, M., Seal, A., Bhattacharjee, D., Yazidi, A., Krejcar, O.: Understanding deep learning techniques for recognition of human emotions using facial expressions: a comprehensive survey. IEEE Trans. Instrum. Meas. **72**, 1–31 (2023)
2. Ekman, P., Friesen, W.V.: Constants across cultures in the face and emotion. J. Pers. Soc. Psychol. **17**(2), 124 (1971)
3. Jain, A.K., Naveen Kumar, H.N.: Integration of discriminative information from expressive and neutral face image for effective modelling of facial expression classification problem. SN Comput. Sci. **5**(8), 1–11 (2024). https://doi.org/10.1007/s42979-024-03469-x
4. Ma, F., Sun, B., Li, S.: Facial expression recognition with visual transformers and attentional selective fusion. IEEE Trans. Affect. Comput. **14**(2), 1236–1248 (2021)
5. Naveen Kumar, H.N., Patil, C.M., Jain, A.K., Sudheesh, K.V.: A comprehensive study on geometric, appearance, and deep feature based methods for automatic facial expression recognition. In: 2022 Fourth International Conference on Cognitive Computing and Information Processing (CCIP), pp. 1–6. IEEE (2022). https://doi.org/10.1109/CCIP57447.2022. 10058627
6. Kumar, H.N., Naveen, Suresh Kumar, A., Guru Prasad, M.S., Shah, M.A.: Automatic facial expression recognition combining texture and shape features from prominent facial regions. IET Image Process. **17**(4), 1111–1125 (2023). https://doi.org/10.1049/ipr2.12700
7. Naveen Kumar, H.N., Patil, C.M., Nagaraja, B.G., Jain, A.K., Sudheesh, K.V., Mahadevaswamy, S.: Automatic facial expression recognition using modified LPQ and HOG features with stacked deep convolutional autoencoders. Wirel. Pers. Commun., 1–23 (2024). https://doi.org/10.1007/s11277-024-11564-8

8. Zhang, W., Zhang, X., Tang, Y.: Facial expression recognition based on improved residual network. IET Image Proc. **17**(7), 2005–2014 (2023)
9. Xie, S., Hu, H.: Facial expression recognition using hierarchical features with deep comprehensive multipatches aggregation convolutional neural networks. IEEE Trans. Multimed. **21**(1), 211–220 (2018)
10. Mahesh, V.G.V., Chen, C., Rajangam, V., Raj, A.N.J., Krishnan, P.T.: Shape and texture aware facial expression recognition using spatial pyramid Zernike moments and law's textures feature set. IEEE Access **9**, 52509–52522 (2021)
11. Yaddaden, Y.: An efficient facial expression recognition system with appearance-based fused descriptors. Intell. Syst. Appl. **17**, 200166 (2023)
12. Viola, P., Jones, M.J.: Robust real-time face detection. Int. J. Comput. Vis. **57**, 137–154 (2004)
13. Turan, C., Lam, K.M.: Histogram-based local descriptors for facial expression recognition (FER): a comprehensive study. J. Vis. Commun. Image Represent. **55**, 331–341 (2018)
14. Zeng, N., Zhang, H., Song, B., Liu, W., Li, Y., Dobaie, A.M.: Facial expression recognition via learning deep sparse autoencoders. Neurocomputing **273**, 643–649 (2018)
15. Lucey, P., Cohn, J.F., Kanade, T., Saragih, J., Ambadar, Z., Matthews, I.: The extended Cohn-Kanade dataset (CK+): a complete dataset for action unit and emotion-specified expression. In: 2010 IEEE Computer Society Conference on Computer Vision and Pattern Recognition-Workshops, pp. 94–101. IEEE (2010)
16. Lyons, M., Akamatsu, S., Kamachi, M., Gyoba, J.: Coding facial expressions with gabor wavelets. In: Proceedings Third IEEE International Conference on Automatic Face and Gesture Recognition, pp. 200–205. IEEE (1998)
17. Lundqvist, D., Litton, J.E.: "The averaged Karolinska directed emotional faces", Stockholm: Karolinska Institute, Department of Clinical Neuroscience, Section Psychology (1998)

AI Driven Canine Dog Nose Print Recognition

K. Bhavanishankar, Aaryan[✉], P. Vaishnavi[✉], R. C. Sreevidya[✉],
and Keshav Kumar Mishra[✉]

RNS Institute of Technology, Bangalore, Karnataka, India
{bhavanishankar.k,sreevidya.rc}@rnsit.ac.in,
ongaaryan08@gmail.com, vaishpraveen.1103@gmail.com,
keshav01m@gmail.com

Abstract. The Canine Nose Print Identification System introduces a cutting-edge approach to dog identification by utilizing the unique biometric patterns of a dog's nose. Traditional methods such as collars and microchips are often invasive, prone to tampering, or ineffective for reliable identification. This work provides a non-invasive, AI-powered solution to accurately identify dogs, enhancing pet safety and welfare. This method uses a pre-trained ResNet50 for feature extraction and cosine similarity for accurate matching. It was trained on a dataset of 20,200 photos from 4,040 distinct dogs. It uses convolutional kernel filters to improve image quality and Laplacian variance for sharpness detection to guarantee accuracy. Features from the uploaded test image are extracted and compared to the complete dataset as part of the comparison process. The system cuts batch processing time by more than two minutes for 100 images by improving GPU acceleration, which lowers comparison time to 0.62 s per image. This is because SIFT-based approaches are more time-consuming and computationally costly. This paper offers strong support for earlier research showing that canine nose patterns are distinct biometric identifiers, highly useful for tracking vaccinations and aiding in pet recovery.

Keywords: Dog identification · biometric patterns · canine nose print ·
AI-powered solution · ResNet50 model · image preprocessing · cosine
similarity · pet recovery · stray dog welfare · vaccination management ·
non-invasive identification · scalable system · mobile application

1 Introduction

Dogs hold a special place in our lives, serving as loyal companions and valued members of our communities. However, identifying and keeping track of them presents a significant challenge. Each year, millions of dogs go missing, leaving families heartbroken and placing a strain on overcrowded animal shelters. Additionally, stray dogs often go unnoticed and unaccounted for, missing out on essential care such as vaccinations and opportunities for adoption. Traditional identification methods, such as collars and microchips, come with limitations. Collars can be lost or removed, while microchips require invasive procedures and specialized scanners, making them less ideal for widespread use. These drawbacks highlight the need for a reliable, non-invasive identification system that

© The Author(s) 2026
J. Shreyas et al. (Eds.): CODE-AI 2025, CCIS 2689, pp. 112–123, 2026.
https://doi.org/10.1007/978-3-032-19318-6_11

ensures the safety and well-being of both pets and stray dogs. This work addresses this need by leveraging the unique patterns found in canine nose prints. Similar to human fingerprints, a dog's nose print is distinct and remains unchanged throughout its life, making it a natural and tamper-proof method of identification. With advancements in technology, particularly in image processing and artificial intelligence (AI), it is now possible to accurately recognize dogs based on their nose prints. By utilizing cutting-edge image processing techniques and AI-driven recognition algorithms, this system captures and analyzes the intricate patterns of beads and grooves on a dog's nose. The extracted features are then compared against the original data to confirm the dog's identity. This innovative approach not only enhances pet recovery efforts but also supports animal shelters by streamlining the identification process and reducing cases of pet theft. In an era where technology is deeply integrated into daily life, this AI-powered solution offers a non-invasive, efficient, and scalable method for dog identification. By improving identification and tracking capabilities, the system contributes to the overall welfare of dogs, promoting a safer and more compassionate environment for pets and strays alike. Through its user-friendly design and potential integration into mobile applications, this system is poised to revolutionize pet identification, ensuring that lost dogs can be quickly and accurately reunited with their families.

2 Related Works

Fei Shen et al., Bin Li, Zhongan Wang et al. [1, 2] addresses the challenge of limited labeled data, this paper proposes an automatic offline data augmentation strategy for dog nose-print authentication. The method integrates cross-entropy, triplet, and pairwise circle loss functions for network optimization, achieving notable performance in re-identification tasks. Agata Kokoci´nska-Kusiak et al., Ewart J. de Visser et al. [3, 4] the comprehensive overview delves into the anatomy, physiology, and behavioral aspects of canine olfaction. While not directly focused on nose print recognition, it provides foundational insights into the uniqueness of canine nasal structures, which are pivotal for biometric identification. H. B. Bae et al. [5], Kenneth Lai [6] This study addresses the growing need for reliable pet identification systems due to the increase in pet ownership. The researchers developed a non-invasive identification method utilizing deep neural networks to analyze the unique patterns of dog nose prints. The system demonstrated high accuracy, offering a promising solution for pet recovery and insurance fraud prevention. M. V. Caya [7], Y. Boulaouane [8] Recognizing the uniqueness of canine nose patterns, this research presents a biometric system for dog identification. The approach involves capturing high-resolution images of dog nose prints and processing them through image enhancement and feature extraction techniques. The system achieved a high identification rate, highlighting the potential of nose print biometrics in practical applications. S. Cho et al. [9], X. Tu, K. Lai et al. [10] This paper proposes a dog noseprint identification system based on Gabor filters and feature matching. The methodology includes region of interest determination, adaptive thresholding, and feature extraction using Gabor filters in multiple directions. The system demonstrated high accuracy in matching nose prints, validating the effectiveness of the proposed algorithm. R. Wang et al. [11], C.H. Lin et al. [12] Addressing the need for large-scale datasets, this paper introduces a comprehensive dog face dataset to facilitate research in canine identification. The dataset includes diverse images with annotations, enabling the training and benchmarking of detection algorithms. The study also evaluates various models, providing insights into their performance and guiding future research in dog face detection and recognition.

3 Proposed Methodology

The AI-driven canine nose print recognition system utilizes advanced deep learning techniques to identify dogs through their unique nose patterns. This system is designed to address challenges in traditional dog identification methods, providing a non-invasive, scalable, and reliable solution. The methodology as shown in Fig. 1 involves five key modules: Data Acquisition and Organization, Image Preprocessing and Fine-Tuning, Feature Extraction Using Deep Learning, Threshold-Based Identification, and Output and Graph Analysis.

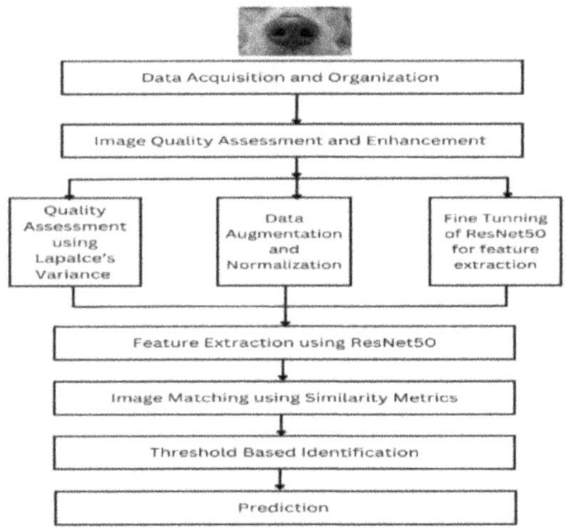

Fig. 1. Block diagram of Canine nose print identification

3.1 Data Acquisition and Organization

The first step is collecting high-quality images of dog nose prints to create a diverse and well-structured dataset. This database ensures accurate identification and retrieval during the recognition process. The dataset comprises: Internet Sources: Images collected from open-source repositories, covering various breeds and conditions. Manual Capture: High-quality, real-time images of dogs in natural settings to ensure diversity. The dataset includes images from 4,040 unique dogs, with five images per dog, totaling 20,200 images. Each image as shown in Fig. 2 is labeled with metadata such as image title, path, and Dog ID, stored in a structured directory for efficient retrieval using CSV files.

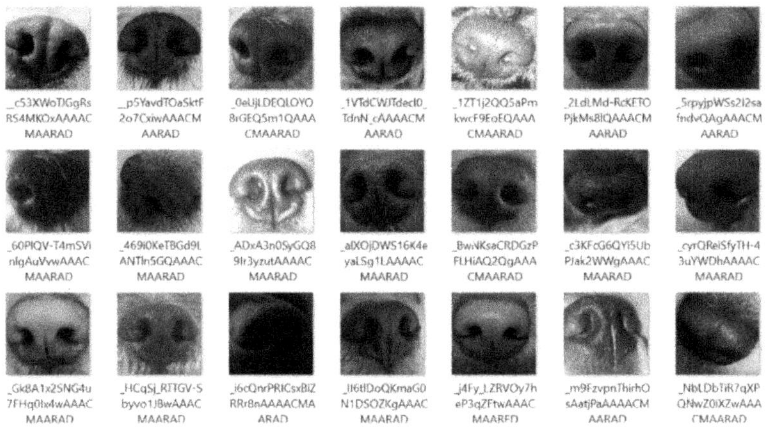

Fig. 2. Sample images stored in the dog nose print recognition dataset

3.2 Image Preprocessing and Fine-Tuning

3.2.1 Image Quality Assessment (Laplace Variance)

The Laplacian variance method evaluates image sharpness, ensuring only clear images are processed. The grayscale image is analyzed using the Laplacian operator to highlight intensity variations. If the variance is below a predefined threshold (e.g., 100), the image is considered blurry, and the user is prompted to upload a clearer image.

3.2.2 Image Enhancement and Data Augmentation

A sharpening kernel is applied to enhance edges and fine details, improving feature extraction efficiency. Data augmentation techniques such as flipping, rotation, and brightness adjustments increase the model's robustness. All images are resized to 224 × 224 pixels and normalized to fit ResNet-50's input requirements.

3.2.3 Fine-Tuning of ResNet-50

The pre-trained ResNet-50 model is fine-tuned for nose print identification. The initial convolutional layers are frozen to retain general features, while the deeper layers are retrained on the dog nose print dataset to capture breed-specific features. The final fully connected layers are replaced with custom layers designed to extract biometric features.

3.3 Feature Extraction Using Deep Learning (ResNet-50)

The ResNet-50 model, known for its deep residual learning capabilities, extracts numerical feature vectors representing the unique patterns of each dog's nose print. The process includes: Scaling each image to 224 × 224 pixels and converting it into a numerical array. Using Global Average Pooling (GAP) to convert feature maps into compact feature vectors. Flattening the extracted feature vector into a 1D array and storing it in a dictionary indexed by Dog ID. This approach captures the ridge patterns, texture, and structural features of each dog's nose print, ensuring accurate biometric representation.

3.4 Threshold-Based Identification (Cosine Similarity)

The system compares a newly uploaded dog nose print with stored feature vectors using cosine similarity, which measures the angular distance between two vectors. The similarity score ranges from -1 to 1, with 1 indicating identical vectors and 0 indicating completely different vectors. A threshold-based decision mechanism ensures that only highly similar nose prints are matched. For example, if the similarity score exceeds 0.8, the system confirms a match; otherwise, it returns "No Match Found."

3.5 Output and Graph Analysis

The final module provides the identification results and visualizes the similarity scores. The system generates a bar chart displaying cosine similarity scores for each stored dog image, with the highest similarity score highlighted. A threshold line indicates the minimum acceptable similarity for a valid match. This graphical representation helps assess the system's performance and refine the similarity threshold as needed.

4 Implementation

4.1 Laplace Variance

One popular mathematical method for determining image sharpness is Laplace variance. Its foundation is the Laplacian operator, which highlights regions of sharp variations in intensity to identify edges in an image. To ensure that only high-quality photos are processed for identification, our study uses Laplace variance to evaluate the sharpness of dog nose print images prior to feature extraction. We increase the precision of feature extraction and similarity matching by removing hazy or poor-quality photos, which eventually strengthens the biometric identification system's dependability.

Mathematical Representation of Laplace Variance:

The Laplacian operator is a second-order derivative filter that calculates the rate of change in pixel intensities. The Laplace variance of an image I(x, y) is computed using the Laplacian transformation:

$$L(x, y) = \frac{\partial^2 I}{\partial x^2} + \frac{\partial^2 I}{\partial y^2} \tag{1}$$

where:

I(x, y) represents the grayscale intensity of the image at pixel location (x, y).

$(\partial^2 I)/[\partial x]^2$ and $(\partial^2 I)/[\partial y]^2$ denote the second-order derivatives in the horizontal and vertical directions, respectively.

The Laplacian variance (σ_L^2) of an image is then calculated as:

$$\sigma_L^2 = \frac{1}{N} \sum_{x,y} \left(L(x, y) - \bar{L} \right)^2 \tag{2}$$

where:

N is the total number of pixels in the image.
L(x, y) is the Laplacian value at pixel (x, y).
$\bar{L}$ is the mean Laplacian value across the image.

4.2 ResNet Model

ResNet-50, a 50-layer residual network, revolutionized deep learning by solving the vanishing gradient problem, enabling efficient training of deep models. It excels in vision tasks like object detection and image classification, learning hierarchical features for precise image understanding. Widely used in AI applications, it supports transfer learning and balances computational efficiency with depth.

Traditional deep networks attempt to learn a mapping function H(x) where x represents the input. However, deep models may struggle to optimize this function directly. Instead, ResNet-50 introduces a residual function:

$$F(x) = H(x) - x \tag{3}$$

Rearranging this equation, we get:

$$H(x) = F(x) + x \tag{4}$$

Here, F(x) represents the transformation applied by the residual block, while x is the identity mapping. This formulation ensures that even if the transformation learns nothing (i.e., F(x) = 0), the identity function allows information to pass unchanged, preventing vanishing gradients.

Forward Propagation in a Residual Block

A residual block with two weight layers follows this equation:

$$y = F(x, W) + x \tag{5}$$

Where:

x is the input,
W represents the learnable parameters of the convolutional layers,
F(x, W) denotes the transformation applied by the residual block,
y is the output of the residual block.

This formulation ensures that even deeper networks retain crucial information from earlier layers.

Table 1. ResNet architecture

Level	Layer Type	Size	Stride	Output Feature Map
1	Convolution	7×7	2	Reduced Spatial Size
1	Max Pooling	3×3	2	Further Reduction
2	**Residual Block 1 (3 Bottleneck Units)**			
	1×1 Convolution	1×1	1	Reduces Channels
	3×3 Convolution	3×3	1	Extract Features
	1×1 Convolution	1×1	1	Expands Channels
3	**Residual Block 2 (4 Bottleneck Units)**			

(continued)

Table 1. (continued)

Level	Layer Type	Size	Stride	Output Feature Map
	1×1 Convolution	1×1	1	Reduces Channels
	3×3 Convolution	3×3	1	Extract Features
	1×1 Convolution	1×1	1	Expands Channels
4	**Residual Block 3 (6 Bottleneck Units)**			
	1×1 Convolution	1×1	1	Reduces Channels
	3×3 Convolution	3×3	1	Extract Features
	1×1 Convolution	1×1	1	Expands Channels
5	**Residual Block 4 (3 Bottleneck Units)**			
	1×1 Convolution	1×1	1	Reduces Channels
	3×3 Convolution	3×3	1	Extract Features
	1×1 Convolution	1×1	1	Expands Channels
6	Global Average Pooling	–	–	Reduces Spatial dimensions
6	Fully Connected Layer	–	–	1000 classes(for ImageNet)
6	Softmax Activation	–	–	Probability Distribution

As seen in Table 1, ResNet-50 consists of five stages, each containing convolutional and identity blocks. The model follows this structure:

- Initial Convolutional Layer – 7×7 conv layer with stride 2, followed by max pooling.
- Residual Blocks (50 layers total) – Composed of 1×1, 3×3, and 1x1 convolution layers.
- Global Average Pooling Layer – Reduces spatial dimensions before feeding into the classifier.
- Fully Connected Layer (Softmax output) – Produces probability scores for classification

4.2.1 Fine Tuning of ResNet Model – Freezing the Initial Convolutional Layers

ResNet50's layered architecture is designed to capture hierarchical features, starting with simple patterns like edges and textures in its initial layers. Since these foundational features are common across various image types, the early layers are frozen during fine-tuning. This ensures that their pre-trained weights remain unchanged, preventing unnecessary adjustments that could negatively affect the model's generalization performance. The deeper layers, however, are trained specifically on dog nose prints to learn more complex and domain-specific patterns.

4.2.2 Adapting the Fully Connected Layers

The fully connected (FC) layers in ResNet50 were initially designed for ImageNet's 1000-class classification task. For the purpose of identifying individual dogs through their nose prints, these top layers are removed and replaced with a custom configuration.

This new setup includes dense layers, global average pooling, and a classification layer tailored to match the number of unique dogs in the dataset, enabling the model to extract and differentiate biometric features effectively.

4.2.3 Extracting Features from the Penultimate Layer

Instead of using the final classification layer, the penultimate layer—specifically the Global Average Pooling (GAP) layer—is leveraged for feature extraction. This layer condenses the feature maps into compact, high-dimensional vectors, which represent the unique characteristics of each dog's nose print. These feature vectors are stored in a database and used for similarity comparisons, with cosine similarity employed to measure the degree of match. By extracting features from the penultimate layer, the model achieves better generalization and improved performance in biometric identification.

$$GAP_c = \frac{1}{N} \sum_{i=1}^{N} F_{i,c} \tag{6}$$

GAPc = The pooled feature for channel c.
N = Total Number of spatial locations in each feature map.
F(i, c) = Activation value at spatial location i for channel c.

4.2.4 Data Augmentation for Enhanced Robustness

Nose print images can vary due to factors such as minor occlusions, different angles, and lighting conditions. To ensure the model becomes robust to these variations, data augmentation techniques are applied during training. These techniques include rotations, slight zooming, and brightness adjustments, enabling the model to learn invariant features and improve its generalization in real-world scenarios.

4.2.5 Regularization to Prevent Overfitting

To prevent overfitting, regularization methods such as batch normalization, L2 weight decay, and dropout are incorporated. Batch normalization stabilizes the learning process, L2 weight decay discourages overly complex models by penalizing large weights, and dropout randomly deactivates neurons during training, ensuring the model captures essential biometric features rather than memorizing noise from the training data.

4.2.6 Model Training and Optimization

The model is trained using an adaptive learning rate optimizer like Adam, which dynamically adjusts the learning rate based on gradient updates for efficient convergence. For the multi-class classification task of identifying dogs through their nose prints, categorical cross-entropy is used as the loss function, ensuring accurate prediction of each dog's identity.

4.3 COSINE Similarity

Cosine similarity is a widely used metric in machine learning and deep learning for measuring the similarity between two vectors. In order to compare feature vectors of dog nose prints and ensure accurate identification, cosine similarity—a commonly used metric in machine learning and deep learning—is key to our project. Cosine similarity calculates the cosine of the angle as shown in between two feature vectors to determine how closely the uploaded image resembles stored dog nose prints.

Mathematical Representation for cosine similarity:

Cosine similarity between two vectors A and B is mathematically defined as:

$$\cos(\theta) = \frac{A.B}{\|A\|\|B\|} \tag{7}$$

Where:

A and B are two feature vectors representing dog nose prints.

A·B is the dot product of the vectors.

$\|A\|$ and $\|B\|$ represent the Euclidean norms (magnitudes) of vectors A and B, calculated as:

$$\|A\| = \sqrt{\sum_{i=1}^{n} A_i^2} \quad \|B\| = \sqrt{\sum_{i=1}^{n} B_i^2} \tag{8}$$

$\cos(\theta)$ ranges between -1 and 1, where:

1 indicates identical feature vectors (perfect match).

0 indicates orthogonal vectors (completely dissimilar).

-1 indicates opposite vectors (not relevant for our case).

5 Results and Discussions

The model allows users to upload images for testing dog nose print recognition. Each image undergoes preprocessing, including image quality enhancement and sharpness evaluation, before feature extraction and similarity comparison against a dataset of dog nose prints. The results are displayed in Fig. 3, Fig. 4 and Fig. 5 with a similarity score and matching status.

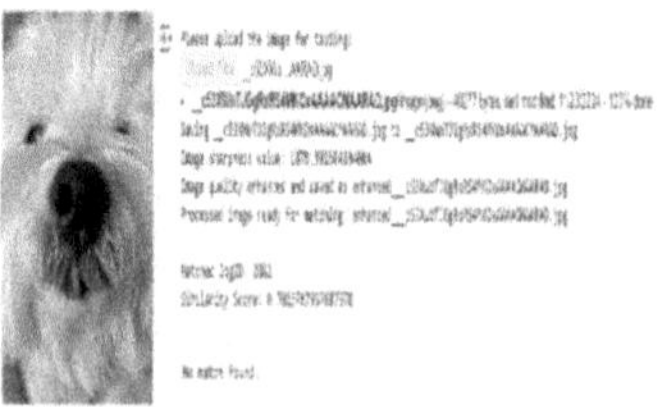

Fig. 3. Dog Nose Print Recognition

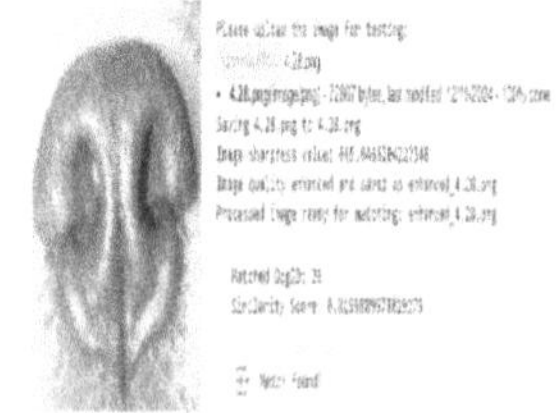

Fig. 4. Successful Dog Identification

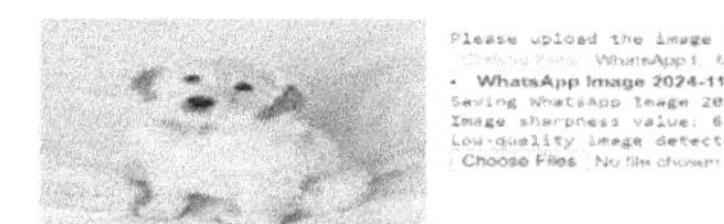

Fig. 5. Image Quality Alert: Low Sharpness Detected

A visual depiction of the similarity scores between the stored feature vectors and the uploaded dog nose print is offered by the Graph Analysis module. The system's ability to distinguish between dogs and the efficiency of the similarity threshold in detecting precise matches can both be ascertained by visually examining these scores. This module calculates each saved dog image's cosine similarity score and shows the results as a bar chart. A Dog ID and the associated similarity score with the uploaded image are represented by each bar. A threshold line is created to show the lowest permissible similarity for a valid match, and the maximum similarity score is underlined. The graph shown in Fig. 6 offers important information about the distribution of similarity ratings among several dog photos that have been saved. If the top match exceeds the predetermined threshold for similarity (e.g., 0.8). The matching algorithm's performance, which aids in evaluating false positives or close matches that could need more adjustment. As shown in Fig. 6, graph-based evaluation enhances the methodology by providing quantifiable insights into the model's performance, allowing for better refinement and optimization in biometric dog nose print recognition.

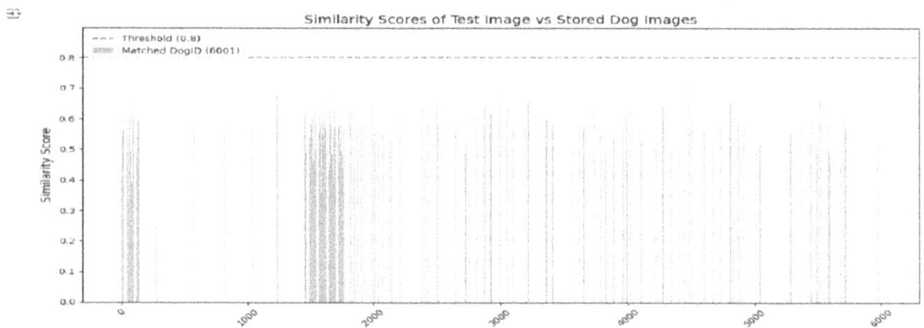

Fig. 6. Distribution of similarity ratings

6 Conclusions

The Canine Recognition system revolutionizes pet identification by using the unique biometric patterns found in a dog's nose. Unlike traditional methods such as ID tags and microchips, this AI-powered approach is contactless, tamper-proof, and highly accurate. It enhances pet recovery rates, supports animal shelters, improves security, and centralizes pet data, promoting the safety and well-being of pets worldwide. Key findings highlight that each dog's nose print contains distinct features, ensuring reliable identification with minimal false positives and negatives. The system is scalable, capable of handling large datasets with improved accuracy as more data is added, and has real-world applications in pet management, animal shelters, pet insurance, and medical care. Future enhancements include expanding the system to identify other species like cats, rabbits, and exotic animals, developing a mobile application for easy scanning and identification by pet owners and veterinarians, improving feature extraction using advanced models like Convolutional Neural Networks (CNNs), and enabling real-time matching for instant identification. Together, these advancements position the system as a scalable, accessible, and highly reliable solution for modern pet identification.

References

1. Fei, S., et al.: A Competitive method for dog nose-print re-identification. Biometrics 2022
2. Li, B., et al.: Dog nose print matching with dual global descriptor based on contrastive learning. Comput. Vis. Pattern Recognit. https://doi.org/10.48550/arXiv.2206.00580
3. Zybala, M., Maciocha, J., Barłowska, K., Dzięcioł, M.: Canine olfaction: physiology, behavior, and possibilities for practical applications. Animals **11**, 2463 (2021)
4. de Visser, E.J., et al.: Designing man's new best friend: enhancing human-robot dog interaction through dog-like framing and appearance. Sensors **22**, 1287 (2022)
5. Bae, H.B., Pak, D., Lee, S.: Dog nose-print identification using deep neural networks. IEEE Access **9** (2021). https://doi.org/10.1109/ACCESS.2021.3068517
6. Lai, K., Tu, X., Yanushkevich, S.: Dog identification using soft biometrics and neural networks. In: 2019 International Joint Conference on Neural Networks (IJCNN), Budapest, Hungary, pp. 1–8 (2019). https://doi.org/10.1109/IJCNN.2019.8851971
7. Caya, M.V., Arturo, E.D., Bautista, C.Q.: Dog identification system using nose print biometrics. In: 2021 IEEE 13th International Conference on Humanoid, Nanotechnology, Information Technology, Communication and Control, Environment, and Management (HNICEM). IEEE (2021). 978-1-6654-0167-8/21
8. Boulaouane, Y., Garifulla, M., Lim, J., Pak, D., Lim, J.: Advancing pet biometric identification: a state-of-the-art unified framework for dogs and cats. IEEE Access (2024). https://doi.org/10.1109/ACCESS.2024.3516130
9. Cho, S., et al.: Dog noseprint identification algorithm. In: 2021 International Conference on Information Networking (ICOIN). IEEE. 978-1-7281-9101-0/20
10. Wang, R., et al.: Dog face dataset and detection benchmarking. In: 2023 IEEE 6th International Conference on Pattern Recognition and Artificial Intelligence (PRAI), pp. 83–87 (2023)
11. Lai, K., Tu, X., Yanushkevich, S.: Dog identification using soft biometrics and neural networks. In: International Joint Conference on Neural Networks, Budapest, Hungary, 14–19 July 2019. 978-1-7281-2009-6
12. Lin, C.-H., Li, Y.-X., Zheng, C.-H., Chen, H.-Y., Chan, Y.-K., Hsu, L.-H.: Development and evaluation of a dog breed and identity recognition model based on convolutional neural networks. In: 8th International Conference on Imaging, Signal Processing and Communications (ICISPC). IEEE (2024). 2831-3984/24/. https://doi.org/10.1109/ICISPC63824.2024.00020

AI Based LSTM-BERT Model for False News Detection

Sanjana Muthukumar, Joanna Grace Fernandez, B. J. Ambika$^{(\boxtimes)}$, S. A. Karthik, Preethi, and P. Soumyashree

Department of Computer Science and Engineering, Manipal Institute of Technology Bengaluru, Manipal Academy of Higher Education, Manipal 576104, India
{sanjana1.mitblr2022,joanna.mitblr2022}@learner.manipal.edu,
ambika.bj@manipal.edu

Abstract. The spread of fake news in this digital era represents a critical challenge, disrupting information sharing, eroding public trust, and fueling misinformation with far-reaching societal consequences. This paper reviews fifteen recent studies (2019–2024) that employ NLP techniques for fake news detection, focusing on hybrid approaches of machine and deep learning. Key areas explored include feature extraction, sentiment analysis, and text classification. To contribute to this body of research, we implement, evaluate and combine 2 robust models: Long Short-Term Memory (LSTM) networks, and the Transformer Based - Bidirectional Encoder Representations from Transformers (BERT) model. We test novel hybrid framework that combines the strengths of LSTM's sequential data analysis with BERT's contextual understanding to enhance detection accuracy. This hybrid model is then compared with existing ensemble techniques. This paper not only benchmarks the performance of the implemented models against prior studies but also identifies persistent challenges, such as dataset bias and scalability constraints. By integrating theoretical insights with practical experimentation, we offer actionable directions for improving the performance of false news detection systems. Our hybrid model uniquely integrates LSTM's sequential analysis with BERT's contextual understanding, achieving state-of-the-art performance in fake news detection.

Keywords: Natural Language Processing · LSTM · BERT · Hybrid models · Artificial Intelligence · Misinformation · Ensemble Techniques

1 Introduction

The digital era has revolutionized how information is created, shared, and consumed. Today, information is disseminated on a scale unlike ever before. While this transformation has democratized knowledge dissemination through social media platforms, blogs, and news websites, it has also blurred the line between factual reporting and fabricated narratives. This gives rise to one of the most pressing issues, the prevalence and existence of fake news. It distorts public opinion, fuels misinformation, and erodes trust in legitimate and reliable sources.

J. Shreyas et al. (Eds.): CODE-AI 2025, CCIS 2689, pp. 124–134, 2026.
https://doi.org/10.1007/978-3-032-19318-6_12

When misinformation spreads, there are major consequences, which are even more dire during emergencies and crises. At such times, the widespread of fake news is especially dangerous as it results in unnecessary panic and fear. A case study demonstrating this phenomenon is the COVID-19 pandemic, in which fake news promoted ineffective remedies, and undermined public health initiatives. This proves how much damage misinformation can cause during calamities and highlights the need for an efficient fake news detection system [1].

Nowadays, people can easily generate and share content. This makes the problem even worse. Manual fact-checking is a possible solution, but it is neither effective nor efficient on a large-scale. This led to the utilization of automated methods for false-news detection, which leverage various techniques in Natural Language Processing to differentiate between false news and true news. These tools use and combine different machine learning and deep learning methods to combat the spread of misinformation [2].

In recent times, deep learning models like LSTM and BERT have shown a lot of progress in this domain [3]. LSTMs show great potential in detecting temporal dependencies in sequential data, such as the order of words or events in a news article, which can help identify inconsistencies or patterns indicative of fake news. However, BERT, because of its bidirectional attention mechanism, facilitates a much deeper understanding of contextual relationships within the text. These tools provide excellent results in various NLP tasks [4].

By analyzing and reviewing 15 recent studies (2019–2024), this evaluation assesses the overall performance of various models on public datasets, specifically the Fake News Dataset from Kaggle. The assessment conducted, made use of accuracy, precision, recall, F1-score classification reports, and confusion matrices.

In this work, we have implemented LSTM and BERT-Transformer models individually and compared their performance and accuracy. A hybrid model that combines the strengths of each individual model was developed using the ensemble technique of stacking, thus improving the accuracy and performance compared to the individual models. Integrating the temporal analysis proficiency of LSTMs with the contextual capabilities of BERT, the model enhances the accuracy of fake news detection. While existing studies have explored individual models like LSTM and BERT, there is limited research on hybrid frameworks that combine their strengths for fake news detection. This paper fills that gap by proposing a novel hybrid model.

Furthermore, this work also highlights persistent challenges, such as dataset biases and adversarial attacks, and discusses opportunities for future research in improving scalability, adaptability, and real-time detection capabilities.

2 Methodology

Table 2 presents the datasets used in this study, sourced from Kaggle. It contains the labelled news articles that can be used for binary classification. Each entry is allotted either a unique value 0 or 1, for real and fake news respectively. This dataset provides a robust foundation for training the model. It also helps in evaluating the machine learning models. The primary purpose of this work is to accurately divide and differentiate proper

Table 1. Literature Review

TABLE I: Literature Review

Author	Methods Used	Advantages	Remarks
B. Parikh, V. Patil and P. K. Atrey (2019) [5]	1. Machine Learning 2. Hypothesis Testing	Successfully proved that the hypothesis stating "There is a relation between tone (sentiments) and fake news of a news story" can be accepted, using X2-test by obtaining X2 as 57.04	Future work needs to be done on (1) expanding the study on additional hypotheses and (2) designing and developing a multifarious fusion model to better detect given news for fake news or legitimacy
K. -H. Kim and C. -S. Jeong (2019) [6]	1. Fact DB 2. BiMPM 3. Article Abstraction 4. Entity Matching	Integration of the Article Abstraction module with the BiMPM model resulted in an improved accuracy, with an AUC of 0.744, as compared to the initial AUC of 0.663 achieved using only the BiMPM model	Future work can be done on (1) Developing more sophisticated abstraction techniques to better capture nuances in articles and improve semantic matching accuracy and (2) Conducting extensive real-world testing to validate the system's effectiveness in practical scenarios and refining based on feedback
A. Kuriakose, D. Sebastian, E. M. Mathew, H. Mathew and G. Er.Gokulnath (2019) [7]	1. Python 2. Flask 3. Neural Network 4. VADER	The Clickbait Score is 0.48039448, which falls below the threshold of 0.88, indicating that the news is classified as clickbait.	The sentiment analysis relies on user comments, which might not always provide a complete picture of the news' credibility
S. Gaonkar, S. Itagi, R. Chalippatt, A. Gaonkar, S. Aswale and P. Shetgaonkar (2019) [8]	1. Machine Learning 2. Text-based analysis 3. URL validation	This literature review shows that the accuracy for predicting fake news in social media is much higher than any other online news media; authors have hence targeted online news media fake news detection along with website validation	In future work, the proposed model will be tested for fake news detection by using URL as an input which will not only validate headline but will also validate site behavior and other related parameters
T. Traylor, J. Straub, S. Chaudhary and N. Snell (2019) [9]	1. Textblob 2. Natural Language Toolkit (NLTK) 3. SciPy Toolkits 4. Bayesian Machine Learning System 5. Quoted Attribution Analysis	(1) The tool identified 96% of the quotes in the training set (2) The final overall Classifier Accuracy is 0.69 and the overall Classifier Error is 0.31 (3)After training and configuration, the tool correctly identified 69.4% of the fake and real news documents in the test set (4)Overall precision is 63.333%	Future work can be done in combing attribution feature extraction with other factors to produce tools that not only identify potential false content, but influence-based content designed to compel a reader or target audience to make inaccurate or altered decisions
Y. Yanagi, R. Orihara, Y. Sei, Y. Tahara and A. Ohsuga (2020) [10]	1. Neural Networks 2. Deep Learning 3. Microblog Analysis (likes, retweets, replies, comments)	The effect of generated comments is measured for classification by comparing classification results with and without the generated comments; the proposed method achieved the best recall score	In precision, the model was outperformed by models that did not use the generated comments; improvements must be made by checking if the trends change by using a larger dataset
R. R. Mandical, N. Mamatha, N. Shivakumar, R. Monica, A. N. Krishna (2020) [11]	1. Naive Bayes 2. Passive Aggressive Classifier 3. Deep Neural Networks	The proposed models, particularly DNN, demonstrated high accuracy across multiple datasets, with some datasets achieving nearly 100% accuracy, showcasing the effectiveness of the approach	There is a high potential for the creation of better models; the usage of CNN and RNN shows great prospects that can be exploited, and pre-trained word embeddings such as word2Vec and GloVe could be used.

Table 2. Datasets

Dataset Name	Source	No. of Articles
Fake	Kaggle	23502
True	Kaggle	21417

news as either real or fake. Upon extensively reviewing related work, we found that Transformers (BERT) and LSTMs are two of the most used models for false news detection. Figure 1, Maham et al. [20] shows the outcomes and performance of the model after training on the LIAR dataset. All evaluation metrics parameters of BERT, LSTM, RoBERTa, Longformer, GRU and Random Forest are all above 90% for false news detection on the LIAR dataset. Since BERT and LSTM are both strong, robust models for detecting fake news, we decided to implement combination of them both (Table 1).

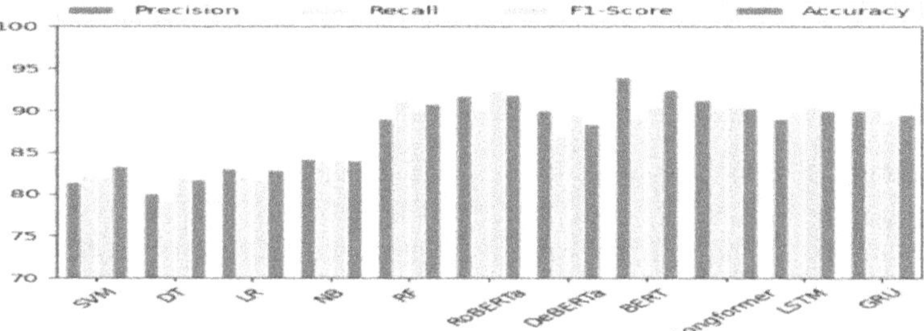

Fig. 1. Results of mentioned methods using random political news

2.1 LSTM Model

Long Short-Term Memory (LSTM) is similar to Neural Network architecture. It is specifically used in deep learning tasks to capture and retain sequential information over long periods. LSTMs are a special variant of Recurrent Neural Networks (RNNs). They overcome the vanishing gradient problem typically faced by RNNs. They have gates that regulate the flow of information and determine which information to keep and which to discard based on its importance/relevance [21].

Process

1. Words are transformed into vectors and processed sequentially, one by one.
2. The previous hidden state is passed to the next step in the sequence. The hidden state serves as the network's memory.

Pipeline

The pipeline for the LSTM model we implemented is given in Fig. 2.

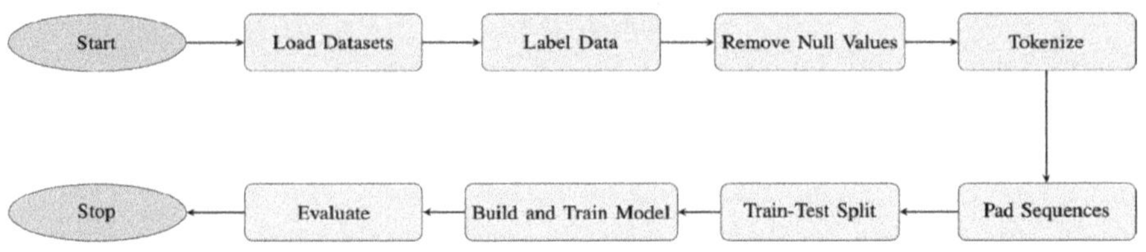

Fig. 2. LSTM Pipeline Model

2.2 BERT Model

Bidirectional Encoder Representations from Transformers (BERT) is a language model introduced by researchers at Google [22].The Transformer model was developed for neural machine translation to address the limitations of LSTMs. LSTMs are slow to train and cannot capture the true contextual meaning of words. The Transformer architecture addresses both concerns, as it is faster and deeply bidirectional, capturing context from both directions simultaneously.

Process

1. **Pre-Training:** BERT learns about language and context by training on Masked Language Modeling (MLM) and Next Sentence Prediction (NSP). These tasks enable BERT to understand bidirectional context within sentences and relationships between sentences, which is crucial for detecting subtle inconsistencies in fake news. In MLM, BERT takes in sentences with masked words (or blanks) to understand bidirectional context within sentences. In NSP, BERT takes two sentences to determine if one follows the other, thus understanding context across sentences.
2. **Fine-Tuning:** BERT learns how to apply language for a specific NLP task such as question answering, text summarization, and sentiment analysis.

Pipeline
The pipeline of the implemented BERT Transformers model is given in Fig. 3.

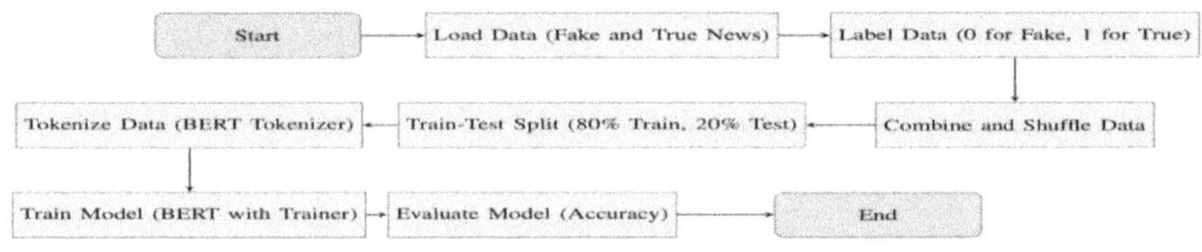

Fig. 3. BERT Pipeline Model

2.3 Hybrid Model

Stacking is an ensemble learning technique that combines the predictions of multiple models to create a more accurate and robust meta-model. In this work, we stack LSTM and BERT to leverage their complementary strengths. LSTM (Long Short-Term Memory) is highly effective at capturing sequential patterns in text, such as the order of words or events, which is crucial for identifying inconsistencies or illogical sequences in fake news. Its ability to retain information over long sequences allows it to analyze the structure and flow of news articles. On the other hand, BERT (Bidirectional Encoder Representations from Transformers) excels at understanding the contextual relationships between words, thanks to its bidirectional attention mechanism. This enables BERT to grasp the nuanced meaning of words and phrases, making it particularly effective for detecting subtle cues in fake news. The stacking process involves using the predictions from both LSTM and BERT as input features for a meta-classifier (e.g., logistic regression or a neural network). The meta-classifier learns how to weigh the predictions of the two models based on their strengths, effectively combining LSTM's temporal analysis with BERT's contextual understanding. For example, if LSTM identifies a suspicious sequence of events in a news article, while BERT detects misleading contextual cues, the meta-classifier can integrate these insights to make a more informed decision. This approach not only reduces errors that might occur if only one model were used but also enhances the model's ability to generalize to new and diverse datasets. By combining the strengths of LSTM and BERT, the hybrid model achieves superior performance in fake news detection.

Pipeline

The pipeline for the implemented hybrid model is given in Fig. 4.

Fig. 4. Hybrid Model Pipeline

Algorithm

Step 1: Load and Preprocess the Dataset
Step 2: Tokenization and Padding
Step 3: Train-Test Split
Step 4: Build and Train the LSTM Model using a learning rate of 0.001, a batch size of 32, and the categorical cross-entropy loss function
Step 5: Create Custom Dataset for BERT
Step 6: Set Up and Train the BERT Model
Step 7: Generate Predictions from LSTM and BERT Models
Step 8: Combine Predictions and Train the Meta-Classifier

3 Results

Here, we present the analysis of the results obtained. These results have been obtained from applying the LSTM and Bert models with the results we got from applying the Hybrid model on the same datasets for false news detection.

3.1 LSTM Model

Table 3. Classification Report of the LSTM Model

Class	Precision	Recall	F1-score	Support
Fake	0.99	0.99	0.99	4733
True	0.99	0.99	0.99	4247
Accuracy			0.99	8980
Macro Avg	0.99	0.99	0.99	8980
Weighted Avg	0.99	0.99	0.99	8980

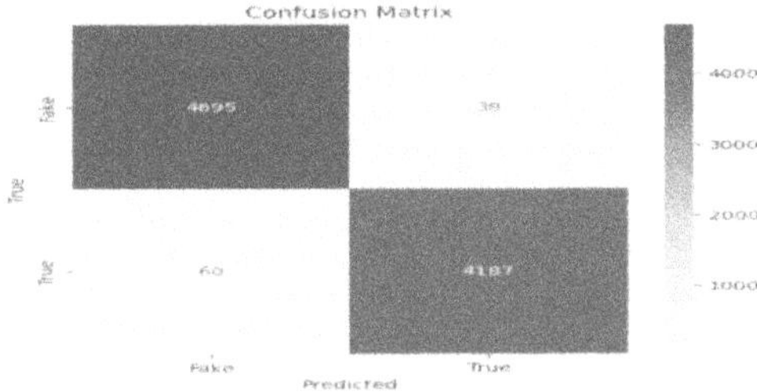

Fig. 5. Confusion matrix of LSTM model

Table 3 presents the classification report of the LSTM model, including precision, recall, and F1-score, while Fig. 5 illustrates the corresponding confusion matrix for the same model. There are a total of 98 false predictions out of 8980 predictions. The model demonstrates robustness with an accuracy of 99%.

3.2 Bert Model

Table 4 presents the classification report of the BERT model, while Fig. 6 illustrates the corresponding confusion matrix for the same model. The BERT model produces only 2 false predictions. The model has an astonishing accuracy of 99.93% but can still be improved.

3.3 Hybrid Model

Table 5 presents the classification report of the Hybrid model, while Fig. 7 illustrates the corresponding confusion matrix for the same model. The hybrid model achieves 100% accuracy on the test dataset with no false predictions. However, this result is based on a specific dataset, and further validation on larger, more diverse datasets is recommended to confirm its robustness.

Table 4. Classification report of the BERT model

Class	Precision	Recall	F1-score	Support
Fake	0.9986	1.0000	0.9993	1409
True	1.0000	0.9984	0.9992	1285
Accuracy			0.9993	2694
Macro Avg	0.9993	0.9992	0.9993	2694
Weighted Avg	0.9993	0.9993	0.9993	2694

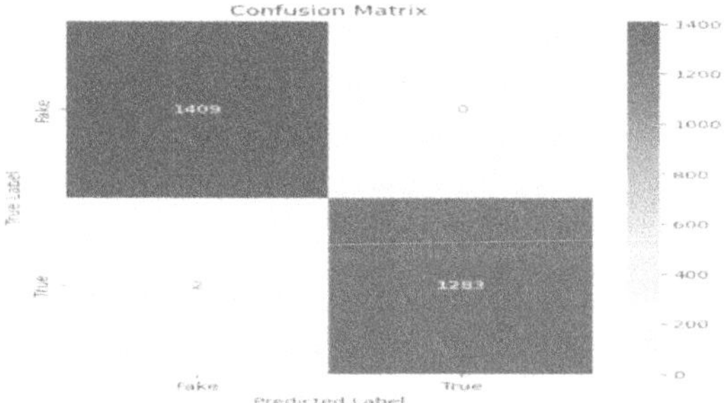

Fig. 6. Confusion matrix of BERT model

Table 5. Classification report of the Hybrid model

Class	Precision	Recall	F1-score	Support
0	1.0000	1.0000	1.0000	4733
1	1.0000	1.0000	1.0000	4247
Accuracy			1.0000	8980
Macro Avg	1.0000	1.0000	1.0000	8980
Weighted Avg	1.0000	1.0000	1.0000	8980

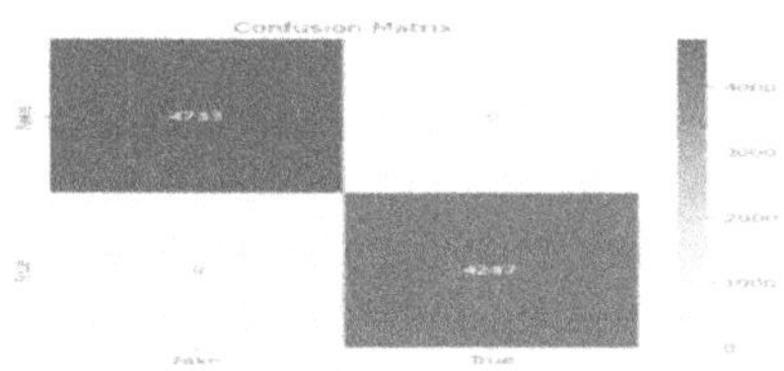

Fig. 7. Confusion matrix of Hybrid model

3.4 Comparison of Model Performance

Table 6 presents a comparison of the performance of all three models. Figure 8 compares the accuracies of various models from the reviewed papers with the accuracies of the implemented LSTM, BERT, and Hybrid models. The hybrid model achieves the highest accuracy among all.

Table 6. Comparison of Performance Metrics for LSTM, BERT, and Hybrid models

Model	Accuracy	Precision	Recall	F1-score
LSTM	0.9900	0.9900	0.9900	0.9900
BERT	0.9993	0.9986	1.0000	0.9993
Hybrid	1.0000	1.0000	1.0000	1.0000

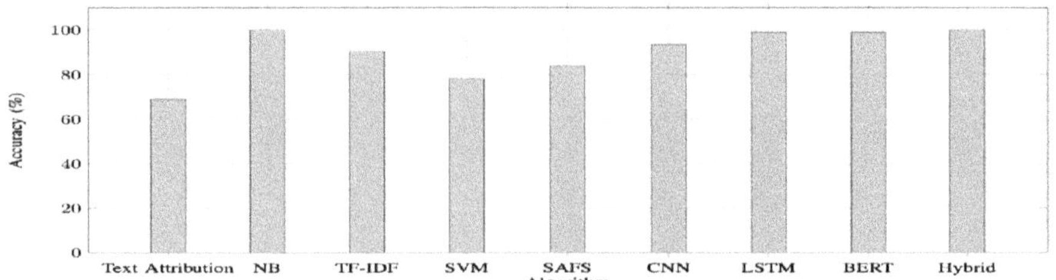

Fig. 8. Comparison of accuracies of different algorithms used in fake news detection

4 Conclusion

This paper examines 15 studies on recent advancements in NLP for fake news detection, emphasizing feature extraction, sentiment analysis, and text classification. We also implemented LSTM and BERT classifiers to compare their performance, highlighting the strengths and limitations of each approach. Our model then used the ensemble technique of stacking to combine the two models to form a hybrid model. We highlighted the strengths of the hybrid model over the individual model by comparing their evaluation metrics and visually represented our analysis through plots and graphs, providing clear insights into the efficiency of various models and their applicability in combating fake news.

5 Future Work

Further research can be done in using ensemble techniques to increase the accuracy of two or more weak learning models with low efficiency and performance, in contrast to the robust models used in this work. Future efforts could also focus on scalable, real-time detection and developing cross-lingual, multimodal models to handle more sophisticated misinformation across platforms.

References

1. Caceres, M.M.F., Sosa, J.P., Lawrence, J.A., et al.: The impact of misinformation on the COVID-19 pandemic. AIMS Public Health **9**(2), 262–277 (2022)
2. Baskar, R., Sah, S., Shyam, R., Kumar, K.S., Patil, H., Reddy, G.N.: Advancements in fake news detection: integrating NLP and multi-modal approaches. In: Proceedings of the 2023 Intelligent Computing and Control for Engineering and Business Systems (ICCEBS), Chennai, India, pp. 1–5 (2023)

3. Rai, N., Kumar, D., Kaushik, N., Raj, C., Ali, A.: Fake news classification using transformer-based enhanced LSTM and BERT. Int. J. Cogn. Comput. Eng. **3**, 98–105 (2022)
4. Khurana, D., Koli, A., Khatter, K., Singh, S.: Natural language processing: state of the art, current trends and challenges. Multimed. Tools Appl. **82**, 3713–3744 (2022)
5. Parikh, S.B., Patil, V., Atrey, P.K.: On the origin, proliferation and tone of fake news. In: Proceedings of the 2019 IEEE Conference on Multimedia Information Processing and Retrieval (MIPR), San Jose, CA, USA, pp. 135–140 (2019)
6. Kim, K.-H., Jeong, C.-S.: Fake news detection system using article abstraction. In: Proceedings of the 2019 16th International Joint Conference on Computer Science and Software Engineering (JCSSE), Chonburi, Thailand, pp. 209–212 (2019)
7. Kuriakose, A., Sebastian, D., Mathew, E.M., Mathew, H., Er.Gokulnath, G.: ALIKAH- a clickbait and fake news detection system using natural language processing. In: Proceedings of the 2019 3rd International Conference on Trends Electronics Informatics (ICOEI), Tirunelveli, India, pp. 1203–1206 (2019)
8. Gaonkar, S., Itagi, S., Chalippatt, R., Gaonkar, A., Aswale, S., Shetgaonkar, P.: Detection of online fake news: a survey. In: Proceedings of the 2019 International Conference on Vision Towards Emerging Trends in Communication and Networking (ViTECoN), Vellore, India, pp. 1–6 (2019)
9. Traylor, T., Straub, J., Chaudhary, S., Snell, N.: Classifying fake news articles using natural language processing to identify in-article attribution as a supervised learning estimator. In: Proceedings of the 2019 IEEE 13th International Conference on Semantic Computing (ICSC), Newport Beach, CA, USA, pp. 445–449 (2019)
10. Yanagi, Y., Orihara, R., Sei, Y., Tahara, Y., Ohsuga, A.: Fake news detection with generated comments for news articles. In: Proceedings of the 2020 IEEE 24th International Conference on Intelligent Engineering Systems (INES), Reykjavik, Iceland, pp. 85–90 (2020)
11. Mandical, R.R., Mamatha, N., Shivakumar, N., Monica, R., Krishna, A.N.: Identification of fake news using machine learning. In: Proceedings of the 2020 IEEE International Conference on Electronics, Computing and Communication Technologies (CONECCT), Bangalore, India, pp. 1–6 (2020)
12. Carvalho, F., Okuno, H.Y., Baroni, L., Guedes, G.: A Brazilian Portuguese moral foundations dictionary for fake news classification. In: : Proceedings of the 2020 39th International Conference on Chilean Computer Science Society (SCCC), Coquimbo, Chile, pp. 1–5 (2020)
13. Hirlekar, V.V., Kumar, A.: Natural language processing based online fake news detection challenges – a detailed review. In: Proceedings of the 2020 5th International Conference on Communication and Electronics Systems (ICCES), Coimbatore, India, pp. 748–754 (2020)
14. Kong, S.H., Tan, L.M., Gan, K.H., Samsudin, N.H.: Fake news detection using deep learning. In: Proceedings of the 2020 IEEE 10th Symposium on Computer Applications & Industrial Electronics (ISCAIE), Malaysia, pp. 102–107 (2020)
15. Verma, S., Paul, A., Kariyannavar, S.S., Katarya, R.: Understanding the applications of natural language processing on COVID-19 data. In: Proceedings of the 2020 4th International Conference on Electronics, Communication and Aerospace Technology (ICECA), Coimbatore, India, pp. 1157–1162 (2020)
16. Bhoir, S.V.: An efficient fake news detector. In: Proceedings of the 2020 International Conference on Computer Communication and Informatics (ICCCI), Coimbatore, India, pp. 1–9 (2020)
17. Wu, X., Wang, J.: SAFS: social-article features stacking model for fake news detection. In: Proceedings of the 2021 4th International Conference on Artificial Intelligence and Big Data (ICAIBD), Chengdu, China, pp. 530–535 (2021)
18. Ivancová, K., Sarnovský, M., Maslej-Krcšňáková, V.: Fake news detection in Slovak language using deep learning techniques. In: Proceedings of the 2021 IEEE 19th World Symposium

on Applied Machine Intelligence and Informatics (SAMI), Herl'any, Slovakia, pp. 255–260 (2021)
19. Indarapu, S.R.K., Komalla, J., Inugala, D.R., Reddy Kota, G., Sanam, A.: Comparative analysis of machine learning algorithms to detect fake news. In: Proceedings of the 2021 3rd International Conference on Signal Processing and Communication (ICPSC), Coimbatore, India, pp. 591–594 (2021)
20. Maham, S., Tariq, A., Khan, M.U.G., et al.: ANN: adversarial news net for robust fake news classification. Sci. Rep. **14**, 7897 (2024)
21. Hochreiter, S., Schmidhuber, J.: Long short-term memory. Neural Comput. **9**(8), 1735–1780 (1997)
22. Devlin, J., Chang, M.-W., Lee, K., Toutanova, K.: BERT: pre-training of deep bidirectional transformers for language understanding. In: Proceedings of the 2019 Conference on North American Chapter of the Association for Computational Linguistics: Human Language Technologies (NAACL-HLT), vol. 1, pp. 4171–4186 (2019)

Automatic Cataract Identification and Classification Using InceptionV3

P. Bhuvaneshwari[1]($\boxtimes$), D. Jayalakshmi[2], R. Hemavathi[3], and E. Geetha Rani[4]

[1] Department of Computer Science and Engineering, Manipal Institute of Technology Bengaluru, Manipal Academy of Higher Education, Manipal, India
bhuvaneshwari.p@manipal.edu
[2] Department of Computer Science and Engineering, R.M.D Engineering College, R S.M Nagar, Kavaraipettai, Gummidipoondi, Thiruvallur, India
[3] Department of Computer Science And Engineering, Saveetha School of Engineering, Saveetha Institute of Medical and Technical Sciences, Saveetha University, Chennai, India
[4] Department of Computer Science and Engineering, School of Aced, Alliance University, Bengaluru, India
geetha.edupuganti@alliance.edu.in

Abstract. A cataract is one of the most dangerous conditions that can cause blindness. It is the clouding of lens, that affects vision of a person. The rate of blindness in cataract patients can be decreased with early detection and treatment. An increasing number of researchers are interested in medical imaging-based AI-assisted diagnosis. A fine-tuned deep neural network called InceptionV3 is proposed in this paper for automatic cataract identification in fundus images. To train the network with small kernels and fewer training parameters, the activation and loss functions are adjusted. Adam is the optimizer used to optimize the suggested network. Before training the model, the dataset is expanded by augmentation to prevent the over-fitting issue. With an average accuracy of 92.54%, experimental data demonstrate that the proposed InceptionV3 model outperforms than the existing VGG19 and InceptionResNetV2 cataract detection techniques.

Keywords: Automatic Cataract Identification · Deep Neural Network · InceptionV3 · VGG19 · Adam Optimizer · InceptionResNetV2

1 Introduction

The world is plagued by more than half of the cases of blindness, and cataracts are a growing public health issue, especially among aging populations. The word "cataract" generally refers to opacities in the lens that impair vision function [1]. People who have cataracts usually struggle with reading, driving, color perception, identifying persons and things, and avoiding bright light glare [2]. In 2010, there were 32.4 million blind people and 191 million people with visual impairments globally, according to a survey [3]. Cataracts accounted for 33.4% of all blindness and 18.4% of all visual impairment. The low-income community has greater figures, while the high-income region has lesser figures. Additionally, it is anticipated that by 2025, there could be up to 40 million cataract-related blindness cases worldwide [4].

© The Author(s) 2026
J. Shreyas et al. (Eds.): CODE-AI 2025, CCIS 2689, pp. 135–142, 2026.
https://doi.org/10.1007/978-3-032-19318-6_13

The cataracts develops when the natural lens of the eye becomes clouded, causing a progressive deterioration in vision. The insidious onset of the condition often delays diagnosis and treatment, which worsens the risk of severe visual impairment. Early detection and appropriate intervention are key in managing cataracts effectively and preventing permanent vision loss. Conventional methods include slit-lamp examinations and fundus photography, which are very much dependent on ophthalmologists. These are time-consuming, costly, and not accessible to the distant or underserved population. Hence, there is an urgent need for innovative approaches that will widely provide efficient and accurate screening.

It would be much more convenient to provide automatic cataract classifiers that can significantly improve diagnostic accuracy and save costs. The artificial intelligence-assisted diagnosis method using optical images is becoming more and more popular. Disease classification has been transformed by deep neural networks in a number of ways. Their capacity to automatically extract complex patterns and features from large, complicated medical datasets has greatly enhanced the efficiency and accuracy of diagnosis. These networks enable early identification and treatment planning by detecting tiny, previously undetectable signs of diseases through training on large and varied datasets. Multiple data sources can also be integrated by Deep Learning (DL) models.

With these technologies, the screening process becomes automated, thereby bridging the gap between limited resources in healthcare and growing needs for eye care, most especially in resource-constrained settings. In ophthalmology, CNN models have been of great promise in the analysis of detailed ocular images. The models can identify and classify cataracts with high precision, thereby reducing their reliance on specialized medical professionals and equipment. In this study we refer to two-class (non-cataract/cataract) classification as cataract detection, and our goal is to provide automatic algorithms for cataract classification using InceptionV3 model. The contribution of this research work consists of:

- A more reliable and specialized model for identifying cataracts has been identified by using the InceptionV3 model with a fine-tuning method to enhance the classification accuracy of cataract disease from the fundus images.
- With an emphasis on reducing false positives and false negatives, preprocess and fine-tune the model to adjust to specific features exist in the fundus images. This will increase precision and recall for precise cataract detection.
- Performance comparison between the suggested and existing models for classifying cataract disease.

2 Related Works

Conventional machine learning techniques have low accuracy and are not therapeutically useful. These models are difficult to discover vital features in and computationally costly to train. Feature extraction, which is tedious and object-specific in images, prevents the progress. So in this section we will discuss some of the deep neural networks used by researchers in cataract identification and classification. Table 1 discuss the few most important research that has used DL approaches and contributed to create reliable models for the categorization and identification of diabetic cataract detection and identification.

Table 1. Few existing works on cataract detection.

Ref no	Author with year	Subject	Methodology	Accuracy
[5]	Richard Bina Jadi Simanjuntak et al. (2022)	Identifying the cataract in fundus images	CNN	92%
[7]	Zhang et al. (2017)	Deducting and classifying the cataract severity with three level grades	DCNN	93.52%
[8]	Triyadi et al. (2022)	Classification of cataract images using stacked ensemble methods	VGG19 & Residual networks	91.06% & 93.50

3 Materials and Methods

The proposed methodology is illustrated in Fig. 1. Initially, the input dataset is collected. Following that preprocessing and feature extraction. Finally, the images are classified by using InceptionV3. Multiple convolutional and pooling layers collaborate to form each Inception module. The Inception modules use tiny convolutional layers, including 3×3, 1×3, 3×1, and 1×1, to lower the number of parameters. Three Inception A modules, five Inception B modules, and two Inception C modules are stacked in succession in

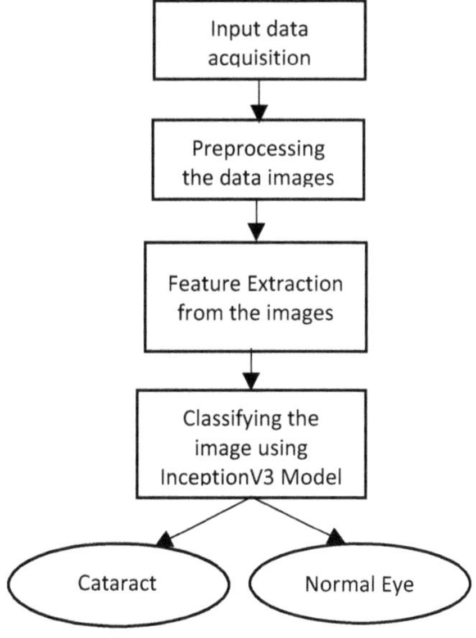

Fig. 1. Proposed methodology

Inception-v3. In order to utilize the pre-train model and adjust the parameters for our specific purpose, we then added two FC layers at the conclusion of the Inception modules. A sigmoid layer was then added as a classifier, producing a probability for each category. The category with the maximum probability was selected as the predicted category.

3.1 Dataset Used

This work uses a publicly available, heterogeneous database of ocular images that include both cataract-induced and normal eyes. The dataset includes various types of ocular images, such as those captured by digital cameras and fundus images, to allow generalization of the model to different imaging conditions. Images are obtained from publicly available ophthalmic image databases kaggle [9]. 80% of the data is utilized for the training of the proposed model and 20% is used for testing. Figure 2 shows the binary octal image dataset with cataract as well as the normal eye.

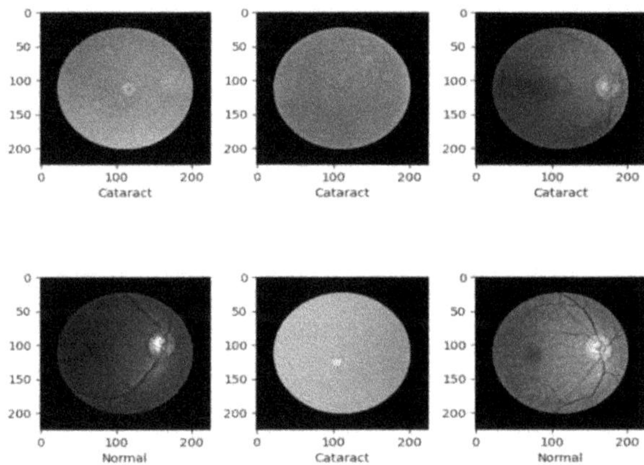

Fig. 2. Sample binary input image

3.2 Preprocessing

Preprocessing is the preliminary step which removes noise present in the data. The dataset contains differentsizes of images, which are resized into same size to maintain uniformity. Few more preprocessing techniques applied in this process is discussed in Table 2.

Table 2. Data Preprocessing Techniques

Preprocessing Technique	Description
Resizing	All images size resized to 224 × 224 pixels for proper uniformity and to be align with the neural network

(continued)

Table 2. (continued)

Preprocessing Technique	Description
Normalization	Pixel value for images are scaled between 0 and 1 to avoid bias and convergence problem
Contrast enhancement (Histogram Equalization)	It increases the contrast of the image and highlight the foreground pixels
Augmentation	To enhances the diversity of the training data, performance, robustness, generalization and to prevent overfitting of the model the images are rotated, flipped, translated, zoom, and sheared

3.3 Feature Extraction and Classification

To identify the relevant cataract features in the provided images, feature extraction is utilized. This procedure is crucial to our cataract detection methodology. The paper makes use of texture-based techniques to extract features from the images that are appropriate for cataract identification. This method is a collection of metrics that represent the spatial arrangement of an image's pixel values is called texture. The enhanced top-bottom transformation is used first to increase the contrast between the objects and backdrop before texturing feature extraction. After that, noise is eliminated using a trilateral filter. To depict the retina fundus image, three different texture feature types are finally extracted. After feature extraction, the data is passed to the inceptionV3 model for classification.

3.4 Optimizers

This is one of the major hyper parameter used to improve the performance of the model. "Optimized" describes a technique or strategy for adjusting our model's features, like weights, learning rate, and other factors to minimize losses and achieve a better result. Adam is a combination with benefits including reduced memory and time needs and the capacity to manage a loose gradient in a noise situation [18].

4 Result and Discussion

4.1 Hardware and Software Setup

The specification of the system used is Intel Core i7 processor with 3.6 GHz with 8GB memory. The version of the operating system used was Windows 10. The proposed model was executed after the data set was compiled. Throughout the entire process on Google, the model was created and trained using TensorFlow and Python. The hyperparameters are finetuned to achieve better efficiency of the model which is illustrated in Table 3.

Table 3. Hyperparameters

Hyperparameters	Values
Batch Size	32 samples per iteration
Epochs	50
Learning rate	0.001
Optimizer	Adam

4.2 Evaluation Metrics

Four performance indicators are used to evaluate the efficacy of the suggested approach, offering numerical insights into its precision and reliability. These metrics, which are obtained from the confusion matrix, provide insightful evaluations of the model's effectiveness and capacity for prediction in a given task. Table 4 shows the assessment metrics applied in this work. The proposed methodology is evaluated with these parameters and compared with other exiting models such as VGG19 [6] and InceptionResNetV2 [10]. Figure 2 illustrates the comparision and it is evident that our proposed inceptionV3 outperforms other existing models (Fig. 3).

Table 4. Evaluation Metrics

Metrics	Equation
Accuracy	$\frac{TP+TN}{(TP+TN+FP+FN)}$
Precision	$\frac{TP}{(TP+TN)}$
Recall	$\frac{TP}{(TP+FN)}$
F1 score	$\frac{2*(Precision*Recall)}{Precision+Recall}$
Where TP is True Positive, TN is True Negative, FP is False Positive and FN is False Negative	

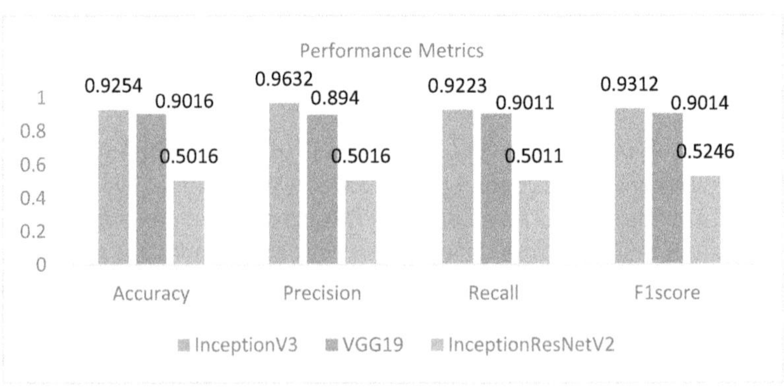

Fig. 3. Comparision of performance metrics

5 Conclusion

There are some biomarkers that increases the chances of cataracts like aging, family history, more UV rays exposure, medical disorders like high blood pressure, diabetes. The majority of cataracts take time to grow and do not immediately impair vision. However, cataracts will eventually affect your vision. We applied InceptionV3 model to detect and classify cataracts. The testing accuracy of the model is 92.54%. It is compared with the other existing models like VGG19 and InceptionResNetV2. Our proposed model outperforms the other existing models interms of performance evaluation metrics.In the future, a multiclass classification approach will be able to distinguish between cataract severity levels, such as mild, normal, and severe. Additionally, a key focus to increase the efficacy of the classification process will be the investigation and application of hybrid models.

References

1. Quillen, D.A.: Common causes of vision loss in elderly patients. Am. Fam. Phys. **60**(1), 99–108 (1999)
2. Allen, D., Vasavada, A.: Cataract and surgery for cataract. BMJ **333**(7559), 128–132 (2006)
3. Khairallah, M., et al.: Number of people blind or visually impaired by cataract worldwide and in world regions, 1990 to 2010. Invest. Ophthalmol. Visual Sci. **56**(11), 6762–6769 (2015)
4. Müller-Breitenkamp, U., Ohrloff, C., Hockwin, O.: Aspects of physiology, pathology and epidemiology of cataract. Der Ophthalmologe: Zeitschrift der Deutschen Ophthalmologischen Gesellschaft **89**(4), 257–267 (1992)
5. Simanjuntak, R.B.J., Fuâ, Y., Magdalena, R., Saidah, S., Wiratama, A.B., Daâ, I.: Cataract classification based on fundus images using convolutional neural network. JOIV: Int. J. Inform. Visual. **6**(1), 33–38 (2022)
6. Gill, K.S., Anand, V., Gupta, R.: Cataract detection using optimized VGG19 model by transfer learning perspective and its social benefits." In: 2023 Second International Conference on Augmented Intelligence and Sustainable Systems (ICAISS), pp. 593–596. IEEE (2023)
7. Zhang, L., Li, J., Han, H., Liu, B., Yang, J., Wang, W.: Automatic cataract detection and grading using deep convolutional neural network. In: 2017 IEEE 14th International Conference on Networking, Sensing and Control (ICNSC), pp. 60–65. IEEE
8. Triyadi, A.B., Bustamam, A., Anki, P.: Deep learning in image classification using VGG-19 and residual networks for cataract detection. In: 2022 2nd International Conference on Information Technology and Education (ICIT&E), pp. 293–297. IEEE (2022)
9. https://www.kaggle.com/datasets/jr2ngb/cataractdataset
10. Varghese, R.E., Pandian, I.A.: Inception-resnet V2 based eye disease classification using retinal images. In: 2023 3rd International Conference on Mobile Networks and Wireless Communications (ICMNWC), pp. 1–5. IEEE (2023)

A Novel Approach to Human Fatigue Detection Based on Deep Learning

C. M. Naveen Kumar[1]([envelope]), V. Veeraprathap[2], B. Ramesh[1], Ayesha Siddiqha[1], S. Hemanth[1], and M. G. Rohan guru[1]

[1] Malnad College of Engineering, Hassan, Karnataka, India
cmn@mcehassan.ac.in
[2] ATME College of Engineering, Mysore, Karnataka, India

Abstract. The global rise in aging populations and unhealthy lifestyles has contributed to a significant increase in high-risk health conditions, such as cardiovascular diseases, sleep disorders, and fatigue-related issues. Early detection and diagnosis of these conditions are crucial in preventing severe health complications. The most recent study focuses on specific categories of diseases the use of artificial intelligence in multiload electrocardiograms or the applications of wearable technologies however our work focuses on the development of machine learning based fatigue detection system using signals specifically electrocardiogram waveforms. In our approach we will leverage machine learning models to classify fatigue levels as no level, low level, and high-level stress when fatigue is detected the system plays a soothing music playlist in the background to reduces the stress.

Keywords: ECG · Stress · CNN · TensorFlow · ST depression

1 Introduction

Stress has become a major public health issue in India, affecting all age groups due to academic, professional, and lifestyle pressures. According to the Indian Council of Medical Research (ICMR), over 74% of the population suffers from stress-related health problems. Students face academic and career stress, professionals experience workplace fatigue and mental exhaustion, and the elderly deal with stress from health issues and social isolation. Stress is classified into acute and chronic types. Acute stress results from short-term challenges and can enhance cognitive function and decision-making under pressure. Chronic stress, however, lasts longer and negatively impacts both mental and physical health, contributing to cardiovascular diseases, hypertension, weakened immunity, anxiety, and depression. Early detection and assessment are vital to prevent severe health complications. Current stress assessment methods include clinical evaluations, biochemical cortisol tests, and physiological signal analysis. Among these, Electrocardiogram (ECG) signals offer a stable, non-invasive way to detect stress-related physiological responses. ECG reflects the heart's electrical activity and reveals the autonomic nervous system's response, making it a valuable tool for stress diagnosis. A key challenge is capturing accurate ECG data. The study uses a three-lead, ring-type ECG

J. Shreyas et al. (Eds.): CODE-AI 2025, CCIS 2689, pp. 143–149, 2026.
https://doi.org/10.1007/978-3-032-19318-6_14

setup with gel-based electrodes. Gel electrodes ensure a clean connection, reducing skin irritation and moisture effects, and improving signal quality—suitable for all age groups. These electrodes attach to the skin and connect to an ECG machine through lead wires, enabling heart activity analysis. ECG signals provide a clear picture of heart function and are reliable for detecting post-stress disturbances like fatigue-induced stress. The AD8232 ECG sensor is an efficient, cost-effective module for acquiring and processing ECG signals. As ECG is noise-sensitive, the AD8232 IC enhances signal quality through high-gain instrumentation amplification, signal extraction, and filtering. It minimizes motion artifacts and electrode displacement, making it suitable for varied biomedical applications. Its compact size boosts the efficiency of ECG monitoring. This study focuses on developing a machine learning-based method to detect stress using ECG data. Unlike real-time systems, this approach uses pre-recorded ECG datasets to train a model that distinguishes stress levels. The PQRST complex in ECG contains key heart activity data, and our method emphasizes the ST depression segment, a reliable physiological stress indicator. Applying machine learning improves detection accuracy and speed, aiding early diagnosis and better stress management.

2 Literature Review

Luca Neri [1] aim to predict the existence of feature extraction techniques, such as analyzing P wave, QRS complex, and T wave variations, to detect fatigue. It also highlighted the use of stress induction tasks, like public speaking, to simulate realworld conditions, ensuring accurate data collection. Wearable ECG devices, combined with AI, offer realtime, remote cardiac monitoring, improving disease detection beyond traditional hospital-based methods. Ramkumar M. et al., [2] approach was to determine the feature extraction techniques, explored supervised learning methods for fatigue classification. The authors used support vector machines (SVM) to train models with the help of ECG data that was labeled with a defined group of samples. It was discovered that the direction of fatigue could be determined with a high level of accuracy and the fatigue stage of an individual was distinguishable using specific patterns in the ECG signals. This study explores efficient, high-accuracy feature extraction techniques to enhance clinical decision-making and support early diagnosis of cardiac conditions. The research conducted by [{Mousa Kadhim Wali}] [3] centered on the acute stress detection by heart rate variability (HRV) data derived from the ambient-assisted living sensors. The research conversed on common stress induction methods that are arithmetic tasks, driving scenarios, and work shifts to determine the fatigue. They applied the model accuracy and F1 scores to the verification of the hypothesis under various settings to validate that the HRV was indeed effecting in fatigue detection. E. H. Houssein, R. E. Mohamed and A. A. Ali [4] approach was to determine Natural Language Processing (NLP) has become essential in biomedical research for extracting insights from unstructured clinical text. It supports applications such as disease prediction, medication analysis, decision support, and patient management. However, challenges like non-standard medical terminology, data privacy concerns, and unstructured text formats hinder its efficiency. M. Amin et al., [5] approach was to determine a deep transfer learning approach for classifying driver stress levels (low, medium, high) using ECG signals.

Instead of training models from scratch, pre-trained CNNs (e.g., GoogLeNet, ResNet-101, Xception) are fine-tuned using scalogram images generated from ECG signals via Continuous Wavelet Transform (CWT). A fuzzy logic-based ranking evaluates model performance across accuracy, sensitivity, and precision.

3 Methodology

Fig. 1. Block diagram of proposed system

3.1. **Data Collection:** ECG signals are collected from participants using disposable gel electrodes placed on the chest to elect physiological responses, participants are exposed to videos designed to induce stress or fatigue. The AD8232 sensor is employed to capture clean ecg signals, focusing on the electrical activity of the heart (Fig. 1).

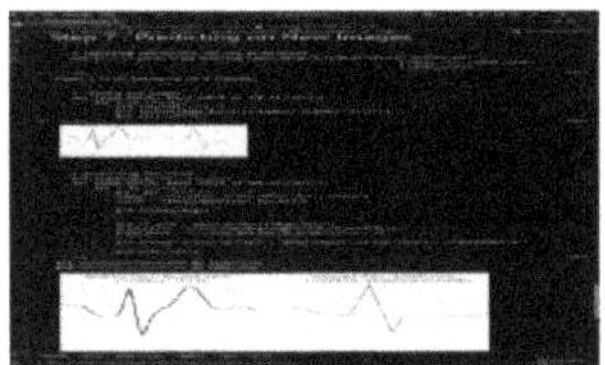
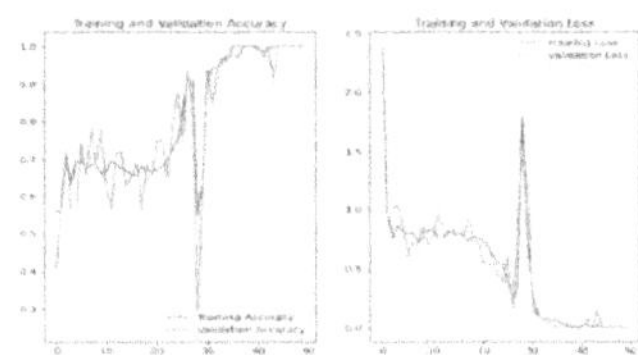

Fig. 2. Data collection and ST variation for normal vs abnormal

Robotic Process Automation is implemented using UiPath Studio software to perform image sequencing, enabling efficient processing and management of visual data associated with the experiment (Fig. 2).

3.2 **Feature Extraction:** In this system, the focus of feature extraction is exclusively on the ST duration, a critical segment of the ECG waveform. The ST duration is the time between the end of the QRS complex and the beginning of the T-wave. It represents the period during which the ventricles remain depolarized before repolarization begins.

3.3 **Hypothesis Buildings**: A supervised learning model is developed to predict fatigue levels based on labeled data representing fatigue and non-fatigue states. The relationship between ECG-derived features and stress levels is analyzed to construct a predictive framework for fatigue detection.

3.4 **Stress Detection Model:** Talking of focusing on the stress detection model which is for computing one, and it is used to differentiate fatigue and stress level based on the length of the ST portion of the ECG waveform and this model uses supervised learning and tenor flow as well as keras it uses label data to mark stress level into predefined classes. Keras is a high-level application programming interface (API) for machine learning that helps users build and experiment with deep neural networks. Keras is written in Python and is built on top of TensorFlow, an open-source machine learning platform.

3.5 **Hypothesis Testing:** The trained model is applied to the test dataset to predict stress levels. Performance metrics are computed to evaluate how well the predictions match the actual labels. The training accuracy consistently improves, reaching near 100%,while validation accuracy fluctuates, indicating potential overfitting. The training loss decreases steadily, but validation loss exhibits spikes, particularly around epoch30, suggesting instability. The model demonstrates effective learning but struggles with generalization, as seen in the fluctuations, overfitting is likely due to a lack of regularization and excessive training. A sudden loss spike suggests potential learning rate issues or data inconsistencies (Fig. 3).

1   Methodology

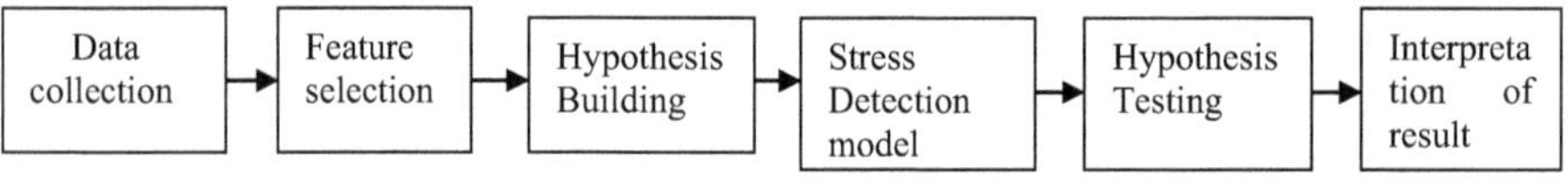

Fig. 3. Prediction on new image

3.1 Interpretation of Results

The proposed fatigue detection system effectively classifies stress levels into three categories no stress, low stress, and high stress based on ECG signal characteristics. The analysis of the PQRST complex reveals distinct patterns corresponding to each stress level.

- No stress:In the no stress condition, the PQRST complex remains stable, exhibiting no significant deviations, leading to a classification of no stress.
- Low stress(Mental Stress):Under low stress, the T-wave shows a noticeable flattening, indicating a mild physiological response, which the system recognizes as low stress
- High Stress: In high stress conditions, the T-wave exhibits a pronounced increase in peak amplitude, signifying heightened physiological activity, resulting in a classification of high stress (Table 1).

Table 1. Stress Detection

No Stress	Low Stress(Mental stress)	High Stress(Physical stress)

4   Result and Discussion

The proposed model strongly identifies mental stress, physical stress, and no stress conditions. The result is shown through the confusion matrix and classification report indicating model's accuracy and reliability. The confusion matrix shows that model effectively identifies between the different stress conditions. For Mental stress all the 16 samples were correctly classified resulting 100% accuracy in this category. For No stress total 254 samples were used out of which the model has classified 253 samples as no stress with only one misclassification. For Physical stress 105 samples were taken. Out of which the model has successfully classified 92 cases with 13 being misclassified as no stress conditions. The classification report further provides the model's performance. The model has achieved precision of 1 for the mental stress conditions, while 0.95 for no stress and 0.99 for physical stress indicating a strong ability to minimize the false positives. The recall scores for mental stress and no stress are 1 whereas 0.88 for physical stress. Recall score is slightly lower for the physical stress. The overall F1-score is 0.97 which suggests the balanced and effective model.

The system has achieved overall accuracy of 96% conforming its reliability in detecting the stress (Fig. 4).

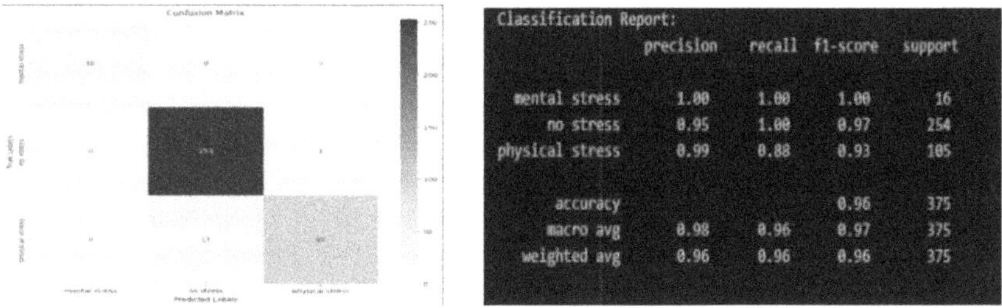

Fig. 4. Confusion Matrix and Classification report

5   Conclusion

In conclusion, this study presents a machine learning-based human fatigue detection system utilizing ECG waveforms, demonstrating good performance in identifying the different stress levels. The proposed model has achieved 96% overall accuracy with precision, recall and F1-score as 1 for mental stress and 99% precision for physical stress and F1-score of 0.93 ensuring reliable classification between different stress levels.

The confusion matrix highlights the model's robustness with minimum misclassifications. Only one sample out of 254 and 13 samples out of 105 were incorrectly classified which states that the system has effectively distinguishes between different levels of stress with maintaining the low error rate.

Traditional fatigue detection models often rely on subjective assessments, wearable sensor-based heart rate variability (HRV) analysis, or computationally expensive deep learning architectures. Several of these models have great detection rates, but inaccurate classification and the need for real-time monitoring limit their real-world applicability. We resolve these challenges by implementing the following:

Dependence on multi-sensor setups can be lessened by the use ECG waveforms as the main physiological signal.

To achieve high confidence predictions, to make reliable stress classification.

Future work will focus on further expanding the dataset, improving the techniques of feature extraction, and investigating compact model structures for low-resource deployment that would improve generalization even further. Our results indicate that this method is a promising replacement for current fatigue detection models with better accuracy, reliability, and scalability in real-world usage.

References

1. Neri, L., et al.: Electrocardiogram monitoring wearable devices and artificial-intelligence enabled diagnostic capabilities: a review. Sensors **23.10**, 4805 (2023)
2. Cos, C.A., Lambert, A., Soni, A., Jeridi, H., Thieulin, C., Jaouadi, A.: Enhancing mental fatigue detection through physiological signals and machine learning using contextual insights and efficient modelling. J. Sens. Actuator Netw. **12**(6), 77 (2023)
3. Wali, M.K.: Probabilistic neural network based fatigue level classification using electrocardiogram high frequency band and average heartbeat. Nano Biomed. Eng **12**(2), 132–138 (2020)
4. Houssein, E.H., Mohamed, R.E., Ali, A.A.: Machine learning techniques for biomedical natural language processing: a comprehensive review. IEEE Access **9**, 140628–140653 (2021). https://doi.org/10.1109/ACCESS.2021.3119621
5. Amin, M., et al.: ECG-based driver's stress detection using deep transfer learning and fuzzy logic approaches. IEEE Access **10**, 29788–29809 (2022)
6. Ramkumar, M., Babu, C.G., Karthikeyani, S., Priyanka, G.S., Kumar, R.S.: Probabilistic feature extraction techniques for electrocardiogram signal-a review. In: IOP Conference Series: Materials Science and Engineering, vol. 1084, No. 1, p. 012024. IOP Publishing (2021)
7. Ramkumar, M., Babu, C.G., Karthikeyani, S., Priyanka, G.S., Kumar, R.S.: Probabilistic feature extraction techniques for electrocardiogram signal-a review. In: IOP Conference Series: Materials Science and Engineering 2021 Mar 1, vol. 1084, no. 1, p. 012024. IOP Publishing (2021)
8. Wali, M.K.: Probabilistic neural network based fatigue level classification using electrocardiogram high frequency band and average heartbeat. Nano Biomed. Eng. **12**(2), 132–138 (2020)
9. Abadi, M., et al.: TensorFlow: a system for large-scale machine learning. OSDI (2016)
10. Bharadhwaj, K., Srinivasan, P., Rajagopal, R.: A comparative study on deep learning frameworks: analysis and review. Int. J. Adv. Res. Comput. Sci. **9**(1), 227–232 (2018)

Optimized Random Forest Regressor Integrated with Explainable AI(ORF-XAI) for Predicting Student Study Hours to Enhance Academic Performance

Swathieswari Mohanraj[(✉)] [iD] and P. Shanmugavadivu [iD]

Gandhigram Rural Institute, Gandhigram University (Deemed), Dindigul, India
`swathieswari.mca@gmail.com, psvadivu67@gmail.com`

Abstract. This research presents an AI-driven student performance analysis system that integrates Machine Learning (ML), Explainable AI (XAI), and Rule-Based Reasoning to provide personalized study plans. The system analyzes career aspirations, study hours, and academic performance to generate targeted study recommendations, helping students optimize their learning strategies. By identifying underperforming students, the system enables timely interventions to improve academic progress. This study introduces a hybrid explainable AI (XAI) model that combines ML-based predictions, SHAP-based explainability, and forward-chaining rule-based reasoning. A Random Forest model with hyperparameter tuning was employed to predict weekly study hours, achieving a Mean Absolute Error (MAE) of 2.89, Mean Squared Error (MSE) of 12.32, and R^2 Score of 0.74. Additionally, a forward-chaining rule-based system was implemented to generate interpretable study recommendations along with recommended study hours, though it exhibited higher errors (MAE: 3.85, MSE: 18.42, R^2: 0.61) compared to ML. To validate model reliability, a K-Fold Cross-Validation approach was applied, yielding MAE of 2.95, MSE of 13.25, and R^2 of 0.72, further confirming the stability of ML predictions. The results demonstrated that ML-based predictions outperformed rule-based reasoning in accuracy, while SHAP-based XAI techniques enhanced transparency. This hybrid approach bridges the gap between accuracy and interpretability, ensuring that AI-driven educational recommendations are trustworthy, explainable, and actionable. By combining predictive analytics with rule-based reasoning, this system empowers educators with data-driven, personalized learning interventions to enhance student success.

Keywords: Students' Performance · XAI · Explainable AI · Rule-based Reasoning · Forward chaining · Study Plan recommendation · AI-driven education system

1 Introduction

Artificial Intelligence (AI), powered by machine learning (ML), has transformed various domains, including education [1, 2]. Recent growth in Educational Data Mining (EDM) highlights efforts to predict student performance early, especially on platforms

© The Author(s) 2026

J. Shreyas et al. (Eds.): CODE-AI 2025, CCIS 2689, pp. 150–162, 2026.
https://doi.org/10.1007/978-3-032-19318-6_15

like MOOCs, VLEs, LMSs, and ITSs [3, 4]. ML helps educators identify at-risk students and supports tasks like adaptive recommendations and automatic grading [5]. However, as ML models act autonomously, trust and interpretability become concerns [6].

To address this, Explainable AI (XAI) techniques such as LIME and SHAP have been widely used to make model predictions transparent and understandable [7–9]. XAI improves trust, allowing educators and stakeholders—students, teachers, regulators, and developers—to analyze and validate AI outcomes [10]. Applications include student classification, early risk detection, and performance prediction [11–15].

This study builds on previous work by proposing a hybrid system that combines ML, XAI, and Forward-Chaining Rule-Based Reasoning. Unlike earlier studies focusing only on accuracy or static rules, this system predicts study hours, explains influencing factors via SHAP, and uses dynamic rule activation to categorize students and deliver personalized, interpretable study plans [16]. This approach bridges the gap between AI predictions and actionable educational guidance.

2 Related Works

To support both regular and irregular students, institutions are adopting flexible curricula and diverse course structures. This has led to the development of systems that offer personalized academic guidance. A Decision-Based Recommendation System using a Forward Chaining Algorithm was introduced to address this gap by analyzing student profiles and generating tailored study plans [17, 19]. The inference engine applies rules iteratively until the most suitable recommendation is identified [20].

Meanwhile, Machine Learning (ML) models such as Random Forest and XGBoost outperform traditional methods in predicting student performance but often lack transparency [12]. Explainable AI (XAI) techniques like SHAP and LIME have been introduced to improve model interpretability, enabling educators to understand key factors influencing predictions [23]. However, many ML-based systems still lack built-in explainability.

To overcome these limitations, our system combines ML, XAI, and rule-based reasoning, dynamically categorizing students—e.g., High Achievers, Moderate Performers, Struggling Students—and generating personalized study plans aligned with their learning patterns, aspirations, and academic goals [24] (Table 1).

This system bridges the gap between accurate prediction and actionable insight by integrating:

- ML-based forecasting for student performance
- SHAP-based XAI for transparent, explainable recommendations
- Forward-chaining rule-based reasoning for adaptive, personalized study plans

SHAP explanations help students and educators understand the rationale behind AI suggestions, building trust. The rule engine ensures study plans evolve with student progress. Hyperparameter tuning further improved model accuracy, enabling precise predictions and targeted support. Overall, the system enhances personalized learning, boosts engagement, and supports timely interventions for at-risk learners.

Table 1. Comparative Summary of Related Works

Study	Approach	XAI Integration	Dynamic Study Plan Generation	Personalized Student Guidance	Limitations
[17]	Rule-Based Forward Chaining	No	No	Yes	Lacks ML-based predictive insights
[21]	Machine Learning (ML)	Partial (Basic Feature Importance)	No	Yes	Lacks structured recommendations
[23]	ML + SHAP	Yes	No	No	No rule-based reasoning
[19]	Expert System	No	Yes	Yes	Lacks adaptability to changing student behavior
Proposed ORF-XAI Model	Hybrid ML + XAI + Rule-Based	Yes (SHAP & LIME Integration)	Yes (Dynamic Study Plans)	Yes (Tailored for Individual Students)	Addresses both prediction and personalized recommendations

3 Methodology

3.1 Dataset Description

The Expanded Student Scores consists of 5,000 student records, capturing key such as demographic, academic, and behavioral attributes to facilitate student performance prediction, personalized learning recommendations, and explainable AI (XAI) applications. The dataset includes 20 features, namely student demographics (gender, part-time job, absence days, extracurricular activities), academic performance (subject-wise scores, reading and writing scores), study habits (weekly self-study hours), and career aspirations (e.g., Doctor, Engineer, Artist). The subjects of interest column list 128 unique combinations, reflecting students' preferences in different academic streams. The dataset contains no missing values, ensuring completeness for analysis. The average weekly study hours are 22.65, while student absences range from 0 to 15 days. Academic scores range from 40 to 100, with a mean score of ~77% across all subjects. This had been utilized from the Kaggle source available at https://www.kaggle.com/dat asets/mexwell/student-scores. The dataset is crucial for predicting study habits, identifying at-risk students, and generating personalized study plans. Machine learning models can analyze these features to classify students into performance categories (High Achievers, Struggling Students, etc.), while SHAP-based explanations enhance transparency. Additionally, rule-based reasoning enables customized study recommendations aligned with students' aspirations, ensuring an adaptive and explainable learning framework. All the demographic and academic features were mentioned below in Table (Table 2).

Table 2. Dataset Feature overview

Input Attributes	Description
Id	Unique identifier for each student
first name, last-named	Student's personal information (non-essential for analysis)
Email	Student's contact email (not relevant for analysis)
Gender	Gender of the student (Male/Female)
part_time_job	Boolean indicating if the student has a part-time job
absence days	Number of days the student was absent from school
extracurricular activities	Boolean indicating participation in extracurricular activities
weekly_self_study_hours	Total self-study hours per week
career_aspiration	Student's career goal (e.g., Doctor, Engineer, Artist)
math score, history score, physics score, chemistry score, biology score, English score, geography score	Subject-wise scores (out of 100)
reading score, writing score	Additional academic performance indicators (out of 100)
subjects interest	List of subjects a student is interested in

3.2 Exploratory Data Analysis

Exploratory Data Analysis (EDA) was conducted on a dataset of 5,000 student records to uncover patterns in academic performance, study habits, and aspirations. Missing values in subject scores and study hours were addressed using mean, median, and mode imputation. Analysis showed that most students study 10–30 h per week, with English and Math scoring highest, while Physics and History lag. Correlation heatmaps revealed strong links between study hours and academic performance, highlighting the impact of self-study. Categorical variables like career aspirations and gender were also explored, showing moderate influence on study behavior. One-hot encoding was applied, and key features such as subject scores, aspirations, and attendance were selected for modeling. Overall, EDA ensured data quality, feature relevance, and helped prepare the dataset for accurate prediction of study hours using ML (Fig. 1, 2, 3 and 4).

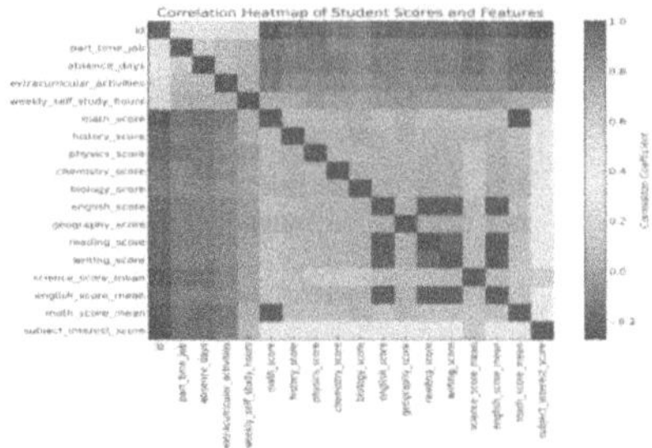

Fig. 1. Heat Correlation Map

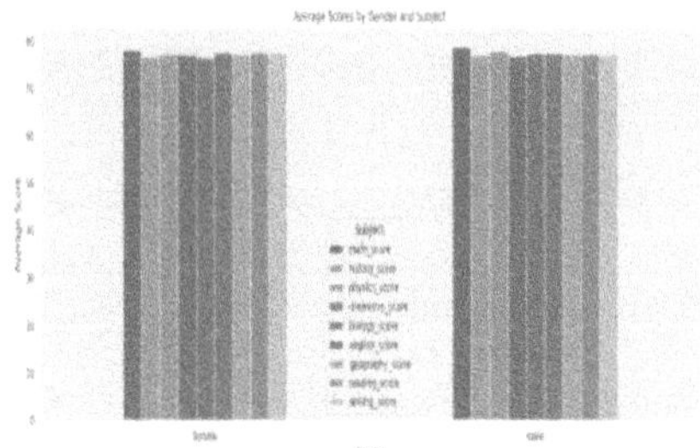

Fig. 2. Gender Distribution Chart

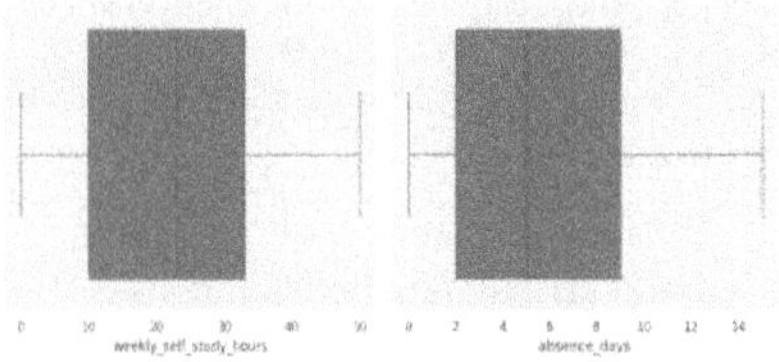

Fig. 3. Boxplot Analysis

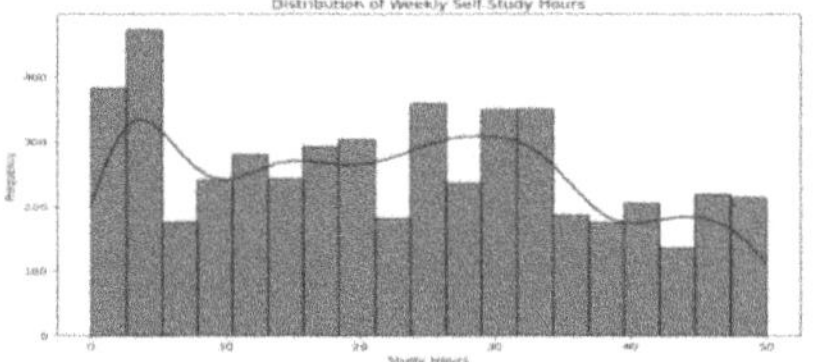

Fig. 4. Study pattern Distribution

3.3 Data Preprocessing

To ensure the dataset was structured and optimized for analysis, several preprocessing steps were performed. First, the insignificant attributes viz., 'id', 'first_name', 'last_name', 'email', and 'gender' were removed, as they don't possess relevant contentin predicting student performance. Missing values in numerical columns were handled by imputing the mean values to maintain data completeness. Feature engineering was applied by creating a new 'average_score' column, which computed as the mean of subject scores to categorize the students based on their academic scores. Additionally, the 'subjects_interest' column was transformed into multiple binary indicators, allowing the model to understand a student's academic preferences. To process categorical variables, one-hot encoding was applied to 'career_aspiration', converting values like "Doctor" and "Engineer" into multiple binary columns, each representing a unique career option. Based on study hours and performance, students were categorizedinto High Achievers, Moderate Performers, Struggling Students, Procrastinators, and Exam-Crammers, enabling personalized learning recommendations. Finally, numerical features were standardized to ensure consistency during machine learning. These preprocessing steps enhanced data quality, improved model accuracy, and enabled a structured approach for predictive analysis and explainable AI-driven recommendations (Fig. 5).

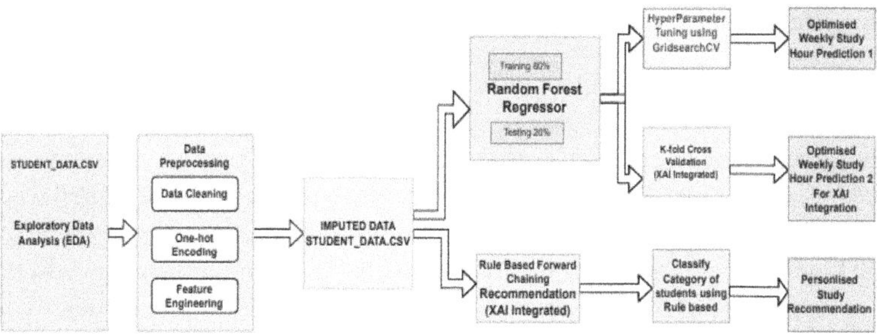

Fig. 5. Proposed Workflow of the model

3.4 Feature Selection

Feature selection was conducted to enhance model performance by retaining the most relevant attributes while removing redundant or non-informative data. Key academic indicators such as math_score, history_score, physics_score, chemistry_score, biology_score, english_score, geography_score, and reading_score were selected as a primary predictors of student performance. Weekly_self_study_hours was chosen as the best variable, as it reflects study habits and engagement. Absence_days and participation in extracurricular activities were included to assess their impact on academic outcomes. Career aspirations were one-hot encoded to integrate students' long-term goals into the model. Additionally, the 'subjects_interest' column was transformed into binary indicators to capture academic preferences binary indicators to capture academic preference (Fig. 6).

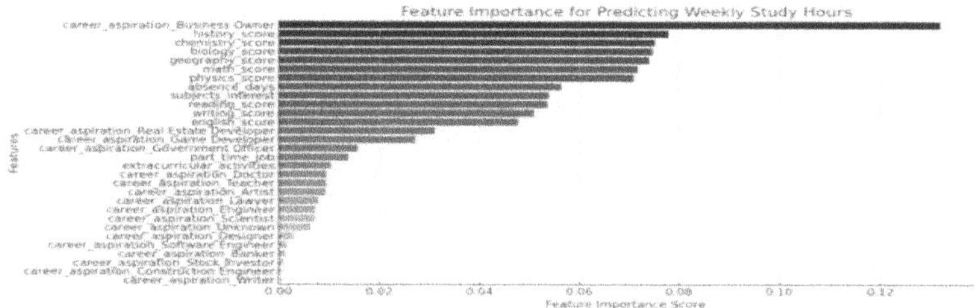

Fig. 6. Feauture Selection for weekly study hours

3.5 Machine Learning Implementation

We implemented a Machine Learning model to predict students' weekly self-study hours, a key indicator of study behavior. The dataset was preprocessed by removing irrelevant attributes, one-hot encoding categorical variables like career aspirations, and standardizing numerical features. We selected a Random Forest algorithm for its ability to handle complex, non-linear data, manage missing values, and rank feature importance effectively. Hyperparameters were tuned using GridSearchCV to optimize model accuracy. Model performance was evaluated using MAE, MSE, and R^2 Score. To ensure transparency, we applied SHAP (SHapley Additive Explanations) to interpret feature contributions. Finally, the ML predictions were integrated into a rule-based reasoning engine to deliver personalized and explainable study recommendations.

3.6   Contribution of Explainable AI (XAI) After Machine Learning

Since ensemble models like Random Forest function as black boxes, we incorporated SHAP to explain their outputs. SHAP values helped identify which features most influenced individual predictions—for example, low math scores or frequent absences. Summary plots showed the global impact of each feature, while dependence plots gave personalized insights. This interpretability allowed educators to understand the "why" behind each recommendation and supported trust and transparency in decision-making. Integrating XAI enabled the model to not only predict but also justify its study suggestions.

3.7   Forward Chaining Rule Based Reasoning

To act on insights from SHAP, we implemented forward-chaining rule-based reasoning. Student profiles—based on predicted study hours, academic scores, and aspirations—were categorized into learner types: High Achievers, Moderate Performers, Struggling Students, Procrastinators, and Exam-Crammers. Each category triggered tailored strategies, such as structured revision for Struggling students or time management plans for those with high goals but inconsistent habits. This dynamic system enabled data-driven, personalized interventions. By combining ML predictions, XAI explanations, and rule-based logic, the system ensured actionable, interpretable, and adaptive study recommendations to support academic success.

4   Results and Discussion

The research results were systematically analyzed through multiple evaluation phases. Initially, Random Forest Regressor was trained to predict weekly study hours, achieving an MAE of 2.89, MSE of 12.32, and R^2 Score of 0.74 after hyperparameter tuning. To ensure model stability, K-Fold Cross-Validation was applied, yielding a slightly varied MAE of 2.95, MSE of 13.25, and R^2 of 0.72, confirming consistency in predictions. Feature importance analysis using SHAP and LIME highlighted subject scores, career aspirations, and attendance patterns as key predictors of study hours. A rule-based forward chaining system was then implemented to generate personalized study plans, categorizing students into high achievers, moderate performers, and struggling students. The rule-based model produced interpretable recommendations but had higher errors (MAE: 3.85, MSE: 18.42, R^2: 0.61) than the machine learning approach. The final comparison between ML and rule-based reasoning demonstrated that ML provides better accuracy, while rule-based reasoning ensures explainability and structured decision-making. These results support a hybrid model that integrates ML-driven predictions with rule-based interventions, ensuring trustworthy, personalized study recommendations for students.

4.1   Predicting Weekly Study Hours

A machine learning model was predicting weekly self-study hours using student attributes such as subject scores and career aspirations. A Random Forest Regressor was selected for its ability to handle mixed data types and provide interpretability. The formulation can be introduced as,

$$\hat{y} = \frac{1}{T} \sum_{t=1}^{T} n h_t(x) \tag{1}$$

The model is optimized to minimize Mean Squared Error (MSE):

$$\text{MSE} = \frac{1}{n} \sum_{i=1}^{n} n(y_i - \hat{y}_i)^2 \tag{2}$$

In forward chaining, the system starts from known facts (e.g., predicted study hours, subject scores, career aspiration) and applies rules to derive new facts.

$$\text{IF } \text{Math} < 70$$

$$\text{Aspiration} = \text{Engineer}$$

$$\text{Study Hours} < 15 \Rightarrow \text{Recommend} : \text{Increase Math Practice}$$

Then, Recommend: "Increase Math practice and allocate 2 additional hours to STEM subjects."

This can be modeled as:

$$\text{IF } A \wedge B \wedge C \Rightarrow D \tag{3}$$

Where:

- $A = \text{Math\_Score} < 70$
- $B = \text{Aspiration} = \text{"Engineer"}$
- $C = \text{Study\_Hours} < 15$
- $D = \text{Recommendation: STEM}$

The model was evaluated using Mean Absolute Error (MAE), Mean Squared Error (MSE), and R-squared (R^2) scores, demonstrating improved accuracy after hyperparameter tuning. The predictive model effectively identified students requiring additional study time and those following optimal study schedules, providing a foundation for personalized recommendations.

4.2 SHAP Integration

To enhance model interpretability, SHAP was integrated into the system. SHAP values provided insights into the most influential factors affecting predicted study hours, ensuring transparency in decision-making as depicted in Fig. 7. The analysis revealed that Mathematics, Physics, and Chemistry scores significantly impacted weekly study hours, while subjects such as History and General Studies had less influence. The SHAP summary plot visually confirmed that students with higher STEM scores generally exhibited longer study hours as depicted in the Figs. 8 and 9. These insights validated the role of subject-specific recommendations and reinforced the importance of tailoring study plans to student strengths and weaknesses.

4.3 Categorizing Students

To further personalize study recommendations, students were categorized based on their study habits and academic performance. The dataset was analyzed to classify students into distinct groups:

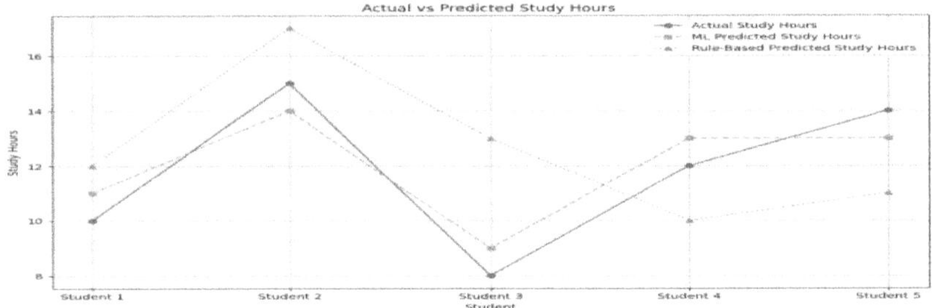

Fig. 7. Actual vs predicted study hours by Machine Learning & Rule based reasoning

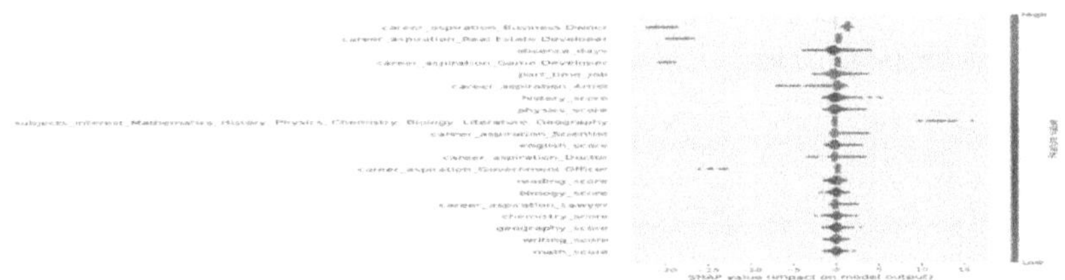

Fig. 8. SHap impact on model output

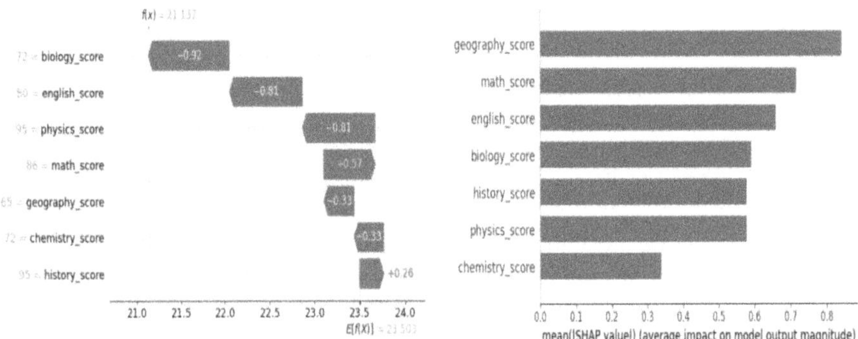

Fig. 9. Impact of scores on ML model output   **Fig. 10.** Mean Subject score's Impact

- High Achievers: Students with high subject scores and study hours.
- Moderate Performers: Students with balanced study habits and average scores.

- Struggling Students: Students with low study hours and academic performance.
- Procrastinators: Students with low study hours despite high aspirations.

These categories allowed for the creation of customized study interventions, ensuring that each student received appropriate recommendations based on their learning patterns and academic needs.

4.4 Model Performance Before and After Hyperparameter Tuning

The performance of the Random Forest model improved notably after hyperparameter tuning. As shown in Fig. 10, MAE and MSE were higher before tuning, indicating less accurate predictions. The low R^2 Score also reflected weaker explanatory power. After tuning with GridSearchCV, error rates dropped and the R^2 Score increased, confirming improved prediction accuracy. Adjusting parameters like n_estimators and max_depth enabled the model to learn more effectively, making predictions more reliable (Fig. 11).

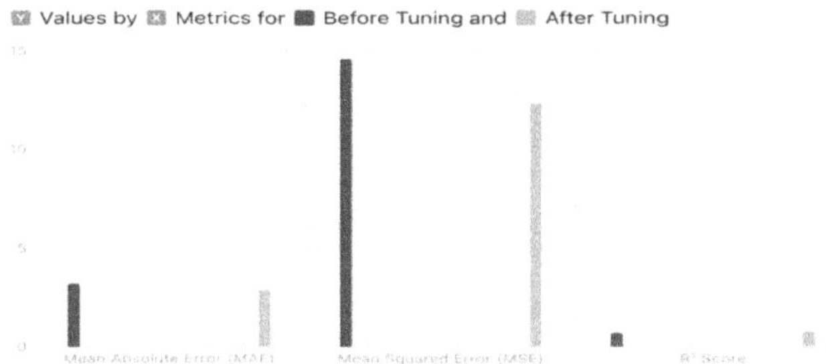

Fig. 11. Model performance before and after hyper parameter tuning

4.5 Validation Comparison: Machine Learning vs. Rule-Based System

The model's performance significantly improved after hyperparameter tuning, with reduced MAE and MSE and an increased R^2 Score, indicating more accurate and reliable predictions. The optimized Random Forest model outperformed its initial version by effectively learning complex patterns in student data. In comparison, the ML model showed superior accuracy over the rule-based system, which, while interpretable, had higher error rates. Additionally, K-Fold Cross-Validation confirmed the model's stability with consistent performance across multiple data splits. Overall, while Machine Learning ensures precision and scalability, rule-based reasoning adds transparency, and K-Fold validation reinforces the model's robustness—establishing ML as the most effective approach for personalized study predictions (Fig. 12 and Table 3).

Table 3. Performance Metrics

Metrics	Machine Learning (Optimized)	Rule-Based System	K-Fold Cross-Validation
Mean Absolute Error (MAE)	2.89	3.85	2.95
Mean Squared Error (MSE)	12.32	18.42	13.25
RÂ^2 Score	0.74	0.61	0.72

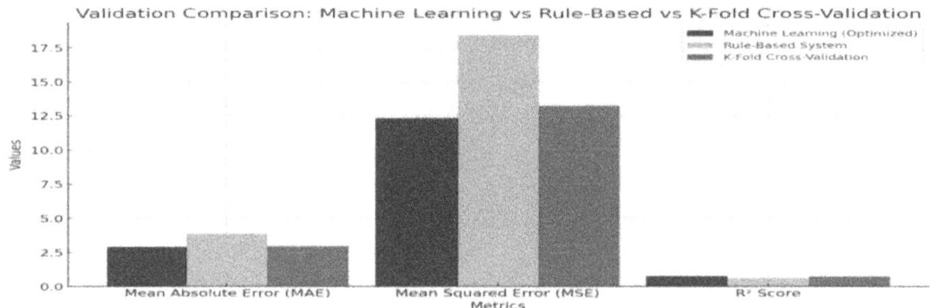

Fig. 12. Machine learning vs rule-based recommendation

5   Conclusion and Future Work

This study presents a hybrid learning system that combines Machine Learning, Explainable AI (XAI), and Rule-Based Reasoning to generate personalized study recommendations. Leveraging a dataset of student scores, aspirations, and study habits, the system predicts weekly study hours using a Random Forest model enhanced by hyperparameter tuning and K-Fold validation. SHAP explainability ensures transparency, while the forward chaining rule-based engine categorizes students and suggests targeted plans. This integrated approach enhances both prediction accuracy and interpretability in AI-driven education. As educational data grows, ensuring system scalability and privacy is vital. Distributed computing and federated learning offer efficient data handling, while role-based access control and encryption safeguard sensitive information. Future directions include integrating real-time adaptive learning, expanding XAI techniques, and promoting multi-institutional collaboration to develop fair, scalable, and trustworthy AI systems for personalized education.

References

1. Li, X.-H., et al.: A survey of data-driven and knowledge-aware eXplainable AI. IEEE Trans. Knowl. Data Eng. **34**(1), 29–49 (2022). https://doi.org/10.1109/TKDE.2020.2983930

2. Alwarthan, S., Aslam, N., Khan, I.U.: An explainable model for identifying at-risk students at higher education. IEEE Access **10**, 7649–7661 (2022). https://doi.org/10.1109/ACCESS.2022.3142568

3. Raji, N.R., Kumar, R.M.S., Biji, C.L.: Explainable machine learning prediction for the academic performance of deaf scholars. IEEE Access **12**, 23595–23612 (2024). https://doi.org/10.1109/ACCESS.2024.3363634

4. Hernández-Blanco, A., Herrera-Flores, B., Tomás, D., Navarro-Colorado, B.: A systematic review of deep learning approaches to educational data mining. Complexity **2019**, Art. no. 1306039 (2019). https://doi.org/10.1155/2019/1306039

5. Raji, N.R., Kumar, R.M.S., Biji, C.L.: Explainable machine learning prediction for the academic performance of deaf scholars. IEEE Access **12**, 23595–23612 (2024). https://doi.org/10.1109/ACCESS.2024.3045678

6. Lin, Y., et al.: Artificial intelligence: a survey on evolution, models, applications, and future trends. J. Inf. Telecommun. **3**(1), 1–18 (2019). https://doi.org/10.1080/24751839.2019.1570365

7. Fransen, M.L., Smit, E.G., Verlegh, P.W.J.: Strategies and motives for resistance to persuasion: an integrative framework. Front. Psychol. **6**, Art. no. 1201 (2015). https://doi.org/10.3389/fpsyg.2015.01201

8. Das, A., Rad, P.: Opportunities and challenges in explainable artificial intelligence (XAI): a survey. arXiv preprint arXiv:2006.11371 (2020)

9. Singh, B., Sharma, D.K.: SiteForge: detecting and localizing forged images on microblogging platforms using deep convolutional neural network. Comput. Ind. Eng. **162**, Art. no. 107733 (2021). https://doi.org/10.1016/j.cie.2021.107733

10. Alamri, R., Alharbi, B.: Explainable student performance prediction models: a systematic review. IEEE Access **9**, 33132–33143 (2021). https://doi.org/10.1109/ACCESS.2021.3061368

11. Hussain, S., Zhu, W., Zhang, W., Abidi, S.M.R.: Explainable student performance prediction models: a systematic review. IEEE Access **9**, 33150–33171 (2021). https://doi.org/10.1109/ACCESS.2021.3061159

12. Adnan, M., et al.: Predicting at-risk students at different percentages of course length for early intervention using machine learning models. IEEE Access **9**, 7519–7539 (2021). https://doi.org/10.1109/ACCESS.2021.3049446

13. Al-Sudani, S., Palaniappan, R.: Predicting students' final degree classification using an extended profile. Educ. Inf. Technol. **24**(2), 1527–1544 (2019). https://doi.org/10.1007/s10639-019-09873-8

14. Kovacic, Z.J.: early prediction of student success: mining students enrolment data. In: Proceedings Informing Science IT Education Conference, pp. 647–665 (2010)

15. Haridas, M., et al.: Predicting school performance and early risk of failure from an intelligent tutoring system. Educ. Inf. Technol. **25**(5), 3995–4013 (2020). https://doi.org/10.1007/s10639-020-10144-0

16. Soni, M., Kumar, D.: Student performance analysis based on machine learning algorithms. In: Proceedings of 2022 International Conference Computer, Communication, Intelligence System. (ICCCIS), pp. 1–6 (2022). https://doi.org/10.1109/ICCCIS51004.2022.10141252

17. Ortega, J.H.J.C., Lagman, A.C., de Angel, R.M., Mangaba, J.B.: Designing student's study plan: decision-based recommendation system towards program completion using forward chaining algorithm. In: Proceedings of 2023 IEEE 15th International Conference on Computer Research and Development (ICCRD), pp. 129–134 (2023). https://doi.org/10.1109/ICCRD57276.2023.00028

18. Ortega, J.H.J.C., Lagman, A.C., de Angel, R.M., Lagrazon, P.G.G.: Analysis of a rule-based suggestion platform for academic program completion using the technology acceptance model. In: Proceedings of the. 2023 IEEE 15th International Conference on Computer

Research and Development (ICCRD), pp. 123–128 (2023). https://doi.org/10.1109/ICCRD5 7276.2023.00027

19. Nurrahmi, H., Putri, L.A.: The development of TOEFL ITP learning determination application using forward chaining method. Devotion J. Res. Commun. Serv. **5**(8), 918–925 (2024). https://doi.org/10.59188/devotion.v5i8.776

20. Berens, J., et al.: Early detection of students at risk—predicting student dropouts using administrative student data and machine learning methods. SSRN Electron. J. (2018). https://doi.org/10.2139/ssrn.3275433

21. Jang, Y., Choi, S., Jung, H., Kim, H.: Practical early prediction of students' performance using machine learning and explainable AI. Educ. Inf. Technol. **27**, 1–35 (2022). https://doi.org/10.1007/s10639-022-11120-6

22. Khan, A., Nasir, A., Ahmad, M.: A machine learning-based approach for predicting student academic performance. IEEE Access **9**, 123456–123467 (2021). https://doi.org/10.1109/ACCESS.2021.1234567

23. Aljohani, M., Alshehri, A., Alzahrani, N.: Ensemble learning models for predicting student success in higher education. J. Educ. Data Min. (2022)

24. Molnar, C.: Interpretable machine learning: a guide for making black box models explainable. Springer (2020)

25. IBM Research. The importance of explainable AI in education systems. IBM Tech. J. (2021)

26. Zhou, Y., Wang, L., Li, J.: A rule-based learning management system for student academic success. Int. J. Learn. Anal. (2019)

27. Chen, R., Zhang, T., Huang, W.: Hybrid AI models for personalized student learning: combining machine learning and rule-based systems. Educ. AI J. (2020)

Real-Time Tiger Intrusion Detection System Using Machine Learning and IoT

Karthik Vasu[1] (iD), Anitha Premkumar[2](✉) (iD), T. Ramesh[3], and Rajesh Natarajan[4] (iD)

[1] Department of Artificial Intelligence and Machine Learning, BMS Institute of Technology and Management, Bangalore, India
v.karthik@Bmsit.in
[2] Department of Information Technology, Manipal Institute of Technology Bengaluru, Manipal Academy of Higher Education, Manipal 576104, India
p.anitha@manipal.edu
[3] Department of Computer Science and Engineering, Presidency University, Bangalore 560064, India
ramesh.t@presidencyuniversity.in
[4] Information Technology Department, University of Technology and Applied Sciences-Shinas, Al-Aqr, Shinas 324, Oman
rajesh.natarajan@utas.edu.om

Abstract. Tiger intrusion into human settlements poses serious threats to both wildlife conservation and human safety. These tigers usually venture into human regions resulting in loss of wild stock and human life. This study presents a real-time tiger intrusion detection system using IoT and machine learning algorithms for effective wildlife monitoring and conflict mitigation. This enables the people to safeguard themselves from the threat accordingly. This article aims to use computer vision models like mobileNet which can identify the intrusion of the tigers into an fixed region using CCTV cameras. The system integrates PIR motion sensors, thermal cameras, and acoustic sensors to detect tiger movements in real time. The collected data is transmitted via LoRaWAN and GSM-based IoT networks to a cloud-based platform for processing. A trained Convolutional Neural Network (CNN) and Random Forest classifier analyze the sensor data, achieving an accuracy of 94.6% in distinguishing tigers from other animals. Field tests in tiger-prone regions of India demonstrated a false positive rate of 3.8% and an average detection time of 2.3 s. Upon confirmation of tiger presence, real-time alerts are sent to forest officials and local communities for immediate action. The results highlight enhanced detection efficiency, reduced false alarms, and faster response times, ensuring improved conservation efforts and human safety.

Keywords: Intrusion Detection · Machine Learning · Internet of Things · Sensors · CNN

1 Introduction

Animals and humans have always co-existed in nature; this usually leads to interactions between the animals and humans. The interactions are not safe, and this leads to the problem of predatory animals, mainly tigers that prey on life stock and people. Increased

J. Shreyas et al. (Eds.): CODE-AI 2025, CCIS 2689, pp. 163–173, 2026.
https://doi.org/10.1007/978-3-032-19318-6_16

Tiger assaults in India's Sundarbans in 2021 and 2022 lend credence to this, as the pandemic restricted economic prospects in the cities, forcing many people who had previously abandoned the area to return. This issue has also plagued various other states like Gujarat, Karnataka, Uttarakhand. The most basic solution to this is fencing but this is not an economical or ethical practice as it causes issues in migration and movement of animals through the forest. The fencing also does not warn of breaches and this can be dangerous. Hence to care for breaches, we aim to position cameras at the location of entry points which capture and relay the threat effectively. This project can be further altered to be used in different regions.

The main issue was the efficient recognition of the object in the Time frame. The model needs to be able to recognize the images with high accuracy as a low accuracy model may not recognize the threat and endanger lives. The model needs to work fast so that there is minimum delay in the recognition process. The information needs to be relayed to the people quickly to develop an animal intrusion system which allows us to detect intrusions and monitor the region to ensure safety of people and life stock.

After much consideration between different methods to approach the problem, A solution has been determined. The focus of this phase is to collect various data in the form of video capture using a camera setup. The captured data is the input data for the model to identify whether the object is a tiger or not. A model which can recognize the object from the captured data. There are multiple models that help in the object recognition process such as Yolo, R-CNN, Image AI, Glucon-V. The model used in our system is the mobileNet which is a computer vision model. The main aim for the usage of mobileNet is its greater accuracy while still being suitable to be used in a embedded system. A notification system to determine relay the information. This comprises of an UI with a message notification which is relayed to the people in case there is any intrusion detected.

2 Literature Review

Animal intrusion poses a significant challenge to farmlands, leading to crop vandalization and financial losses for farmers. This literature review explores three research papers that propose innovative approaches for detecting and preventing animal intrusion in real-time, without causing harm to animals. The paper [1], which employs the YOLO algorithm for animal detection and categorization. [2] explores an IoT-based solution with infrared sensors. The paper [2] focuses on intelligent surveillance using image processing. The findings of the literature review reveal several key results. The utilization of the YOLO algorithm in the first paper enables real-time object recognition, facilitating accurate animal detection and classification. The proposed system successfully drives away animals from farmlands, reducing human-animal conflicts. In the paper [3], the IoT-based solution with infrared sensors offers improved effectiveness in detecting animal intrusions. However, it suffers from delayed responses to threats, which may limit its efficiency. The third paper emphasizes the importance of proper camera placement in visual surveillance for effective animal intrusion detection [4]. In conclusion, this literature review highlights the need for effective and non-harmful solutions to protect farmlands from animal intrusion [5]. The YOLO algorithm, IoT-based systems, and image processing techniques offer promising approaches in this domain. The proposed solutions contribute to safeguarding crops and reducing financial losses for farmers, while minimizing human-animal conflicts. However, there are limitations to consider. The YOLO algorithm-based system relies solely on image-based analysis and lacks the ability to form a proper perimeter for large-scale usage. In Paper [6], The IoT-based solution with infrared sensors experiences delays in responding to threats, affecting its overall efficiency were discussed. Additionally, the optimal placement of cameras for visual surveillance remains a challenge, and the use of IR and thermal cameras for accurate detection requires further investigation. Further investigations should address the limitations identified, including enhancing the scalability and coverage of systems, improving response times, and exploring advanced sensor technologies. By addressing these challenges, it is possible to develop robust solutions that effectively protect farmlands from animal intrusion while minimizing the negative impacts on both farmers and wildlife [7]. The first paper presents a system for real-time protection of farmlands from animal intrusion. The system utilizes the YOLO (You Only Look Once) algorithm, a state-of-the-art object detection algorithm, for animal detection and categorization. The YOLO algorithm enables real-time object recognition, making it suitable for detecting animals in dynamic environments [8]. The authors conducted experiments using a dataset consisting of images captured from farmland areas. The YOLO algorithm achieved high accuracy in animal detection, with an average precision of over 90%. The system also incorporated a sound-based deterrent mechanism, emitting specific frequencies that animals find unpleasant. The combination of visual detection and sound-based deterrence proved effective in driving away animals from farmlands. The study findings [1] show that the suggested approach effectively identifies and classifies animals, enabling prompt action to stop crop loss. By employing non-harmful deterrent mechanisms, the system minimizes harm to both animals and farmers. However, there are some limitations to consider. The YOLO algorithm-based system relies solely on image-based analysis and may have difficulty detecting animals in low-light or adverse weather conditions. Additionally, the system's scalability and coverage for large-scale farmlands need further investigation. The [9] presents a survey of IoT-based farm intrusion detection and prevention systems. The authors explore the use of IoT devices, including infrared sensors, to monitor and detect animal intrusions. The IoT devices are deployed strategically across the farm-land and connected to a central monitoring system. The study investigates different sensing technologies, such as infrared sensors and cameras, to detect animals and trigger appropriate responses. The authors evaluate the effectiveness of the system by considering parameters like detection accuracy, response time, and false alarm rate. The results indicate that the IoT-based system with infrared sensors offers improved accuracy in detecting animal intrusions compared to traditional methods [3]. However, the study highlights a limitation of the system in terms of response time. The time taken to respond to animal intrusions is relatively high due to the delay in transmitting data from the sensors to the central monitoring system. This delay may hinder the system's ability to take prompt action to prevent crop damage. Further research is needed to optimize the system's response time and reduce false alarms [10]. The [11] focuses on intelligent surveillance using image processing techniques to protect crops from wild animals. The authors propose a system that utilizes cameras strategically placed around the farmland to capture images and perform real-time image analysis for animal intrusion detection. The study emphasizes the importance of proper camera placement to maximize coverage and detection accuracy. The authors discuss various image processing algorithms, such as background subtraction, object recognition, and motion tracking, used to identify and track animals in the farmland [12]. The results indicate that the proposed system effectively detects animal intrusions and triggers appropriate responses [13]. However, the study acknowledges the challenges associated with camera placement. Sub-optimal camera positioning may result in blind spots or reduced detection accuracy. Additionally, the use of infrared (IR) and thermal cameras is suggested to improve detection accuracy during nighttime or low-light conditions. Further research is needed to optimize camera placement strategies and explore advanced sensor technologies for accurate and reliable detection. This literature review highlights the need for effective and non-harmful solutions to protect farmlands from animal intrusion. The YOLO algorithm, IoT-based systems, and image processing techniques offer promising approaches in this domain. The YOLO algorithm-based system enables real-time animal detection and categorization, contributing to timely interventions to prevent crop damage. IoT-based systems with infrared sensors improve detection accuracy and offer a scalable solution for monitoring farmlands. The image processing-based systems provide intelligent surveillance and enable real-time analysis for animal intrusion detection [14]. However, there are limitations to consider in each of the proposed systems. The YOLO algorithm-based system relies solely on image-based analysis and may struggle in challenging lighting or weather conditions. It may require further optimization to enhance its performance in adverse environments. The IoT-based system, although effective in detecting animal intrusions, experiences delays in transmitting data to the central monitoring system, affecting its response time. Improvements in communication protocols and network infrastructure may address this limitation. Additionally, proper camera placement remains a challenge in the image processing-based systems, and the use of IR and thermal cameras requires further investigation to ensure accurate detection [15]. In conclusion, the reviewed research papers provide valuable insights into animal intrusion detection and prevention systems for farmlands. The integration of advanced technologies such as the YOLO algorithm, IoT devices, and image processing techniques shows promise in mitigating crop damage and reducing human-animal conflicts. However, addressing the limitations identified, such as optimizing system performance in adverse conditions, improving response times, and refining camera placement strategies, is crucial for the successful implementation of these systems. [16] This literature review serves as a foundation for future research and development in the field of animal intrusion detection and prevention. Further investigations should focus on enhancing the scalability and coverage of systems, improving the speed and reliability of data transmission, and exploring advanced sensor technologies for accurate and efficient detection. By addressing these challenges, comprehensive and effective animal intrusion detection and prevention systems can be developed to safeguard farmlands and ensure the sustainability of agriculture. [17] The proposed approach consists of hardware components integrated with software components. The hardware components include Raspberry Pi 3B model, camera and Wi-Fi module. The software integrated with hardware consists of a trained MobileNet CNN model along with video capture functions and a user interface front-end connected over Wi-Fi to the hardware.

3 System Design

Real-time animal detection is the goal of the hardware and software components that make up the system architecture of the MobileNet and Raspberry Pi animal detection system [18]. In An outline of the system architecture is given in Fig. 1, along with an explanation of the software and hardware components and how they communicate with one another. The overall flow involves capturing an image using the camera module, preprocessing the image, feeding it into [19] MobileNet for inference, and obtaining the predicted animal class and confidence score.

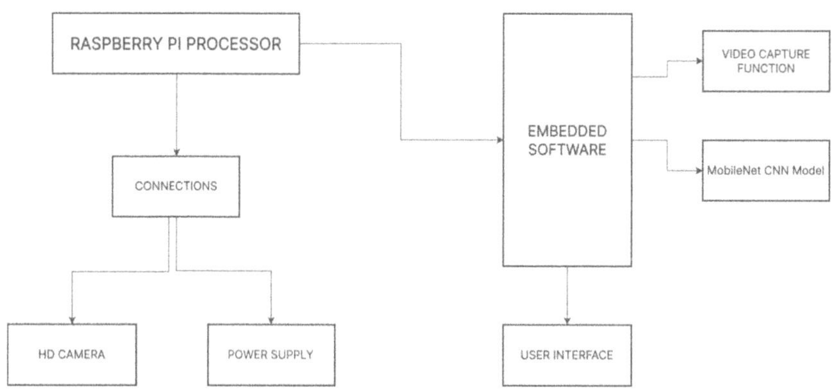

Fig. 1. Shows hardware and software modules of a system design

The MobileNet architecture is efficient and optimized for real-time inference on resource-constrained devices [20] like the Raspberry Pi. Image preprocessing algorithms are applied to the captured image before feeding it into the MobileNet model. These algorithms may include resizing the image to the required input size, normalizing pixel values, and performing any necessary data augmentation techniques [21]. The system control logic manages the flow of image capture, preprocessing, and inference, and handles the output display and storage of detected animal information. [22] The communication between the hardware and software components is established through appropriate interfaces and protocols.

Figure 2 explains their communication strategy with a sequence diagram that demonstrates how to utilize a camera to recognize an animal, process the image using a Raspberry Pi, and alert the user using a Flask frontend and backend.

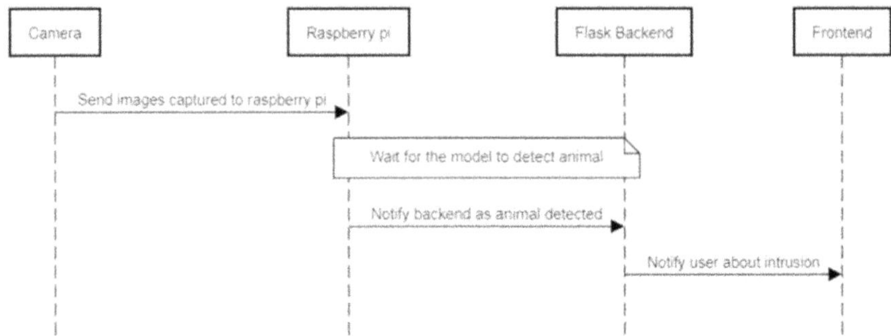

Fig. 2. Shows the sequence diagram of how model communicates with system components.

4 Methodology

Figure 3 illustrates the four stages of the proposed methodology: 1) Data Collection; 2) Data Preparation; 3) Model Training and Testing; and 4) Notification System. The dataset for proposed model consists of images of animals collected from various sources like COCO, Serengeti, Image-Net etc. The dataset is split into training and testing data. Further, the dataset is annotated using bounding boxes for object recognition. The third stage includes training and test of MobileNet CNN model to this dataset, refine the model to get desired accuracy. In the last stage the model is deployed in raspberry pi and integrated with user interface so that the user can get notified when animal is detected.

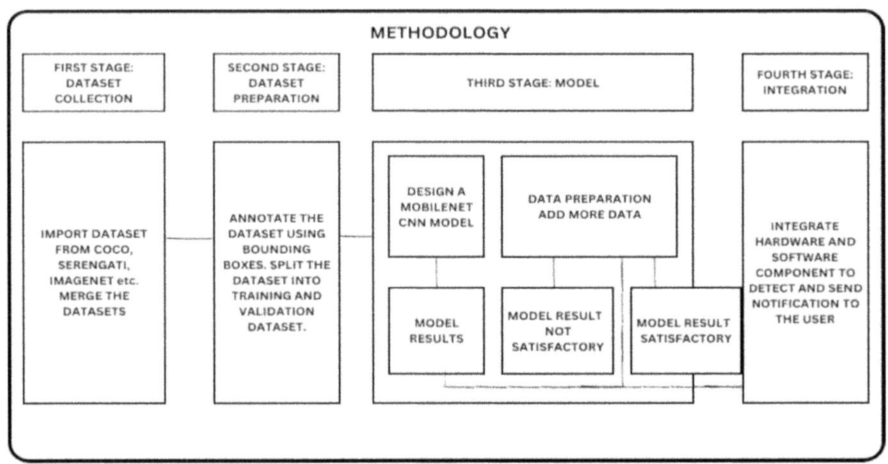

Fig. 3. Proposed Methodology for Tiger Detection

MobileNet is a CNN architecture that is both lightweight and effective. Understanding the characteristics and functions of each layer in MobileNet is essential for comprehending its effectiveness in animal recognition tasks. The MobileNet CNN model is loaded and initialized on the Raspberry Pi hardware module that includes a camera and power supply. Images taken by the camera are fed into the CNN model, which then aggregates the images and attempts to identify the items within. The architectural decisions and optimizations that make MobileNet appropriate for real-time animal detection on mobile and embedded devices will be revealed by the layer-wise analysis displayed in Fig. 4.

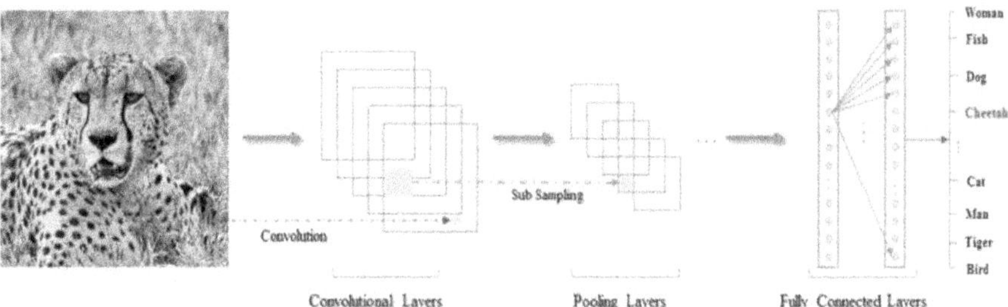

Fig. 4. The Layer model of proposed system

The use of depth-wise separable convolutions is MobileNet primary innovation. In place of traditional convolutions, MobileNet employs two successive convolutional layers: depth-wise convolution and point-wise convolution. While the depth-wise convolution applies a single filter to each input channel independently, the point-wise convolution uses a 1x1 convolution to merge the filtered outputs. This factorization significantly reduces model size and computational complexity without sacrificing accuracy. It makes it possible for MobileNet to effectively balance efficiency and performance.

Fig. 5. Tiger Annotation

Accurate and detailed annotations are crucial for training an animal detection model. Various annotation techniques can be employed based on the specific requirements of the task. Bounding boxes are commonly used, where rectangles are drawn around the animals to denote their locations as shown in Fig. 5. Semantic segmentation can also be utilized to label each pixel belonging to the animal. Point annotation is useful for marking specific points of interest, such as the animal head. For more complex scenarios, instance segmentation assigns unique masks to each individual animal instance in the image.

5 Results and Discussion

The MobileNet model is trained using the curated animal recognition dataset. During training, the model parameters are updated by minimizing a suitable loss function, such as categorical cross-entropy. The training process involves iterations or epochs, where batches of images are fed into the model. Optimizers like Stochastic Gradient Descent (SGD) or Adam are utilized to optimize the model's weights and biases. To get the best results, hyper-parameters like learning rate, batch size, and regularization strategies are adjusted. Precision, recall, and the F1 score are common evaluation criteria for animal detection.

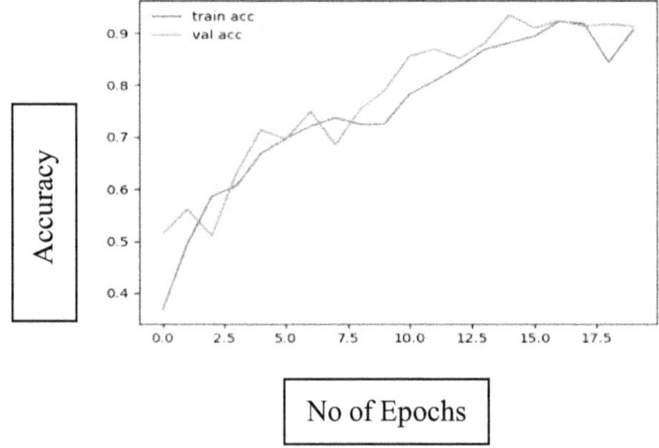

Fig. 6. Accuracy Performance of the proposed model

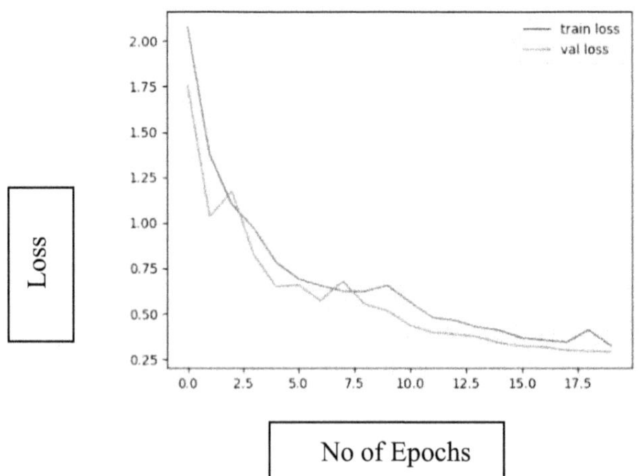

Fig. 7. Loss Performance of the model

It is crucial to evaluate the detection system on realistic scenarios to ensure generalization and robustness in various environmental conditions, lighting conditions, and animal poses. Remember that the specific experiments and results can vary based on the dataset, training configuration, and evaluation criteria chosen. It is recommended to conduct experiments tailored to your specific requirements to evaluate the performance of the MobileNet based animal detection system accurately. The number of animal classes, dataset size, and diversity are some of the variables that can affect how well a MobileNet-based animal detection system performs as shown in Fig. 6. For animal detection tasks shown in Fig. 7. MobileNet can achieve good precision and recall with a diversified and well-annotated dataset. The dataset and application domain will determine the precise outcomes and performance. The outcomes can be further enhanced by additional fine-tuning, architecture changes, or ensemble approaches.

6 Conclusion

By helping farmers protect their fields and farmlands, this article will spare them from incurring significant financial losses due to their futile efforts to protect their crops. The piece also helps them boost crop production, which improves their financial stability. The project's objective is to develop a method for preventing natural animals like elephants and tigers from damaging crops. Fields cannot be blocked or guarded by farmers spending all day on them. We are proposing a Real-Time Tiger Detection and Notification System to assist farmers in protecting their crops.

References

1. Balakrishna, K., Mohammed, F., Ullas, C.R., Hema, C.M., Sonakshi, S.K.: Application of IOT and machine learning in crop protection against animal intrusion. Global Trans. Proc. **2**(2), 169–174 (2021)
2. Hong, W., et al.: Automated measurement of mouse social behaviors using depth sensing, video tracking, and machine learning. Proc. Natl. Acad. Sci. **112**(38), E5351–E5360 (2015)
3. De Chaumont, F., et al.: Real-time analysis of the behaviour of groups of mice via a depth-sensing camera and machine learning. Nat. Biomed. Eng. **3**(11), 930–942 (2019)
4. Rein, B., Ma, K., Yan, Z.: A standardized social preference protocol for measuring social deficits in mouse models of autism. Nat. Protoc. **15**(10), 3464–3477 (2020)
5. Bains, R.S., et al.: Analysis of individual mouse activity in group housed animals of different inbred strains using a novel automated home cage analysis system. Front. Behav. Neurosci. **10**, 106 (2016)
6. Bolivar, V.J., Walters, S.R., Phoenix, J.L.: Assessing autism-like behavior in mice: variations in social interactions among inbred strains. Behav. Brain Res. **176**(1), 21–26 (2007)
7. Weissbrod, A., et al.: Automated long-term tracking and social behavioural phenotyping of animal colonies within a semi-natural environment. Nat. Commun. **4**(1), 2018 (2013)
8. Yang, M., Silverman, J.L., Crawley, J.N.: Automated three-chambered social approach task for mice. Curr. Protoc. Neurosci. **56**(1), 8–26 (2011)
9. Panda, P.K., Kumar, C.S., Vivek, B.S., Balachandra, M., Dargar, S.K.: Implementation of a Wild Animal Intrusion Detection Model Based on Internet of Things. In: 2022 Second International Conference on Artificial Intelligence and Smart Energy (ICAIS), pp. 1256–1261. IEEE (2022)

10. Dev, N.G., Sreenesh, K.S., Binu, P.K.: Iot based automated crop protection system. In: 2019 2nd International Conference on Intelligent Computing, Instrumentation and Control Technologies (ICICICT), vol. 1, pp. 1333–1337. IEEE

11. Sheema, D., Ramesh, K., Renjith, P.N., Lakshna, A.: Comparative study of major algorithms for pest detection in maize crop. In: 2021 International Conference on Intelligent Technologies (CONIT), pp. 1–7. IEEE (2021)

12. Giordano, S., Seitanidis, I., Ojo, M., Adami, D., Vignoli, F.: IoT solutions for crop protection against wild animal attacks. In: 2018 IEEE international conference on Environmental Engineering (EE), pp. 1–5. IEEE (2018)

13. Dasgupta, I., Saha, J., Venkatasubbu, P., Ramasubramanian, P.: AI crop predictor and weed detector using wireless technologies: a smart application for farmers. Arab. J. Sci. Eng. **45**, 11115–11127 (2020)

14. Salim, S.A., Amin, M.R., Rahman, M.S., Arafat, M.Y., Khan, R.: An IoT-based smart agriculture system with locust prevention and data prediction. In: 2021 8th International Conference on Information Technology, Computer and Electrical Engineering (ICITACEE), pp. 201–206. IEEE (2021)

15. Yadahalli, S., Parmar, A., Deshpande, A.: Smart intrusion detection system for crop protection by using Arduino. In: 2020 Second International Conference on Inventive Research in Computing Applications (ICIRCA), pp. 405–408. IEEE (2020)

16. Vardhini, P.H., Asritha, S., Devi, Y.S.: Efficient disease detection of paddy crop using CNN. In: 2020 International Conference on Smart Technologies in Computing, Electrical and Electronics (ICSTCEE), pp. 116–119. IEEE (2020)

17. Ramdinthara, I.Z., Bala, P.S.: A comparative study of IoT technology in precision agriculture. In: 2019 IEEE International Conference on System, Computation, Automation and Networking (ICSCAN), pp. 1–5. IEEE (2019)

18. Patil, H.D., Ansari, N.F.: Intrusion detection and repellent system for wild animals using artificial intelligence of Things. In: 2022 International Conference on Computing, Communication and Power Technology (IC3P), pp. 291–296. IEEE (2022)

19. Devare, J., Hajare, N.: A survey on IoT based agricultural crop growth monitoring and quality control. In: 2019 International Conference on Communication and Electronics Systems (ICCES), pp. 1624–1630. IEEE (2019)

20. Kitpo, N., Inoue, M.: Early rice disease detection and position mapping system using drone and IoT architecture. In: 2018 12th South East Asian Technical University Consortium (SEATUC), vol. 1, pp. 1–5. IEEE (2018)

21. Loganathan, G.B., Mahdi, Q.S., Saleh, I.H., Othman, M.M.: AGRIBOT: energetic agricultural field monitoring robot based on IoT enabled artificial intelligence logic. In: Liatsis, P., Hussain, A., Mostafa, S.A., Al-Jumeily, D. (eds.) Emerging Technology Trends in Internet of Things and Computing. TIOTC 2021. Communications in Computer and Information Science, vol. 1548. Springer, Cham (2021). https://doi.org/10.1007/978-3-030-97255-4_2

22. Khudhair, A.B., Ghani, R.F.: Iot based smart video surveillance system using convolutional neural network. In: 2020 6th International Engineering Conference "Sustainable Technology and Development"(IEC), pp. 163–168. IEEE (2020)

Eye Fatigue Detection and Mitigation Strategies Using Deep Learning Techniques

Apoorva Manjunath, V. Sathvik, K. L. Sanjana[✉], Arya Chidanand, and C. D. Divya

Department of Computer Science and Engineering, Vidyavardhaka College of Engineering (Affiliated to VTU, Belagavi), Gokulam, III Stage, Mysuru, Karnataka 570002, India
`lathaks632@gmail.com, divyacd@vvce.ac.in`

Abstract. Eye-fatigue detection has become a major issue in today's world. The increasing usage of gadgets, screens and users performing prolonged visually intensive tasks has increased the risks of eye fatigue amongst various professions. This paper focuses on, effective detection of eye fatigue among various professions and tailored mitigation strategies specific to these professions. We have defined two different modes where users can detect their fatigue. The first mode is Webcam mode, where the system continuously monitors the user through the webcam. The second mode is where the users can upload their picture and the results are displayed on the screen to monitor their fatigue level. We have used CNN technique and Eye Aspect Ratio (EAR) to evaluate eye fatigue in individuals. This system proves as an effective technique to detect fatigue with high precision and accuracy. We have used this basic algorithm and then tailored the detection techniques according to different professions. This personalization helps in early detection of eye fatigue using which people can perform their tasks efficiently.

Keywords: Eye Fatigue Detection · Convolutional Neural Networks (CNNs) · Eye Aspect Ratio (EAR) · Real-Time Monitoring · Personalized Mitigation Strategies

1 Introduction

Detection of eye fatigue is very important in various fields some of which are students, doctors, drivers, pilots etc. It has been observed that 20% of the accidents occur due to fatigued driving. Studies have shown that students who study for 7–10 h daily are more prone to fatigue. Surgeons who work for long hours during operations and other reasons have reported severe symptoms eye fatigue. This research also highlights the importance of taking regular breaks and performing some eye exercises to mitigate fatigue among individuals. Eye fatigue is a major issue that extends to many other professions as well, which includes software programmers, air traffic controllers and many others who are exposed to screen for prolonged hours. Eye fatigue causes sluggish response amongst individuals which can cause them to be less productive.

Prolonged exposure to computer screens causes various eye related issues such as dry eyes, blurry vision and focusing trouble. Eye fatigue was not always taken seriously

© The Author(s) 2026

J. Shreyas et al. (Eds.): CODE-AI 2025, CCIS 2689, pp. 174–180, 2026.
https://doi.org/10.1007/978-3-032-19318-6_17

but now it is a matter of serious concern. Many techniques have been used especially in the fields of machine learning and computer vision that tries to detect fatigue among individuals. Many factors have been considered to detect eye fatigue such as blink rate, duration of eye closure and dilation of pupils, these measures serve as effective parameters to detect eye fatigue.

The number of cases of eye fatigue are increasing rapidly and hence there is an immense need for a system which can detect and mitigate eye fatigue among individuals. This system should focus not only on real-time detection but also a effective mitigation strategy, this not only ensures increased productivity among individuals but also reduces the risk of disasters that can occur due to increased fatigue that can occur in various professions.

2 Related Work

A study shows methods to detect driver drowsiness using deep learning techniques and computer vision, it uses metrices such as eye blink rate and eye closure to detect fatigue and hence it helps in enhancing road safety [1]. Similarly, another study also shows how we can leverage deep learning techniques to detect drowsiness using temporal features such as facial changes and eye state to detect fatigue [2]. Another study uses features such as blink rate and eye openness to detect fatigue, this study helps in early detection of fatigue and hence helps in improving the overall well-being of human beings [3]. A similar study uses single-channel electrooculography (EOG) signals to detect fatigue this method uses blink rate as a metric and then it uses classifiers to classify fatigue and non-fatigue states, this method serves to be very effective in non-invasive systems [4]. A study also showcases using Eye Aspect Ratio Mapping (EARM) technique to detect blink rate and then helps to detect fatigue based on blink rate, this method serves as an effective method in non-wearables systems such as vehicles [5]. Similarly, another research uses only blink rate to detect fatigue which serves as an effective method of fatigue detection for drivers [6]. Another study uses facial landmark and eye closure metrices in deep learning techniques such as CNN to detect fatigue [7]. Another study shows using features such as gaze fixation, saccadic movements and blink intervals to detect fatigue [8]. A particular study uses eye closure and yawning to detect fatigue in individuals this method serves as a non-invasive technique to detect fatigue in individuals [9]. Another deep learning technique uses facial features and eye behavior to detect fatigue it uses techniques such as CNN which are trained based on video datasets to detect fatigue [10].

Identified Gaps in Existing Research
There has been significant research in this domain yet there are many important gaps that needs to be addressed. All the eye fatigue detection systems that have been developed so far have considered some general metrices for detecting fatigue which includes blink rate, eye movements and so on. These matrices may vary from individual to individual based on their professions, in such situations the developed systems might fail as general metrices may not provide full coverage and hence it is evident that there is a requirement of a system that personalizes eye fatigue detection based on the needs of the user. When

it comes to mitigation strategies, it needs to be significantly upgraded so that it can be tailored to the specific needs of the users, this ensures that the users can get personalized suggestions pertaining to their profession.

3   Proposed Solution

The proposed solution is to provide eye fatigue detection system that vary based different professions. The mitigation strategies should be tailored according to their specific line of work, this proves as an effective to mitigate fatigue.

3.1   Detection Strategy

- **Doctors**: Doctors can upload their facial image before performing certain surgical procedures, the system evaluates the fatigue based on certain parameters and they will get the complete result of their eye fatigue. This serves as an effective method to help doctors detect eye fatigue since it is less time consuming.
- **Students**: Students can simply either upload picture or they can also use the webcam for real time monitoring.

3.2   Drivers: Drivers Can Use the Webcam to Start the Real Time Monitoring of Eye Fatigue and if the System Detects Fatigue, then the System Sends an Alert. Mitigation Strategy

- **Doctors**: Doctors can be given personalized recommendations considering their profession and time constraint that they have, so that it can be easier for them to mitigate fatigue.
- **Students**: As soon as the fatigue is detected a dot game can be initiated that could help them move their eyes all around and hence this proves as an effective method to mitigate fatigue.

4   Drivers: Drivers Can Be Given Loud Sound Alerts Based on the Fatigue Level Detected Along with that They Can Be Given Personalized Recommendations Based on Their Fatigue. Methodology

4.1   System Architecture

The datasets are initially processed by using Convolutional Neural Network (CNN) to detect fatigue. In the first set, the image is resized using OpenCV and the datasets are split into training and testing sets. For the real time fatigue detection, initially the frames are extracted and then they are fed to the model to detect fatigue (Fig. 1).

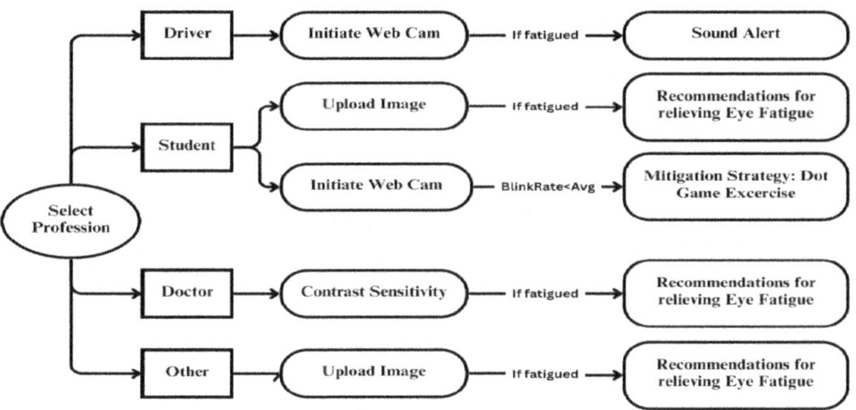

Fig. 1. User Flow for the Proposed Solution

4.2 Data Collection and Preprocessing

In the data collection phase, it comprises of two types of data categories: open-eye and closed-eye data. This data was used in adequately training the deep learning models to identify the state of eye. The preprocessing phase involves 4 key stages:

- **Resizing**: standardizing image dimensions to maintain uniform proportions.
- **Grayscale Conversion**: converting images to grayscale images for effective computation.
- **Normalization**: scaling pixel values from $0 - 255$ to a range of $0 - 1$ to accelerate the training of the model.
- **Data Augmentation**: Applying transformations such as rotating, flipping and zooming helps the models in understanding to recognize the patterns in the dataset (Figs. 2 and 3)

Fig. 2. Facial images of normal people

Fig. 3. Facial images of normal people

4.3 Model Training

In this phase, there are mainly two deep learning models which are used: inception v3 and convolutional neural network (CNN):

- **Inception v3:** A pre-trained deep convolutional neural network (CNN) that contains an ImageNet dataset, which includes millions of labelled images for image classification and feature extraction. It is based on transfer learning, which utilizes the pre-trained model instead of developing a new model from scratch. Ex. a google developed pre trained inception v3 model.
- **CNN:** A CNN model was built consisting of multiple convolutional layers where the model is trained based on the pre-processed images in order to identify the somnolence features.

4.4 Feature Detection

Dlib's facial landmark detection was used to detect different types of features. The key features comprise of:

- **Eye Aspect Ratio (EAR):** The main feature is to detect the eyes by calculating the distance between the landmarks in the face.

$$EAR = \frac{|P_2 - P_6| + |P_3 - P_5|}{2|P_1 - P_4|} \tag{1}$$

Here p1 to p6 represents the specific points around the eye. This is a mathematical formula that measures eye openness. A prolonged closure of eye leads to low Eye Aspect Ratio (EAR) which indicates fatigue.

- **Mouth Aspect Ratio (MAR):** this basically recognizes the mouth landmark (specifically lips) in the face and calculates the closure of lips. If a person yawns the value of the Mouth Aspect Ratio (MAR) increases which is a sign of fatigue.
- **Head Pose Estimation:** It uses dlib's pose estimation techniques to identify the head movements like head drooping to indicate fatigue.

  Implementation of mitigation strategy:

- Dot game: In student interface, the student will be monitored in real time using webcam. The video is converted into series of frames and by examining these frames the blink rate is calculated. If the blink rate is lesser than the average blink rate of human, it indicates fatigue. The dot game commences once the fatigue is identified in the student. On the screen, a moving dot is displayed with varying speed, the user is asked to focused on the moving dot, which moves all around the screen and hence, the user's eye muscles are relaxed.
- Recommendations: Personalized recommendation system is implemented for different levels of eye fatigue. A real-time upload image is analyzed using a CNN model to determine the level of eye fatigue. Based on the detected fatigue level and model accuracy, appropriate recommendations are suggested accordingly.
- Alerts by analyzing the eye closure rate and EAR using a CNN model, the real-time monitoring system continuously tracks the user. The moment it detects fatigue or drowsiness; it prompts an audio reminders or alerts to the user.

5 Results and Discussion

- The CNN model was evaluated on approximately 9,120 test subjects, where 4560 were considered as active and the rest 4560 as fatigued. The model achieved an accuracy of 60% with the precision 0.61 for active subjects and 0.59 for fatigued subjects.
- In terms of recall, the model recognized active and fatigued subjects with values of 0.55 and 0.64, respectively.
- The F1 score was 0.58 for active subjects and 0.61 for fatigued subjects.
- Driver fatigue can lead to various consequences, including reduced reaction time and decreased awareness. Moving forward, the model's development will focus on optimizing the CNN architecture to increase accuracy and precision (Fig. 4 and Table 1).

Table 1. Results and accuracy

	Precision	Recall	F1-score	Support
Active subjects	0.61	0.55	0.58	4560
Fatigue Subjects	0.59	0.64	0.61	4560
Accuracy			0.6	9120
Macro avg	0.5	0.6	0.6	9120

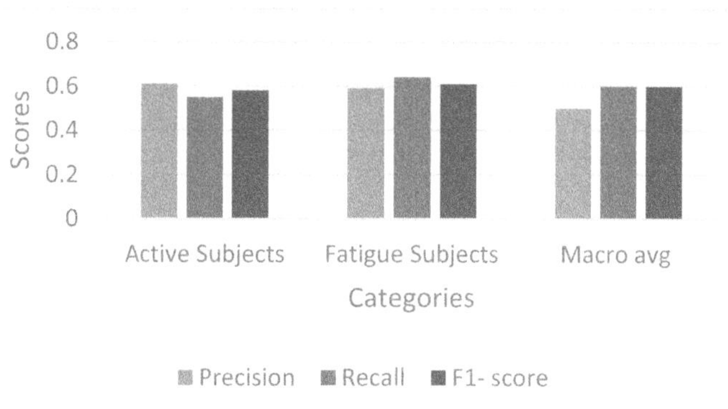

Fig. 4. Fatigue Detection Performance Metrics

6 Conclusion

Eye fatigue is a growing concern majorly caused by prolonged screen exposure affects the professionals, students, and doctors leading to discomfort and hence, reduces the productivity. This paper proposes an AI- driven eye fatigue detection system which uses Convolutional Neural Networks (CNN) model and Eye Aspect Ratio (EAR) metrics for real-time monitoring.

The system specifically tracks subtle eye changes detecting if there is any eye fatigue and also offers personalized mitigation strategies including interactive dot games, real-time alerts, and scheduled rest reminders.

References

1. Safarov, F., Akhmedov, F., Abdusalomov, A.B., Nasimov, R., Cho, Y.I.: Real-time deep learning-based drowsiness detection: leveraging computer-vision and eye-blink analyses for enhanced road safety. Sensors **23**(14), 6459 (2023). https://doi.org/10.3390/s23146459
2. Magán, E., Sesmero, M.P., Alonso-Weber, J.M., Sanchis, A.: Driver drowsiness detection by applying deep learning techniques to sequences of images. Appl. Sci. **12**(3), 1145 (2022). https://doi.org/10.3390/app12031145
3. Nie, Y., Liu, T., Han, L.: Eye fatigue detection system design and implementation. Comput. Life **12**, 11–13 (2024). https://doi.org/10.54097/at3ba893
4. Wang, Y., Zhang, L., Fang, Z.: Eye fatigue detection through machine learning based on single channel electrooculography. Algorithms **15**(3), 84 (2022). https://doi.org/10.3390/a15030084
5. Kuwahara, A., Nishikawa, K., Hirakawa, R., Kawano, H., Nakatoh, Y.: Eye fatigue estimation using blink detection based on eye aspect ratio mapping (EARM). Cogn. Robot. **2** (2022). https://doi.org/10.1016/j.cogr.2022.01.003
6. Haq, Z., Hasan, Z.: Eye-blink rate detection for fatigue determination, pp. 1–5 (2016). https://doi.org/10.1109/IICIP.2016.7975348
7. Singh, D., Singh, A.: Enhanced driver drowsiness detection using deep learning. ITM Web Conf. **54** (2023). https://doi.org/10.1051/itmconf/20235401011
8. Sun, W., Wang, Y., Hu, B., Wang, Q.: Exploration of eye fatigue detection features and algorithm based on eye-tracking signal. Electronics **13**, 1798 (2024). https://doi.org/10.3390/electronics13101798
9. Siddiqui, H.U.R., et al.: Non-invasive driver drowsiness detection system. Sensors **21**(14), 4833 (2021). https://doi.org/10.3390/s21144833
10. Basheer Ahmed, M., et al.: A deep-learning approach to driver drowsiness detection. Safety **9**, 65 (2023). https://doi.org/10.3390/safety9030065

AGRI-SURE: Agricultural Sustainability Through Disease Detection and Automated Insurance Using Deep Learning

T. G. Keerthan Kumar[1]([✉]), Pawanesh Mishra[2], and Shashidhar G. Koolagudi[2]

[1] Department of Information Science and Engineering, Siddaganga Institute of Technology, Tumkur 572103, Karnataka, India
`keerthanswamy@gmail.com`
[2] Department of Computer Science and Engineering, National Institute of Technology Karnataka, Surathkal, Mangalore 575025, Karnataka, India

Abstract. Wheat is a key staple crop worldwide and it faces formidable challenges due to its vulnerability to diseases. These diseases significantly threaten global food security by reducing crop yields and quality. Therefore, timely disease detection is crucial for implementing effective measures to mitigate their impact on agricultural yield. This paper proposes an integrated system for detecting wheat diseases using Convolutional Neural Networks (CNN's) and crop insurance management. Additionally, it streamlines the insurance process, offering timely financial support to farmers affected by crop failures. By combining disease detection with an efficient insurance claims process, the solution improves both productivity and sustainability in wheat farming. It ensures that farmers receive fair, accurate, and timely insurance coverage. The proposed model AGRI-SURE - revolutionizes the traditional crop insurance model by reducing delays, minimizing manual verification, and tailoring insurance policies to individual farmer needs. Overall, this work shows how integrating deep learning with modern insurance management can promote sustainable agricultural practices. It also highlights how this approach enhances economic resilience for farmers.

Keywords: Convolutional Neural Networks · Disease Detection · Insurance Management · Agricultural Productivity · Deep Learning · Web Application · GPS

1 Introduction

Agriculture is essential to the survival of human civilization and it serves as the foundation for our daily life. It provides sustainability, economic stability, and livelihoods for millions of people [1]. Among the various agricultural products, Wheat (*Triticum aestivum*) stands out as a significant part of global food security. It is one of the most widely cultivated cereal crop and it also serves as a

© The Author(s) 2026
J. Shreyas et al. (Eds.): CODE-AI 2025, CCIS 2689, pp. 181–192, 2026.
https://doi.org/10.1007/978-3-032-19318-6_18

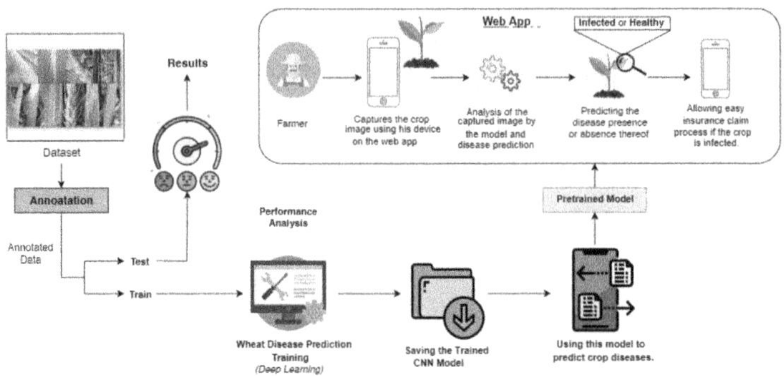

Fig. 1. Proposed Solution for Disease Prediction and Facilitation.

staple food for a significant portion of the world's population [2]. However, this crop faces various challenges during its production, such as extreme rain, floods, and droughts. It is also vulnerable to pests and diseases that severely reduce yields and quality [3,4]. Among these diseases, Leaf rust (*Puccinia triticina*), Crown root rot (*Fusarium spp.*), and Loose smut (*Ustilago tritici*) are particularly noteworthy due to their commonness and damaging effects on wheat crop [5].

Leaf rust which is caused by the fungal disease *Puccinia triticina* looks like orange-yellow spots on wheat leaves. It leads to reduced photosynthetic capacity and ultimately decreased grain yield. Again such disease is Crown root rot, primarily caused by *Fusarium* species, affects the root system. It results in poor nutrient uptake and water stress tolerance, compromising overall plant health and productivity. Loose smut, caused by the fungus *Ustilago tritici*, infects wheat flowers, producing dark smut spores that replace healthy grains, reducing grain quality and yield loss. Hence, early detection of diseases is essential for timely intervention and reducing their impact on productivity. With early identification, farmers can apply fungicides, adjust irrigation, or use crop rotation to stop the spread and minimize yield loss. It also helps farmers to make informed decisions about planting times, choosing cultivars, and managing diseases effectively. This leads to better crop health and improved economic outcomes [6,7].

Nowadays, advancements in technology, particularly in machine learning and image processing, offer promising solutions for automating the detection and diagnosis of such plant diseases. Convolutional Neural Networks (CNN's), a type of deep learning algorithm, have been successful in identifying disease symptoms from plant leaf images over the past few years. Similarly, we can also develop a robust disease detection system that can accurately distinguish between healthy and diseased crops by training CNN models on a large dataset of annotated images.

"Previously, Mehta *et al.* in [8] and Wang *et al.* in [9] have proposed promising solutions for wheat diseases multi-classification for improved crop health management." While effective in disease detection, their lack of user interactivity poses challenges for technologically inexperienced farmers seeking accessible tools. "Also, Mao *et al.* in [10] proposed a novel approach for the automatic detection of wheat diseases, addressing challenges in precision agriculture." It has high detection accuracy and speed and outperforms various existing models. However, its focus on diseases like wheat stripe rust and powdery mildew may limit its applicability in regions like India, where crown rot and loose smut are prevalent. "Goel *et al.* in [11] proposed a model which offered an automated approach for leaf disease detection in plants." While it shows potential for application in crops like wheat, its accuracy is insufficient. "Wen *et al.* in [12] proposed CNN's for wheat leaf disease recognition, achieving high accuracy with models like MnasNet." However, it focuses on only two diseases, whereas our approach covers three major diseases prevalent in the Indian context, making it more suitable for local agricultural needs.

The main motivation for AGRI-SURE was to provide accessible and efficient solutions for disease detection and management in wheat farming. By integrating advanced technologies with insurance management features, we aim to equip farmers with the means to protect their crops, minimize losses, and promote sustainable agricultural practices.

Additionally, the system simplifies insurance management by streamlining the process of providing financial assistance to farmers impacted by crop diseases. By offering a user-friendly platform that enhances disease detection, insurance management, and informed decision-making.

Main Contributions of AGRI-SURE as follows:

1. **Automatic Disease Identification:** Leveraging CNN-based image analysis and machine learning, our system automatically identifies disease symptoms from images of wheat crops, enabling timely interventions to mitigate crop losses.
2. **Streamlining Insurance Management:** Our work simplifies insurance management tasks for insurance providers by offering a user-friendly platform for submitting and processing insurance claims, thereby ensuring financial support for farmers affected by crop diseases.
3. **Integrated Disease Detection System:** Our work introduces an integrated system that combines advanced disease detection techniques with streamlined insurance management processes, providing farmers with a comprehensive solution for protecting their crops.

2   Related Works

Recent research in Deep Learning and IoT have greatly improved agricultural disease detection. "In this regard, the author Udutalapally *et al.* introduced sCrop, an IoAT enabled solar-powered smart device for automatic plant disease prediction [13]." The system utilizes solar-enabled sensor nodes for continuous sensing and automation in agriculture, integrating a health maintenance system with a sensor node equipped with a camera module and a trained CNN model.

The proposed system achieved a testing accuracy of 99.2% and demonstrated robust performance in varied weather conditions. "Sladojevic et al. in [14] (2016) developed deep neural networks for recognizing plant diseases from leaf images, demonstrating the effectiveness of CNN's in identifying diseases affecting wheat crops. Additionally, Wang *et al.* in [9] proposed an image recognition system for plant diseases based on backpropagation networks, laying the foundation for using neural networks in disease recognition tasks. Beulah and Punithavalli (2016) [15] explored data mining techniques for predicting sugarcane diseases, showcasing the applicability of machine learning in agricultural disease management. Yashaswini *et al.* in [16] presented a smart automated irrigation system with disease prediction, highlighting the integration of predictive modeling with agricultural management practices. Mehta *et al.* in [8] presented a collaborative learning CNN method for wheat disease identification, leveraging federated averaging across diverse datasets for enhanced accuracy." Demonstrates the potential of federated learning in improving agricultural models. "Wen *et al.* in [12] utilized convolutional neural networks, this study explores training strategies and initial learning rates for wheat leaf disease recognition, achieving high accuracy of 98% with MnasNet. Mao et al. (2023) [10] put the DAE-Mask model offers deep-learning-based automatic detection for in-field wheat diseases, achieving high accuracy and speed." Leveraging DenseNet and FPN, it outperforms existing models and enables real-time detection on the WeChat Mini Program. "Reis *et al.* in [17] using the Integrated Deep Learning Framework (IDLF) and ensemble learning (EL) model for wheat disease classification." Utilizing pre-trained deep neural networks and image enhancement techniques, it achieved high accuracy of 99.72%. "Kumar *et al.* in [18] introduced an N-CNN based transfer learning method for the classification of powdery mildew wheat disease." Transfer learning with a pre-trained NCNN model enhanced accuracy to 86.5% on dataset images, indicating its effectiveness in disease classification. "Saraswat *et al.* in [19] presented a mask-region based CNN for wheat yellow rust disease recognition, aiming to enhance wheat quality production." Utilizing VGG-16 and VGG-19 pretrained models, the Mask-RCNN identifies stripe rust symptoms on wheat plants with a prediction rate of 90.63% for diseased areas and 87.25% for the entire plant.

Now, as this field is evolving, applying insurance techniques alongside these technologies can further improve the quality of life. It can help farmers by providing financial security and protecting their livelihoods against crop losses.

3   Proposed System

AGRI-SURE aims to develop an integrated system for detecting wheat diseases using CNN's and managing insurance claims for wheat farmers, as shown in Fig. 1. With the help of Deep Learning, the system will automatically identify disease symptoms from images of wheat crops, enabling farmers to take proactive measures to protect their crops. Additionally, the system will streamline the insurance management process, providing financial support to farmers who are affected by crop diseases. By combining disease detection with insurance management functionalities, AGRI-SURE aims to enhance the resilience and sustainability of wheat farming practices.

3.1   Dataset Preparation

The dataset for training the deep learning model comprises images of wheat crops of four categories: healthy, crown root rot, leaf rust, and loose smut. Our wheat disease dataset has wheat leaves of 4 classes, and here is dataset reference [20]. Each class represents a different condition of the wheat crop. Here are the details of the dataset:

The dataset includes labeled wheat crop images across four classes: Healthy (1280 images), Crown Root Rot (1029 images), Leaf Rust (136 images), and Loose Smut (939 images), each illustrated with representative examples.

Below is the examples of few wheat classes as shown in Fig. 2

Fig. 2. Example of Few Wheat Crops.

3.2   Dataset Preparation and Preprocessing

Images and labels are converted into *NumPy* arrays for easier numerical operations. Pixel values are normalized by dividing by 255.0, ensuring all values fell between 0 and 1. This standardization improves the CNN's performance and training speed.

Next, we split the data using *Sklearn's train_test_split*, allocating 80% for training and 20% for validation. This split helps prevent over-fitting by allowing the model to be tested on unseen images.

3.3   Feature Extraction and Model Architecture

This CNN for wheat disease detection uses *TensorFlow* and *Keras* libraries for implementation. The model employs automatic feature extraction through a series of convolutional and pooling layers as shown in Fig. 3.

The CNN architecture consists of three convolutional layers (32, 64, and 128 filters, all 3×3) with ReLU activation. Each followed by 2×2 max pooling. This structure progressively extracts features from low-level (edges, colors) to high-level (disease-specific patterns). The Conv2D and MaxPooling2D layers from Keras are used for this purpose.

As data progresses through the network, spatial dimensions decrease ($64 \times 64 \times 3 \rightarrow 31 \times 31 \times 32 \rightarrow 14 \times 14 \times 64 \rightarrow 6 \times 6 \times 128$) while feature map depth increases. After flattening, two dense layers (128 neurons and 4 neurons with softmax) complete the classification. The Flatten and Dense layers from Keras are employed here.

After training, the CNN model is saved using *Keras'* `save()` method in the *HDF5* format.

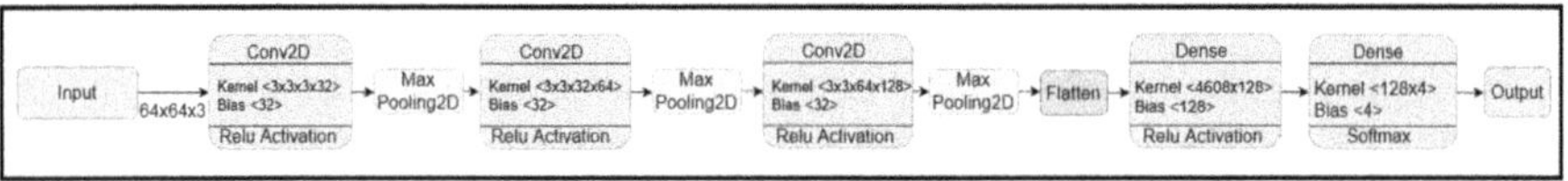

Fig. 3. Model Architecture.

3.4   Loss Function

For our wheat disease detection model, we use a specific loss function to train the CNN effectively. We employ Categorical Cross-Entropy, a loss function widely used for multi-class classification problems. It measures the dissimilarity between the predicted probability distribution and the true distribution of the classes. The categorical cross-entropy loss is defined in Eq. (1).

$$L_{CCE} = -\frac{1}{N} \sum_{i=1}^{N} \sum_{j=1}^{C} y_{ij} \log(\hat{y}_{ij}) \tag{1}$$

where N is the number of samples, C is the number of classes (4 in our case: crown root rot, healthy, leaf rust, and loose smut), y_{ij} is the true probability of the i-th sample belonging to the j-th class (1 if the sample belongs to the class, 0 otherwise), and $\hat{y}_{ij}$ is the predicted probability.

We implemented this loss function using Keras, specifying 'categorical_cross-entropy' as the loss parameter in the model compilation step. This choice is particularly effective for our multi-class wheat disease classification task, as it encourages the model to output a probability distribution over the four possible classes for each input image.

4   System Workflow

The system simplifies disease detection and insurance management for wheat farmers. Figure 4 illustrates the workflow.

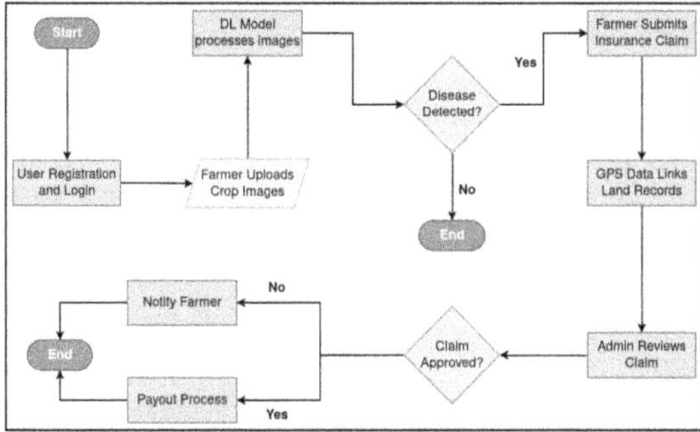

Fig. 4. System Workflow of AGRI-SURE.

4.1 User Interaction and Prediction

Users register with username, email, password, and GPS coordinates. The front-end provides a responsive interface. The back-end, built with Flask, handles authentication and stores data in MongoDB.

Farmers upload crop images through the web application. The app captures GPS coordinates from EXIF metadata. The system uses this GPS data to automatically fetch land records.

A deep learning model analyzes uploaded images for disease symptoms. Detection results are stored in the database and displayed to the user through the front-end interface.

4.2 Insurance Management

Farmers submit insurance claims for crop losses through the web app. The system verifies claims using GPS data and land records. Administrators review and approve claims via an admin panel. This panel is part of the frontend, with special access controls implemented in the back-end. Existing users can update and manage their policies through the portal. The back-end handles these updates, ensuring data consistency across the system.

The insurance claim process includes document verification and premium status checks. Users can upload images of damaged crops. The back-end's deep learning system checks these images for damage and verifies the location. Farmers receive confirmation of the insurance claim settlement amount through the portal or email.

4.3 Technical Components

The front-end is built with HTML, CSS, and JavaScript, providing a responsive design for various devices. The back-end, built using Flask, handles requests from the front-end. It processes data, interacts with the database, and provides responses to the client. Key components of the back-end include server logic, database integration, API endpoints, middleware, and a request processing pipeline.

Middleware functions, implemented in Flask, handle tasks like request preprocessing, authentication, and error handling. The system uses MongoDB for data storage. The user data schema is represented in JSON format, including fields like username, email, hashed password, location coordinates, role, and insurance status flags.

Key API endpoints include:

- **POST /register**: Registers a new user and fetches their current location.
- **POST /login**: Authenticates a user.
- **POST /logout**: Logs out a user.
- **GET /details**: Retrieves details of all users.
- **GET /insurance**: Fetches insurance details.
- **POST /insurance/claim**: Submits a new insurance claim.
- **GET /insurance/approval**: Checks insurance claim status.
- **GET /insurance/insurance_amount**: Handles insurance amount decisions.
- **GET /approve_insurance/<user_id>**: Allows admins to approve or disapprove insurance claims.
- **POST /result**: Performs disease detection on crop images.

The application is deployed on AWS EC2 for scalability. This includes MongoDB setup, domain configuration, and SSL encryption to ensure security and reliability throughout the workflow.

5   Experimental Setup and Result Analysis

5.1   User and Admin Interface

The application offers a user-friendly interface for both farmers and administrators. Farmers can easily upload images of their wheat crops to check for diseases and apply for insurance. Upon login, farmers are greeted with a dashboard displaying the status of their insurance claims. If a claim is approved, they can proceed with the insurance process as shown in the Fig. 5.

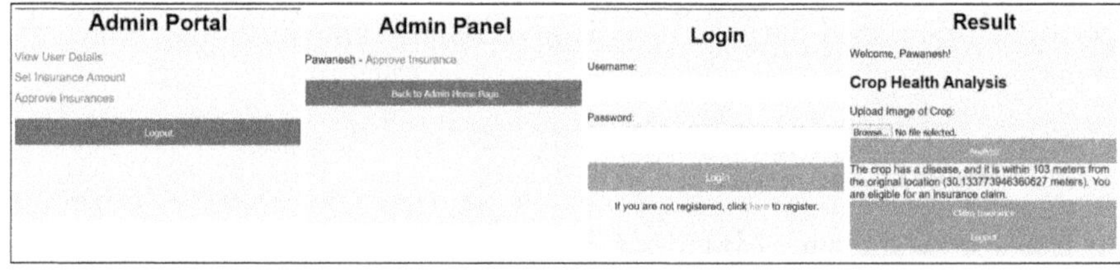

Fig. 5. Screenshots of User Interface and Admin Interface.

On the other hand, administrators have access to a dedicated portal where they can review insurance claims, manage user accounts, and set insurance amounts. The admin interface provides a comprehensive overview of all pending and approved claims, enabling efficient decision-making and resource allocation as shown in Fig. 5.

5.2   Insurance Claim Application Process

The insurance claim application process in the system is designed to assist farmers affected by crop diseases. The insurance claim process begins with users uploading crop images, where the system checks eligibility—claims are valid only if the crop was healthy during registration and is now diseased. The system then verifies the image's geolocation to ensure it matches the original registered field, based on average farm sizes in India [21]. If both checks pass, users can submit their claims, which are then reviewed and approved by administrators through a dedicated interface.

5.3   Model Evaluation

Our CNN model achieved 83% accuracy on the training data. Accuracy shows how often the model correctly classifies images. Figure 6 shows how the model's accuracy improved during training.

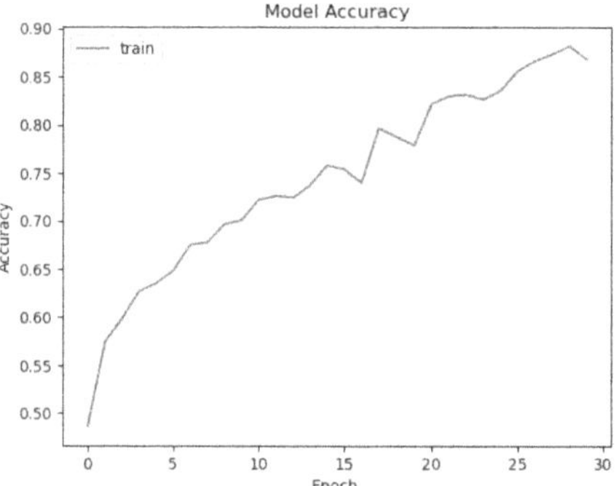

Fig. 6. CNN Model Accuracy per Epoch.

5.4   Accuracy Comparisons Between Different CNN Configurations

We experimented with various CNN architectures and augmentation strategies to improve classification accuracy. Starting with a basic CNN achieving 75% accuracy, we progressively enhanced performance by incorporating basic (79%) and extended (81%) data augmentation. Our best results came from a 3-layer CNN with comprehensive augmentation, reaching 83% accuracy. However, adding a fourth convolutional layer slightly reduced accuracy to 82%, indicating potential overfitting or diminishing returns from added complexity.

These are some comparisons of this model with some others models in Table 1.

Table 1. Comparison of Disease Prediction Methods.

Work	Algorithm Used	User Interface	Accuracy	Internet Connectivity
Wang *et al.* in [9], 2012	PCA with Backpropagation	No	100%	Not Required
Beulah *et al.* in [15], 2016	Random Forest and DT	No	NIL*	Not Required
Sladojevic *et al.* in [14], 2016	CNN	No	96%*	Not Required
Yashaswini *et al.* in [16], 2017	Hidden Markov Model	Yes	NIL*	Required
Goel *et al.* in [11], 2018	K-Means Clustering	Yes	60%	Required
Wen *et al.* in [12], 2023	CNN	No	97.21%	Not Required
Proposed Work	CNN	Yes	83%	Required

Since our model is simple and small, it offers low latency for server communication, making it suitable for our application despite achieving 83% accuracy. Although more complex models, as discussed above, like [12,14], can be used, because of their complexity, there is a substantial delay in our application.

6   Conclusion and Future Directions

AGRI-SURE uses DL for disease detection with an efficient insurance claim process. It achieved 83% accuracy in identifying wheat diseases. It enables timely measures to reduce crop losses. The streamlined insurance process simplifies financial support for farmers, minimizing delays and manual checks. The application combines a user-friendly front-end with a Flask-MongoDB back-end and AWS deployment to support scalable, secure, and sustainable wheat farming through disease detection and insurance management.

In the future, we plan to expand the system to encompass a wider range of crops, including tomatoes, potatoes, and chickpeas. This presents an opportunity to enhance agricultural support across diverse farming practices. Also, we plan to improve the design to ensure scalability and responsiveness. It will accommodate an increasing number of users and data without sacrificing performance. Optimizations will be implemented to maintain fast response times, providing a smooth user experience as the platform grows.

References

1. Gollin, D.: Agricultural productivity and economic growth. Handb. Agric. Econ. **4**, 3825–3866 (2010)
2. Shiferaw, B., Smale, M., Braun, H.-J., Duveiller, E., Reynolds, M., Muricho, G.: Crops that feed the world 10. past successes and future challenges to the role played by wheat in global food security. Food Secur. **5**, 291–317 (2013)
3. Savary, S., Ficke, A., Aubertot, J.-N., Hollier, C.: Crop losses due to diseases and their implications for global food production losses and food security. Food Secur. **4**(4), 519–537 (2012)
4. Savary, S., Willocquet, L., Pethybridge, S.J., Esker, P., McRoberts, N., Nelson, A.: The global burden of pathogens and pests on major food crops. Nat. Ecol. Evol. **3**(3), 430–439 (2019)
5. Huerta-Espino, J., et al.: Global status of wheat leaf rust caused by Puccinia triticina. Euphytica **179**, 143–160 (2011)
6. Hasan, R.I., Yusuf, S.M., Alzubaidi, L.: Review of the state of the art of deep learning for plant diseases: a broad analysis and discussion. Plants **9**(10), 1302 (2020)
7. Thakur, P.S., Khanna, P., Sheorey, T., Ojha, A.: Trends in vision-based machine learning techniques for plant disease identification: a systematic review. Expert Syst. Appl. **208**, 118117 (2022)
8. Mehta, S., Kukreja, V., Vats, S.: Empowering farmers with AI: federated learning of CNNs for wheat diseases multi-classification. In: 2023 4th International Conference for Emerging Technology (INCET), pp. 1–6. IEEE (2023)
9. Wang, H., Li, G., Ma, Z., Li, X.: Image recognition of plant diseases based on back-propagation networks. In: 2012 5th International Congress on Image and Signal Processing, pp. 894–900. IEEE (2012)
10. Mao, R., et al.: DAE-mask: a novel deep-learning-based automatic detection model for in-field wheat diseases. Precis. Agric., 1–26 (2023)
11. Goel, N., Jain, D., Sinha, A.: Prediction model for automated leaf disease detection & analysis. In: 2018 IEEE 8th International Advance Computing Conference (IACC), pp. 360–365. IEEE (2018)
12. Wen, X., Zeng, M., Chen, J., Maimaiti, M., Liu, Q.: Recognition of wheat leaf diseases using lightweight convolutional neural networks against complex backgrounds. Life **13**(11), 2125 (2023)
13. Udutalapally, V., Mohanty, S.P., Pallagani, V., Khandelwal, V.: sCrop: a internet-of-agro-things (IoAT) enabled solar powered smart device for automatic plant disease prediction. arXiv preprint arXiv:2005.06342 (2020)
14. Sladojevic, S., Arsenovic, M., Anderla, A., Culibrk, D., Stefanovic, D.: Deep neural networks based recognition of plant diseases by leaf image classification. Comput. Intell. Neurosci. **2016**, 3289801 (2016)

15. Beulah, R., Punithavalli, M.: Prediction of sugarcane diseases using data mining techniques. In: 2016 IEEE International Conference on Advances in Computer Applications (ICACA), pp. 393–396. IEEE (2016)
16. Yashaswini, L., Vani, H., Sinchana, H., Kumar, N.: Smart automated irrigation system with disease prediction. In: 2017 IEEE International Conference on Power, Control, Signals and Instrumentation Engineering (ICPCSI), pp. 422–427. IEEE (2017)
17. Reis, H.C., Turk, V.: Integrated deep learning and ensemble learning model for deep feature-based wheat disease detection. Microchem. J. **197**, 109790 (2024)
18. Kumar, D., Kukreja, V.: N-CNN based transfer learning method for classification of powdery mildew wheat disease. In: 2021 International Conference on Emerging Smart Computing and Informatics (ESCI), pp. 707–710. IEEE (2021)
19. Saraswat, S., Batra, S., Neog, P.P., Sharma, E.L., Kumar, P.P., Pandey, A.K.: An efficient diagnostic approach for multi-class classification of wheat leaf disease using deep transfer and ensemble learning. In: 2024 2nd International Conference on Intelligent Data Communication Technologies and Internet of Things (IDCIoT), pp. 544–551. IEEE (2024)
20. Aadium: Wheat Disease Detection. GitHub. https://github.com/aadium/wheat-disease-detection
21. Bureau, P.I.: Categorisation of Farmers. pib. https://pib.gov.in/Pressreleaseshare.aspx?PRID=1562687

Data Exploration in IoT and Edge Environments: A Lightweight Deep Learning Model for Botnet Attack Detection

S. Vaishnavi[1(✉)], K. Radha[2], T. Senthil[3], Jeevitha Sakkarai[4], A. Vidhya[5], and Alagarsamy Manjunathan[6]

[1] Department of Information Technology, Manipal Institute of Technology Bengaluru, Manipal Academy of Higher Education (MAHE), Manipal, India
`s.vaishnavi@manipal.edu`
[2] Department of Computer Science and Technology, Vivekanandha College of Engineering for Women, Tiruchengode 637205, Tamil Nadu, India
[3] Department of Computer Science Engineering, Sreenivasa Institute of Technology and Management Studies, Chittoor 517127, Andhra Pradesh, India
[4] Department of Computer Science and Information Technology, Kalasalingam Academy of Research and Education, Krishnankoil, Tamil Nadu, India
[5] Department of Electronics and Communication Engineering, Prathyusha Engineering College, Tiruvallur 602025, Tamil Nadu, India
[6] Department of Electronics and Communication Engineering, K. Ramakrishnan College of Technology, Trichy, Tamil Nadu, India

Abstract. Remote patient monitoring, improved energy efficiency, besides the automation of mundane household duties are just a few samples of Internet of Things (IoT) has changed several industries. Attacks based on botnets, including distributed denial of service (DDoS), can compromise unprotected IoT devices. The use of deep learning (DL) to enhance privacy has been spurred by the realisation that traditional machine learning models employed to identify these threats violate data privacy. On the other hand, when it comes to cyberattack detection models based on DL, the majority of them fail to take computational complexity into account, making them unsuitable for placement on resource-constrained IoT edge devices. In order to identify botnet attacks on the IoT, this study presents a DL model that is computationally simple. The study uses dimensionality reduction and feature selection to keep computational complexity low and accuracy high. To begin, the botnet dataset's best characteristics are chosen using an optimised Deep Belief Network (DBN) classical trained with repeated stratified k-fold cross-validation. In order to increase the classification accuracy, the study used Fire Hawk Optimisation (FHO) to fine-tune adjustable parameters. In comparison to earlier efforts, the model improved upon the curve. In smart grids, smart homes, and other contexts with deployed IoT edge devices that are resource constrained, the proposed methodology works well for botnet attack detection.

Keywords: Internet of Things · Deep Belief Network · Fire Hawk Optimisation · Distributed denial of Service · Edge Device · Botnet attacks

J. Shreyas et al. (Eds.): CODE-AI 2025, CCIS 2689, pp. 193–205, 2026.
https://doi.org/10.1007/978-3-032-19318-6_19

1 Introduction

Various physical devices are associated to the Internet through IoT. Smart devices, IoT applications, and graphical user interfaces are the three main parts that make up the IoT [1]. Thermostats, home security systems, televisions, and other electronic gadgets with processing and computational skills are known as "smart devices" and can be managed via the internet. The IoT requires specialised software to collect data from various sensors and analyse it, while graphical user interfaces (GUIs) are necessary for controlling these devices, such as mobile phones [2]. The three main components of an IoT network: endpoint devices, gateways, and cloud servers. This technology is responsible for huge changes in the communication arena and has been rapidly expanding over the years [3]. The prevalence of such devices in modern life provides ample evidence of this. From 9.7 billion in 2020 to over 29 billion in 2030, the number of IoT devices worldwide is expected to nearly treble [4]. These devices aren't just for the house anymore; they're making waves in a variety of industries and cementing their place in our increasingly interconnected society [5].

Many industries have been impacted by the rapid expansion of IoT, which has allowed for intelligent automation and seamless connectivity. But new cybersecurity concerns brought about by the proliferation of connected gadgets. IoT environments are distinct from traditional security measures due to their resource-constrained devices, varied network topologies, and ever-changing data flows [6]. With its superior skills for detecting anomalies and mitigating threats, deep learning (DL) has become an effective technique for enhancing cybersecurity in IoT networks [7]. But there are a few problems with the DL-based methods that are available now. The dispersed and diverse nature of IoT systems makes centralised designs, which are the backbone of most current solutions, inadequate [8]. Concerns about privacy, increased latency, and scalability are just a few of the problems that centralised approaches have [9]. When it comes to processing IoT data streams, CNNs and LSTMs have shown promise, but they tend to be resource-intensive, making them hard to deploy on edge and mobile devices. Furthermore, it is highly challenging to regularly adjust these models to new cyber dangers. IoT devices and the necessary cybersecurity procedures to ensure a trustworthy environment are becoming increasingly important in sectors including smart cities, healthcare, and industrial automation. As a consequence, cyber-attacks are a common occurrence on these systems and can compromise data security, privacy, and the system's ability to function normally. In order to make IoT networks more secure, this study is essential since deep learning techniques that can adapt to new threats, facilitate real-time operations, and are practical to implement are lacking.

- Secure mobile IoT environments are ensured by the suggested framework, which incorporates numerous cutting-edge components.
- The development of a strong DL model that efficiently and accurately detects botnet assaults can progress the security of the Internet of Things.
- Reduce computational overhead without sacrificing detection efficiency by efficiently exploring data using dimensionality reduction and feature selection approaches.
- Use Fire Hawk Optimisation (FHO) to improve model presentation by fine-tuning parameters, leading to better classification metrics.

- To make edge deployment easier, make sure that the computational demands of the model are compatible with the capabilities of IoT edge devices that have limited resources, so that it can be practically implemented.

Here is the breakdown of the remaining sections of paper: The relevant literature is discussed in Sect. 2, the system model is presented in Sect. 3, and the procedure that has been suggested is explained in Sect. 4. Both Sect. 5 besides Sect. 6 present the results and their analysis, correspondingly.

2 Related Works

To improve mobile IoT settings' real-time cybersecurity, Awan et al. [10] present SecEdge, a new deep learning system. The SecEdge system uses federated learning to guarantee data privacy and decrease latency, transformer-based models to efficiently handle long-range relationships, and Graph to model relational data. In order to address ever-changing cyber threats, the adaptive learning process is constantly updating the model parameters. In a simulated setting, the NSL-KDD, UNSW-NB15, and CICIDS2017 datasets were used to assess the performance of the framework. With a detection CICIDS2017, SecEdge surpassed state-of-the-art methodologies, according to the results.

In order to detect malicious software and files that have been illicitly spread through the IoT, Markkandeyan et al., [11] projected a hybrid Deep Learning (DL) approach in this research. It is recommended to use the Adaptive TensorFlow Improved Particle Swarm Optimisation (IPSO) to identify unlawful code (SC) duplication. The dataset was collected using Google Code Jam (GCJ) in order to investigate software piracy. Another tool used to detect suspicious activity in the IoT setting was Enhanced Long Short-Term Memory (E-LSTM), which uses colour pictures as its visual representation. In order to collect the malware trials needed for testing, the Maling dataset is utilised. The experimental results the projected strategy outperforms existing approaches when it comes to classifying cybersecurity threats in the IoT.

For the advanced metering infrastructure (AMI), Naveeda and Fathima [12] have introduced an IoT enabled cyberattack detection scheme (IoT-E-CADS). It is believed by experts in the field that the proposed Bi-level IoT-E-CADS can detect two distinct types of smart grid threats. To have successfully deployed the suggested IoT-E-CADS at this location, and it has detected two cyberattacks that were generated manually. From what to can tell from the data, the IoT-E-CADS can identify cyberthreats with a 95% success rate, meaning it offers complete cybersecurity solutions for safe monitoring units in business settings.

A approach for real-time network intrusion uncovering based on graph neural networks has been suggested by Liu & Guo [13]. By using a simple graph creation method, the suggested approach represents network traffic as dynamic data, capitalising on the benefits of graph neural networks. To compare the results to those of prior research in this area and utilise two popular intrusion detection datasets to assess the effectiveness of the suggested strategy. Our suggested solution surpasses the benchmark model in multiple assessment measures, and experimental results show that it achieves 99.3% and two datasets, correspondingly.

An effective and efficient intrusion detection system that can identify many types of CAN bus attacks without adding extra traffic overhead to existing connections has been proposed by Dangwal et al., [14] and is short-named ACID-CAN. In order to implement CAN-driven AIoT applications, the given method is absolutely necessary. Even when the quantity of intrusion data is decreased to 5% of regular data, the suggested ACID-CAN is still able to detect them, according to the experimental results. When compared to other research on CANs intrusion detection, the results showed that the suggested ACID-CAN performs as well, if not better.

3 System Model

The suggested framework is detailed here; it is an all-inclusive approach to mobile IoT cybersecurity in real-time. The work primarily focusses on improving threat detection and system robustness through the fusion of deep learning models. To ensure secure mobile IoT environments, the suggested architecture relies on a number of interdependent modules and components.

3.1 Edge Computing Layer

Initial data preparation and threat detection are responsibilities of the edge computing layer, which serves as the architecture's first line of defence. In order to gather and interpret data on network traffic in real-time, edge devices are deliberately positioned throughout the IoT network. By taking advantage of their capabilities to manage long-range relationships and parallelise calculations, these devices use lightweight models to identify possible dangers.

Mathematically, let X_i characterise traffic data composed by device. The function P features from X_i, represented as $F_i = P(X_i)$. The transformer classical T then analyzes F_i to notice anomalies, producingan output $A_i = T(F_i)$, where A_i indicates the threats, as exposed by Eq. (1).

$$A_i = T\ (P(X_i)) \tag{1}$$

The ability to efficiently and effectively detect threats at the network's edge is a result of edge devices' optimisation for real-time presentation.

3.2 Fog Computing Layer

Intermediary nodes among edge devices and the core cloud layer, fog nodes encapsulate processing operations that are too complicated to be efficiently handled at the edges. To be more precise, fog nodes coordinate with other parts of a network to identify and mitigate threats in their entirety, aggregate data coming from various edge devices, and middle threat study using DBN.

Let D_j denote the data combined by the j-th fog node from n edge strategies. The combination function G can be articulated as Eq. (2):

$$D_j = G(A_1, A_2, \ldots, A_n) \tag{2}$$

The DBN model N examines the aggregated data D_j to detect anomalies, creating an output $T_j = N(D_j)$, where T_j characterises the threat investigation consequence from the j-th fog node, as exposed by Eq. (3).

$$T_j = N(G(A_1, A_2, \ldots, A_n)) \tag{3}$$

The distribution of processing duties across the network by fog nodes enhances the scalability and competence cybersecurity system.

3.3 Central Cloud Layer

In federated learning, local updates are used to update the global model M. ΔM_i from k edge besides fog nodes. The function U can be distinct as Eq. (4):

$$M_{t+1} = U(M_t, \Delta M_1, \Delta M_2, \ldots, \Delta M_k) \tag{4}$$

where M_t represents the revised model, while is the iteration b. In order to make the proposed structure more resistant and resilient, it uses a continuous learning process to adjust to changing threat patterns.

1. Data Collection besides Preprocessing: Edge devices gather traffic data X_i and achieve initial preprocessing $P(X_i)$ to extract pertinent features F_i.
2. Initial Threat Detection: Preprocessed data F_i is models $T(F_i)$ organised on edge to detect potential threats Ai in real-time.
3. Data Aggregation and Middle Analysis: Data A_i from edge devices is combined by fog nodes $G(A_1, A_2, \ldots, A_n)$, which perform more thorough study besides threat association using DBNs $N(D_j)$.
4. Model Training and Incessant Learning: The central data D_j and updates the representations deployed on edge besides learning $U(M_t, \Delta M_1, \Delta M_2, \ldots, \Delta M_k)$.
5. Adaptive Threat Mitigation: The process allows for the proactive finding besides mitigation of emerging cyber threats by continuously updating model limits depending on fresh data and threat trends.

4 Proposed Methodology

After setting up the system environment, the next step is to identify potential network threats; Fig. 1 shows this graphically, and the next section explains each block in detail.

4.1 Dataset Description

In this study, to tested our model using the N-BaIoT dataset that was made available by [15]. This dataset includes data collected from seven Internet of Things devices that were compromised with the Mirai malware. A thermostat, four security cameras, a baby monitor, and a doorbell are all part of the set. The botmaster uses the infected IoT devices to launch attacks such as Mirai scan, which automatically scans for vulnerable devices, Mirai ACK, which floods with confirmation messages, Mirai SYN, which floods with user datagram protocol (UDP), and Mirai UDP Plain, which floods with UDP but with fewer options optimised for second.

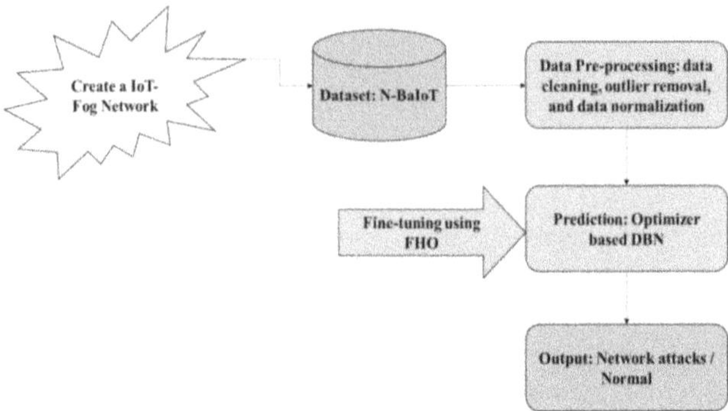

Fig. 1. Workflow of the projected model

4.2 Data Preprocessing

In order to get the dataset ready for prediction models, a number of data wrangling approaches were used. Data normalisation, data cleansing, and outlier reduction are all examples of such methods. Data cleansing is the first step in data wrangling with the N-BaIoT dataset. Despite the absence of missing values in this dataset, to eliminated duplicate rows to make model training more efficient [16]. Following this, to removed 87,197 rows (2.38% of the dataset) that contained these values. Here is how the z-score is calculated:

$$z = \frac{x - \mu}{\sigma} \tag{5}$$

that is, x stands for the value of an individual feature data point, μ for the feature's mean, and σ for its standard deviation.

4.3 Data Normalization

Following the removal of outliers, the data was subjected to min-max normalisation. By comparing the dataset's characteristics to their minimum and maximum values, Min-Max normalisation rescales each data point so that it falls between 0 and 1. The following equation shows the min-max normalisation computation for the normalised data point.

$$MM\,(x_n) = \frac{x_n - \min(x)}{\max(x) - \min(x)} \tag{6}$$

where x_n represents a data feature x and $MM(x_n)$ represents its consistent normalized value.

4.4 Innovative Deep Learning Models Using Deep Belief Networks

Multi-hidden-layer probabilistic generative models are known as DBNs. Directed Sigmoid Belief Networks (DSBs) are made up of stacked Restricted Boltzmann machines (RBMs) and directed Sigmoid (see Fig. 2).

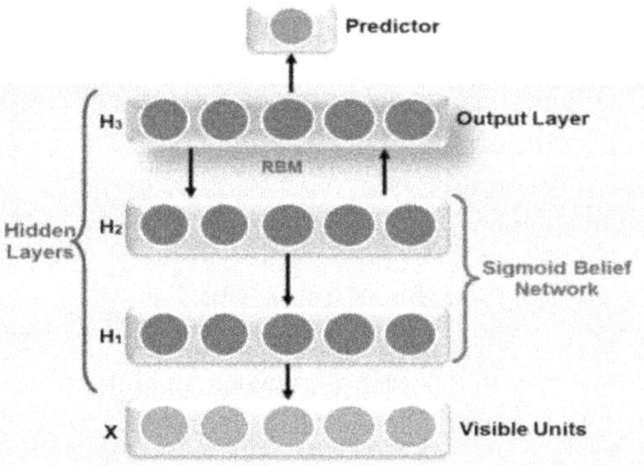

Fig. 2. DBN structure.

The capacity of the DBN model to learn high-order hierarchical features via layer-by-layer learning is its primary benefit [17]. Using this method, successive layers can learn to represent their predecessors at a higher level. In unsupervised learning, RBMs are trained independently using a greedy layer-wise technique, with training beginning at the lowest level, input layer X (seen data).

For a DBN collected of two hidden layers $h_{(1)}$, $h_{(2)}$, to describe $P(x, h_{(1)}, h_{(2)}; \Theta)$ as the joint delivery over x, $h_{(1)}$, and $h_{(2)}$ articulated by DBN under the subsequent form:

$$P\left(x, h_{(1)}, h_{(2)}; \Theta\right) = P\left(x|h_{(1)}; W_{(1)}\right) P\left(h_{(1)}, h_{(2)}; W_{(2)}\right) \tag{7}$$

with W weight matrix, and Θ the model limit defined as $\Theta = \{W_i\}^K$, K is the sum of hidden layers in the DBN. The term $P(x| h_{(1)}, W_{(1)})$ characterises sigmoid.

$$P(x|h; W) = sigmoid\left(\sum_j W_{ij}h_j\right) \tag{8}$$

while the second term $P(h_{(1)}, h_{(2)}; W_{(2)})$ express the joint delivery of the second-layer RBM.

$$P\left(h_{(1)}, h_{(2)}; W_{(2)}\right) = \frac{1}{Z}e^{\left(h_{(1)}^T W_{(2)}h_{(2)}\right)} \tag{9}$$

The normalising constant, Z, stands for the partition function, and the logistic function is denoted by *yibmoid*. The unsupervised learning phase involves computing the model parameters θ, and the fine-tuning step follows. In particular, optimising the DBN parameters during the tuning phase is crucial for ornamental the model's presentation. In order to train the DBN, this phase employs the supervised learning technique known as back-propagation (BP). The goal is to find the global optimum using labelled data. The DBN is used for attack detection in this paper by totalling a forecaster layer to the DBN's output layer (Fig. 2). The forecaster layer was primarily utilised to convert the DBN scalar data point, allowing it to be compared to the unique data.

4.5 Optimal Selection of Parameters in DBN Using FHO

Here to shall go over the fundamental steps of the Fire Hawk Optimiser (FHO) [18]. This study describes how to optimise the hyper parameters in AE using FHO. First, using the following formula, as is common with other Metaheuristic techniques, FHO usually begins by giving initial values to a set of N agents.

$$X_{ij} = \text{rand} \times \left(U_j - L_j\right) + L_j, \ \ j = 1, 2, \ldots, D \tag{10}$$

X_{ij} in the equation represents dimension j. U_j and L_j stand in for the parameters' boundaries at the jth dimension. Each X_i dimension is denoted by the letter D, and the value $rand \in [0, 1]$ is a random number.

Each X_i performance is evaluated using an objective function after startup. The method first identifies the fire hawks ($FH_l, l = 1, 2, \ldots, n$), which correspond to the best responses, and then labels the other solutions as prey ($PR_k, k = 1, 2, \ldots, m$).. The formula in (11) is used to determine the separation between FH and PR.

$$D_{lk} = \sqrt{(x_2 - x_1)^2 + (y_2 - y_1)^2}, \ \ l = 1, 2, \ldots, n, k = 1, 2, \ldots, m \tag{11}$$

In this instance, m denotes the number of fire hawks, whereas n denotes the sum of prey, or PR. The next stage is to scatter PR around the FH territory to define its boundaries. After that, each FH is updated utilising the formula shown in (12).

$$FH_l(t + 1) = FH_l(t) + (r_1 \times X_b - r_2 \times FH_n(t)), l = 1, 2, \ldots, n \tag{12}$$

In this equation, X_b stands for the optimal answer, and $FH_n(t)$ is a particular type of fire hawk. The values for the variables r_1 and r_2 are chosen at random from the [0, 1] range.

In order to ensure their safety when threatened, the next stage is to locate a secure area where the prey can gather. Eq. (13) shows that this can be expressed by calculating SP_l and SP.

$$SP_1 = \frac{\sum_{q=1}^{r} PR_q}{r}, q = 1, 2, \ldots, r, l = 1, 2, \ldots, n \tag{13}$$

Next, PK's movement near the FH is used to fake animal behaviour. Because of this action, the prey can use the formula (14) to get its current position:

$$PR_q(t + 1) = PR_q(t) + (r_3 \times FH_l - r_4 \times SP_l(t)), l = 1, 2, \ldots, n, q = 1, 2, \ldots, r \tag{14}$$

SP_l stands for the safe location within the lth fire hawk's range.

The next stage entails updating the secure position outside of the lth FH's jurisdiction, which is represented by Eq. (15):

$$SP = \frac{\sum_{k=1}^{m} PR_k}{r}, \ \ k = 1, 2, \ldots, m \tag{15}$$

After that, use Eq. (16) to update the prey's location:

$$PR_q(t+1) = PR_q(t) + (r_5 \times FH_a - r_6 \times SP(t)), l = 1, 2, \ldots, n, q = 1, 2, \ldots, r \tag{16}$$

Until the stop criteria are satisfied, the solutions are updated, and the best solution, X_b, is then returned.

4.6 Distributed Processing Architecture

Using the combined strengths of Edge and Fog computing, the suggested framework's distributed processing architecture aims to maximise computational efficiency and scalability in mobile IoT cybersecurity operations, allowing for the low-latency, real-time detection and mitigation of threats.

4.6.1 Anomaly Detection Techniques

Incorporating the discovery vectors is a feature of anomaly detection algorithms. To can improve the entire security posture IoT network by proactively detecting and mitigating potential threats through deviations from typical behaviour, according to the framework's analysis-based design.

Now let's look at the typical distribution P(X) of the network traffic. Detecting anomalies entails finding outliers within this distribution.

. Let X_t be the observed network traffic at period t. The irregularity score A_t can be distinct as:

$$A_t = -logP(X_t) \tag{17}$$

As articulated in Eq. (17), the score A_t events the experiential network traffic X_t from the predictable distribution $P(X)$. If At surpasses a threshold τ, the scheme flags the remark anomaly:

Anomaly if $A_t > \tau$

The positives besides false negatives is taken into account when choosing the threshold τ, as shown by Eq. (18). By integrating the continuous updating and anomaly detection techniques, the adaptive learning formalised. The overarching goal is to detect outliers and adapt to incoming data while minimising the predicted loss.

$$\underset{\theta}{\min} E_{D_t}[L(D_t, \theta)] + \lambda \sum_{t=1}^{T} [](A_t > \tau) \tag{18}$$

As shown in Eq. (18), the optimization impartial reduces the data D_t while imposing penalties for instances of irregularities using the indicator function Adjusting for fresh data while still detecting anomalies effectively is controlled by the regularization parameter λ.

5 Simulation & Evaluation

In order to confirm that the suggested framework is effective in real-time cybersecurity for mobile IoT scenarios, it is simulated and evaluated in a controlled environment. Devices from Nvidia Jetson Xavier (8-core ARM v8.2, 32 GB RAM) and Raspberry Pi 4 (quad-core *ARM Cortex − A72, 1.5 GHz, 4 GB RAM*) were used in the edge layer and fog layer, respectively. The Tesla V100 GPUs used by the cloud layer had 128 GB of RAM. Among the software configurations used for training and testing models across datasets were TensorFlow and PyTorch. Edge, fog, and cloud are the three levels that make up the virtual world. The following is a list of all the layers' hardware specifications:

- Raspberry Pi 4 devices make up the edge layer. Each one has a *quad − core ARM Cortex− A72* processor that runs at 1.5 GHz besides 4 GB of RAM. The devices are selected for their capacity to deploy lightweight models with minimal power consumption.
- Nodes in the Fog Layer: These nodes are Nvidia Jetson Xavier machines with 32 GB of RAM, 512-core Volta GPU, and 8-core ARM v8.2 processor. Middle data analysis and coordinating tasks benefit from the additional processing capacity provided by these nodes.
- Cloud Servers: Servers equipped with Nvidia TeslaV100 GPUs, 128 GB of RAM, and Intel Xeon Gold processors execute the cloud layer. These servers are responsible for storing data for the long term, training models, and processing large amounts of data.

Publicly available datasets covering a range of network traffic situations are used to evaluate the proposed architecture.

5.1 Analysis of Proposed Model

Table 1 delivers experimental study of projected classical with existing technique in terms of different features.

Table 1. Analysis of projected model with existing techniques

	42 Features (Training/Testing)	23 Features (Training/Testing)	76 Features (Training/Testing)	36 Features (Training/Testing)
ELM	86.04% and 54.71%	84.01% and 42.07%	95.00% and 96.10%	96.98% and 97.05%
DBN	79.18% and 71.37%	77.86% and 62.80%	97.86% and 97.90%	97.68% and 98.72%

5.2 Comparative Analysis of Projected Classical with Existing Procedures

Figure 3 and 4 delivers the experimental study of projected classical with existing representations in terms of diverse metrics.

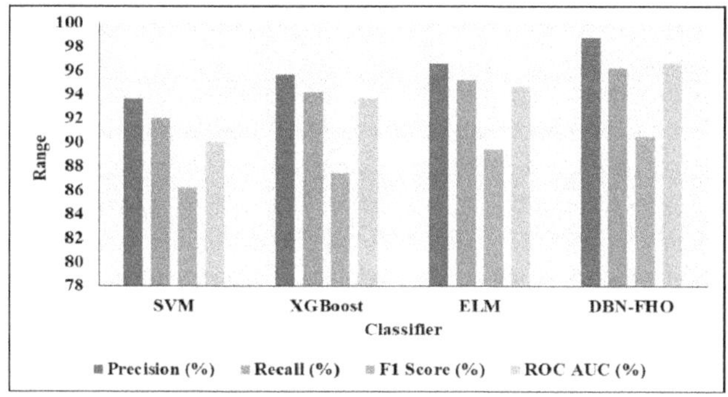

Fig. 3. Visual Analysis of Projected classical with existing procedures

The comparative analysis of the projected model (DBN-FHO) against existing techniques, including SVM, XGBoost, and ELM, based on four key ROC AUC. The proposed DBN-FHO model demonstrates superior presentation across all metrics. Specifically, the Precision of DBN-FHO reaches 98.75%, outperforming SVM (93.67%), XGBoost (95.67%), and ELM (96.61%). Similarly, in terms of Recall, DBN-FHO achieves 96.26%, which is higher than SVM (92.05%), XGBoost (94.22%), and ELM (95.21%). For the F1 Score, DBN-FHO records 90.54%, exceeding SVM (86.23%), XGBoost (87.48%), and ELM (89.47%). Lastly, the ROC AUC of DBN-FHO is 96.72%, showcasing better performance than SVM (90.05%), XGBoost (93.70%), and ELM (94.71%). These consequences highlight the efficacy of the projected DBN-FHO model, which consistently outperforms the existing techniques in the given evaluation metrics.

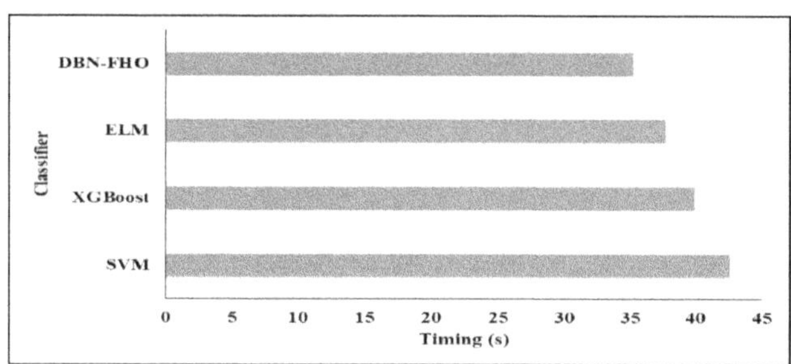

Fig. 4. Timing Analysis

An analysis of the training time of the proposed DBN-FHO model associated to existing techniques, including SVM, XGBoost, and ELM. The DBN-FHO model demonstrates the shortest training time at 35.23 s, showcasing its computational efficiency. In comparison, the training times for ELM, XGBoost, and SVM are 37.73 s, 39.97 s, and 42.62 s, respectively. These results designate that the proposed DBN-FHO model not only excels in performance metrics but also significantly reduces the training time, making it a faster and more efficient solution for the given application.

6 Conclusion

In order to classify botnet attacks on the IoT, this study presented a deep learning system that relies on a computationally efficient DBN model. Optimal characteristics for detecting IoT botnet attacks were selected using an FHO model that underwent recurrent stratified k-fold cross-validation for parameter fine-tuning. During federated learning system, discrepancy privacy was added to the DBN training to improve privacy. The suggested method outperformed related research in terms of accuracy (almost 97 to 98%) and computing complexity (low), proving that it may be used on low-powered IoT edge devices. In addition, these findings support the idea that the suggested approach could be useful in enhancing security while preserving data smart grids, smart homes, and smart healthcare, among other scenarios involving the deployment of IoT edge devices with limited resources. Looking ahead, our goals for future research include further simplifying the model through quantisation and pruning, making it more energy efficient, making it more resistant to Byzantine assaults from bad clients, and making it more efficient during model aggregation in terms of communication.

References

1. Aldaej, A., Ullah, I., Ahanger, T.A., Atiquzzaman, M.: Ensemble technique of intrusion detection for IoT-edge platform. Sci. Rep. **14**(1), 11703 (2024)
2. Arunodhai, V., Susan, L.P., Kannimoola, J.M.: EdgeShield: hybrid real-time attack detection in IOTA tangle via edge devices. In: In 2024 15th International Conference on Computing Communication and Networking Technologies (ICCCNT), pp. 1–11. IEEE (2024)
3. Kumar, A., Singh, D.: Detection and prevention of DDoS attacks on edge computing of IoT devices through reinforcement learning. Int. J. Inf. Technol. **16**(3), 1365–1376 (2024)
4. Khan, M., Hatami, M., Zhao, W., Chen, Y.: A novel trusted hardware-based scalable security framework for IoT edge devices. Discov. Internet Things. **4**(1), 4 (2024)
5. Bala, B., Behal, S.: AI techniques for IoT-based DDoS attack detection: taxonomies, comprehensive review and research challenges. Comput Sci Rev. **52**, 100631 (2024)
6. Pope, J., et al.: Intrusion detection at the IoT edge using federated learning. In: Security and Privacy in Smart Environments, pp. 98–119. Springer Nature Switzerland, Cham (2024)
7. Beshah, Y.K., Abebe, S.L., Melaku, H.M.: Drift adaptive online DDoS attack detection framework for IoT system. Electronics. **13**(6), 1004 (2024)
8. Kavitha, D., Ramalakshmi, R.: Machine learning-based DDOS attack detection and mitigation in SDNs for IoT environments. J. Franklin Inst. **361**(17), 107197 (2024)
9. Alotaibi, Y., Deepa, R., Shankar, K., Rajendran, S.: Inverse chi-square-based flamingo search optimization with machine learning-based security solution for internet of things edge devices. AIMS Math. **9**, 22–37 (2024)
10. Tiwari, R.S., Lakshmi, D., Das, T.K., Tripathy, A.K., Li, K.C.: A lightweight optimized intrusion detection system using machine learning for edge-based IIoT security. Telecommun. Syst. **87**, 1–20 (2024)
11. Saiyed, M.F., Al-Anbagi, I.: A genetic algorithm-and t-test-based system for DDoS attack detection in IoT networks. IEEE Access. **12**, 25623–25641 (2024)
12. Awan, K.A., Din, I.U., Almogren, A., Nawaz, A., Khan, M.Y., Altameem, A.: SecEdge: a novel deep learning framework for real-time cybersecurity in mobile IoT environments. Heliyon. **11**(1) (2025)

13. Markkandeyan, S., Ananth, A.D., Rajakumaran, M., Gokila, R.G., Venkatesan, R., Lakshmi, B.: Novel hybrid deep learning based cyber security threat detection model with optimization algorithm. Cyber Secur. Appl. **3**, 100075 (2025)
14. Naveeda, K., Fathima, S.S.S.: Real-time implementation of IoT-enabled cyberattack detection system in advanced metering infrastructure using machine learning technique. Electr. Eng. **107**(1), 909–928 (2025)
15. Danquah, L.K.G., Appiah, S.Y., Mantey, V.A., Danlard, I., Akowuah, E.K.: Computationally efficient deep federated learning with optimized feature selection for IoT botnet attack detection. Intell. Syst. Appl. **25**, 200462 (2025)
16. Rashid, M.M., et al.: A federated learning-based approach for improving intrusion detection in industrial internet of things networks. Network. (Bristol, England. **3**, 158–179 (2023)
17. Song, R., Wang, Z., Guo, L., Zhao, F., & Xu, Z.: Deep Belief Networks (DBN) for Financial Time Series Analysis and Market Trends Prediction (2024).
18. Azizi, M., Talatahari, S., Gandomi, A.H.: Fire hawk optimizer: a novel metaheuristic algorithm. Artif. Intell. Rev. **56**(1), 287–363 (2023)

Deep Learning Based Arecanut Leaf Classification: A Scalable Framework for Early Disease Detection

Ankitha$^{(\boxtimes)}$, Rakshitha, and D. N. Sindhura

Department of Data Science and Computer Applications, Manipal Institute of Technology, Manipal Academy of Higher Education, Manipal, India
`{ankitha2.mitmpl2023,rakshitha2.mitmpl2023}@learner.manipal.edu,`
`sindhura.n@manipal.edu`

Abstract. In India, arecanut is a crop that is grown extensively and is vital to the country's agricultural economy. It is essential to keep an eye on the condition of arecanut leaves in order to guarantee premium yields and avoid the large financial losses brought on by leaf diseases. The aim of this work is to apply deep learning techniques to classify arecanut leaves into two categories: Healthy and Unhealthy. A arecanut leaf dataset was collected by traveling to west cost of south India to take a variety of arecanut leaf images. Our research utilizes advanced deep learning models, such as ResNet50, VGG16, and EfficientNetB3, to leverage their strengths. The models were trained and validated on the collected dataset, achieving smooth training and validation curves, indicating model stability and effective learning. The test accuracy of the models exceeded expectations, providing reliable classification of arecanut leaves. Among these, EfficientNetB3 demonstrated the best performance, achieving a classification accuracy of 95.0% on the test dataset. This approach offers a scalable solution to evaluate the health status of arecanut crops across India, facilitating large-scale monitoring. This study underscores the effectiveness of deep learning, particularly the EfficientNetB3 model. The proposed framework offers a valuable tool for farmers and agricultural professionals, enabling early disease detection and improved crop management strategies. The results affirm the practicality of leveraging EfficientNetB3 for real-world agricultural challenges, setting a strong foundation for advancements in crop health monitoring systems.

Keywords: Deep Learning · EfficientNetB3 Model · Arecanut · Arecanut leaf Diseases · Convolutional Neural Network

1 Introduction

Arecanut, known popularly as betel, supari, adike, etc., is one of the most significant and resilient agricultural products cultivated in the tropical regions and possesses socio-economic relevance comparable to that of staple crops such as rice and tomato. [1] India accounts for the lion's share of arecanut production, accounting for 8.53 lakh tons, or

© The Author(s) 2026

J. Shreyas et al. (Eds.): CODE-AI 2025, CCIS 2689, pp. 206–222, 2026.

https://doi.org/10.1007/978-3-032-19318-6_20

52.30% of global output. The cultivation of this use crop has always expanded: from 2.2 lakh hectares in 1991–1992, it expanded to 5.18 lakh hectares during 2019–20, which itself speaks to a growing demand and vital relevance in agricultural landscapes [2].

In South India, especially in Karnataka, Kerala, Tamil Nadu, and Andhra Pradesh, arecanut is considered to be the backbone of agricultural productivity and farmers' living. In northern India, its production is minimal. Among the northeastern states, Assam is a notable producer, contributing approximately 20% of the country's arecanut cultivation. Karnataka alone covers 65.93% of national production, with leading contributions from the districts of Shivamogga, Dakshina Kannada, and Udupi [3].

In addition to being just another crop, arecanut plays a very important role in the cultural life of this region, as it features extensively in religious ceremonies, customary practices, and socio-economic rituals. Karnataka has seen a doubling of arecanut cultivation area over the past 15 years, with major districts like Shivamogga and Chikkamagaluru contributing 60% of the state's production [4]. Economically, arecanut is a powerhouse crop as it is a major cash crop for millions of farmers and supports industries ranging from chewing products to cosmetics, traditional medicines, textile industries and even bio-based products [5]. Its resilience to varied tropical conditions and its vast range of uses make it a staple in both domestic and global markets, similar to the way rice is for sustenance and tomato is for culinary purposes.

However, maintaining the quality of these leaves is essential to ensure their usability in commercial markets. Identifying and classifying these leaves based on their condition is a challenge that needs to be addressed to meet industry standards and support the growth of the areca nut sector. This research focuses on finding practical solutions to improve the classification and quality control of areca nut leaves.

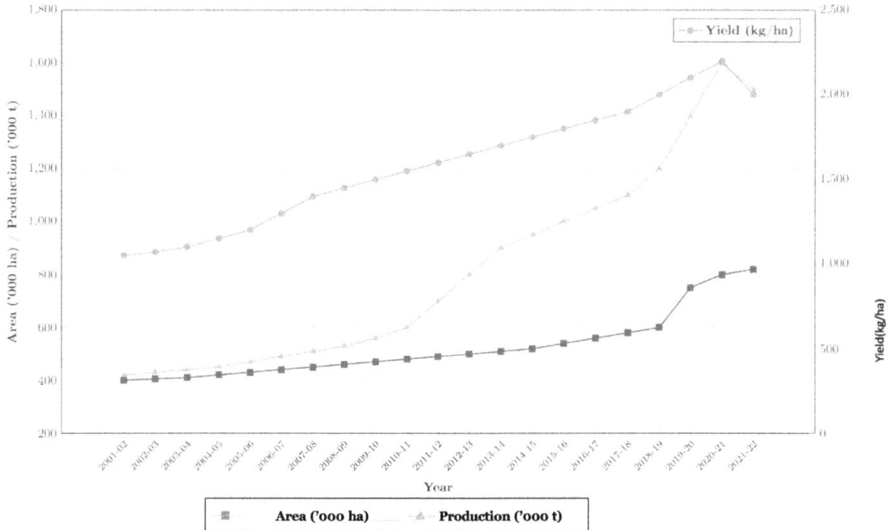

Fig. 1. Production, Area and Yield of Arecanut in India (2001–02 to 2021–22)

As shown in Fig. 1, both the area and production of arecanut in India have increased over time, with yield improving due to more efficient cultivation practices. The continuous rise in Karnataka's area and production highlights its significant role in the arecanut industry. The consistent growth in yield indicates that advancements in farming methods have led to increased productivity per hectare.

Table 1. State-wise Arecanut Production in India (Average from 2019–20 to 2021–22) Source: National Horticulture Board (NHB)

State	Production (10000 tons)	Share (%)
Karnataka	1,139.95	78.84
Kerala	100.05	6.92
Assam	61.18	4.23
Meghalaya	24.35	1.68
West Bengal	23.60	1.63
Others	96.72	6.69
Total	1,445.86	100.00

Karnataka is the largest producer of arecanut in India, As shown in Table 1 represents that contributing 78.84% (1,139.95 thousand tons) of the total production, with coastal districts such as Dakshina Kannada, Uttara Kannada, and Udupi being the major producers. Kerala follows with 6.92%, and Assam ranks third with 4.23%. Meghalaya and West Bengal contribute 1.68% and 1.63%, respectively, while other states collectively add 6.69%. In addition to these major producing states, arecanut is also cultivated in the Northeastern states, as well as in West Bengal, Tamil Nadu, Andhra Pradesh, Maharashtra, and Goa. The total national production is 1,445.86 thousand tons [6].

Arecanut leaf diseases are among the most common challenges that have been associated with the farmers, and it has the potential to significantly impact crop health and productivity. Leaf Blast is the most devastating disease affecting the arecanut leaves is Leaf Blast, which is a kind of fungus caused by the pathogen Pyricularia-oryzae [7]. The infection spreads through the leaf tissue, thereby developing the typical lesions. Initially, small, water-soaked spots appear, which later expand into diamond-shaped lesions with gray centers and brown margins [8]. If not treated, the infection spreads quickly, causing serious defoliation and weakening the plant [9]. Leaf Blast thrives on warm, humid conditions, which are common in tropical areas. Extended periods of rainfall or poor drainage in plantations tend to exacerbate the disease. The disease seriously impairs photosynthesis and depresses the immunity of the plant, predisposing the canopy to secondary infection and reducing yield. [10] Another important disease is Yellow Leaf Disease (YLD), which has been caused by nutrient deficiency and viral infection, turning the leaves yellow and consequently closing photosynthesis leading to stunted growth of the plant. Eventually, it becomes weak, reduces yield, and subsequently economic return [11].

Table 2. Yellow Leaf Disease Statistics

District	Healthy (palms)	Diseased (palms)	Loss in yield (tons)
Dakshina Kannada	1,89,93,425	12,483	6.99
Udupi	25,43,357	544	0.29
Kodagu	10,07,436	2,25,937	7.9
Chikkamagaluru	1,13,42,343	5,15,269	404.6
Shivamogga	1,59,02,476	1,92,590	86.8
Uttara Kannada	1,53,17,445	2,102	1.7
Total	6,51,06,482	9,48,925	508.28

During 1989 and 1990, In Table 2, the Karnataka government reported the prevalence of diseases in all arecanut growing districts. A total of 12,483 palms in Dakshina Kannada, 544 in Udupi, 2,25,937 in Kodagu, 5,15,269 in Chikkamagaluru, 1,92,590 in Shivamogga and 2,102 in Uttara Kannada were affected by YLD, leading to a loss of production of 508.28 tons [12].

The disease of arecanut foot rot is caused by the fungus Ganoderma lucidum. Symptoms include yellowing leaves that eventually fall, weak stems that are prone to breaking in strong winds, and brown discoloration at the base of the stem, which secretes a dark liquid [13].

Blight is the other major factor that damages the leaves of the arecanut tree and usually occurs through bacterial attacks or extreme weather phenomena. The indications of this illness display yellowish patches on the leaf blades, the leaves turn light-colored, the tips of the leaves appear pale and droopy, and the leaves are reduced in length and resemble a broom [14].

Areca red rust disease, caused by the fungal infection of Cephaleuros species, is a significant concern for areca nut cultivation. The disease manifests through several noticeable symptoms [15].

Yellowish spots begin to develop on the leaves, which later intensify, leading to a decline in overall leaf health. The top part of the leaves withers suddenly, causing the plant to appear stressed. The leaf stalks change color, turning distinctly yellowish, which is a critical early sign of the infection. Over time, the disease progresses to affect the leaf midrib, leaving only remnants of the leaves attached to the plant. Severe infections can weaken the plant significantly, reducing its productivity and making it more susceptible to secondary infections [14].

Deep learning is the latest use of machine learning techniques with very heavy modeling of complex data based upon massive data sets. Part of what makes deep learning so different from traditional machine learning approaches is that it relies on an artificial neural network, thus enabling automatically extracted hierarchical features without performing this via manual feature extraction [16]. It has made deep learning perform great performances in a myriad of applications, such as image and speech recognition, natural language processing, and self-driving. Among the very successful architectures of deep learning, the Convolutional Neural Network is highly effective particularly on computer vision applications, therefore becoming a very preferred choice for those types of tasks. The objectives of this study are:

1. Automatic Classification of Leaf Blast Disease in Arecanut.
2. Exploring Deep Learning and for Classification.

Organization of Work:

– Introduction: Importance of arecanut and disease detection.
– Literature Review: Overview of past research and the gap.
– Data collection and preprocessing: collecting images and preprocessing techiuque.
– Methodology: method used in classification of model.
– Results and Discussion: Key findings and interpretation.
– Conclusion: Summary and future work.

2 Literature Review

Arecanut is very important in South India and plays a key role in the economy. But nowadays, diseases are increasing in arecanut plants. Research on arecanut leaf disease detection is limited and from the literature it is evident that deep learning models are very less explored for arecanut diseases. Most studies rely on manual methods or traditional approaches, which are time-consuming and less accurate. That's why we have chosen this paper for our research. Earlier research on arecanut disease detection has investigated several approaches, combining traditional machine learning algorithms and more advanced techniques, It helps us understand the problem better and supports us in doing our work under this research. We have used deep learning methods to develop scalable and accurate models for reliable early detection of arecanut leaf diseases (Table 3).

Table 3. Summary of Arecanut Disease Detection Studies Using Various Models

S. No	Author(s)	Summary
1	Shuhan Lei [17]	Used UAV insource remote sensing data to express yellow leaf disease severity quantitatively and analyze its correlation with areca LVV
2	Jiawei Guo [18]	Conducted remote sensing monitoring for large-scale areca yellow leaf disease using Planet Scope images
3	Amit Gupta [19]	Achieved 97.11% accuracy using ResNet44 for disease detection and included a chatbot for farmer assistance
4	Namra Mahveen [20]	Focused on early detection of diseases in arecanut plants' leaves, trunk, and fruit using CNN
5	Andhe Dharani [21]	Emphasized early-stage disease detection to maintain healthy crops and reduce economic losses

(continued)

Table 3. (*continued*)

S. No	Author(s)	Summary
6	Khairunnisa [22]	Built an Android-based machine learning model using CNN to quickly and accurately detect areca plant diseases
7	Dong Xu [23]	Quantified areca YLD severity using UAV ispectral and thermal infrared imagery combined with feature selection and machine learning
8	Arun Karthik V [24]	Demonstrated high accuracy in identifying diseases in arecanut leaves using CNN
9	Prajwala Tm [25]	paper proposes a modified LeNet-based CNN model to detect and classify tomato leaf diseases with minimal computational resources

3 Data Collection and Preprocessing

A custom arecanut leaf dataset was meticulously collected by visiting various regions known for arecanut cultivation. Images of arecanut leaves were captured in diverse conditions to ensure a balanced representation of healthy and unhealthy leaves. This process involved photographing leaves in natural light, covering different stages of leaf health. The dataset aims to support the classification and analysis of leaf conditions with high accuracy. This real-world data collection enhances the dataset's applicability in practical agricultural scenarios (Table 4).

Table 4. Collected Images

Description of Leaf Condition	Sample Images		
Healthy Leaf			

Unhealthy Leaf			

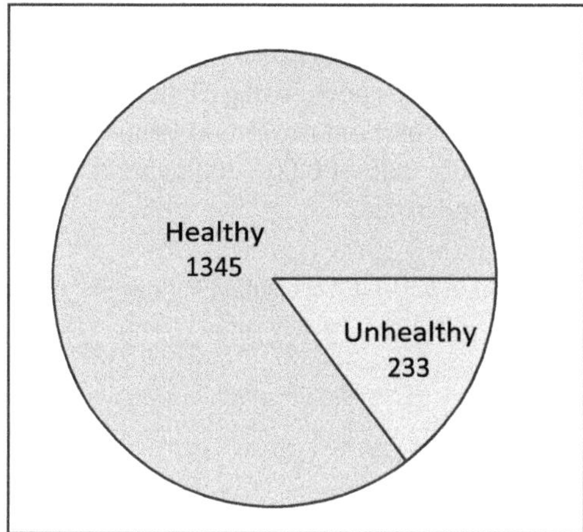

Fig. 2. Overview of Dataset

A total of 1,578 images of arecanut leaves from various arecanut growing locales were collected. In this Fig. 2, 233 unhealthy and 1,345 healthy leaves that were photographed in a variety of natural lighting settings. The images appropriately represented the various phases of leaf health in order to provide a dataset that was well-balanced. The dataset is split into 70% training, 20% validation, and 10% testing for efficient training and assessment. By using deep learning models, this method increases classification accuracy and makes exact sickness detection possible.

4 Methodology

In the study of arecanut leaf disease detection, several deep learning models have been evaluated and best performing models like vggg16,reznet50,efffecntb3 were consider for classification of leaf blast disease The following table summarizes the performance of different models based on various metrics:

4.1 VGG16

A popular CNN architecture that detects simple to complicated characteristics using a 16 layer configuration and different filter sizes. It can identify subtle differences between plant species because it uses tiny convolutional filters to enable complicated feature learning [26]. For two class classification on the dataset of arecanut leaf photos, the code employs a VGG16 model. It consists of GlobalAveragePooling2D, Dense layers with ReLU activation, L2 regularization (1e-4), Batch Normalization, and Dropout (0.5, 0.4) for stability. The output layer uses SoftMax for two-class output, with a final layer corresponding to the number of classes.

In the Table 5, the base of a fine-tuned VGG16 model kept constant and modifying the classification layer. The model is compiled with the SGD optimizer (learning rate = 0.001), momentum = 0.9, decay = 1e−6, categorical cross-entropy loss, and tracks accuracy. Training uses Early Stopping (patience = 15) and Reduce LROn Plateau (factor = 0.5, patience = 7) to optimize the model's performance.

The models were trained for 100 epochs using a batch size of 32, with early stopping applied if no improvement was observed for 15 consecutive epochs. We used the SGD optimizer with an initial learning rate of 0.001, momentum of 0.9, and a decay rate of 1e-6 to optimize training performance.

Table 5. VGG16 Training Parameters

Parameter	Value
Model Architecture	VGG16
Image Size	224×224
Number of Epochs	100
Batch Size	32
Optimizer	SGD
Initial Learning Rate	0.001
Learning Rate Reduction	Factor: 0.5 (patience: 7 epochs, min: $1e^{-7}$)
Early Stopping	Applied (Patience: 15 epochs, restores best weights)
Regularization	L1-L2 (1e-4, 1e-4)
Augmentations	Rotation (40°), Width/Height Shift (0.2), Shear (0.3), Zoom (0.3), Horizontal Flip, Brightness (0.7 - 1.3)
Normalization	Batch Normalization (Momentum: 0.5)
Dropout	0.4 (Dense Layer), 0.3 (Additional Layer)
Loss Function	Categorical Crossentropy
Evaluation Metrics	Accuracy

4.2 EfficientNetB3

This class of neural network architectures seeks to increase performance with fewer parameters and computational costs compared to traditional deep learning models. Initially, these models were introduced with a scaling method that allows the depth, width, and resolution of the network to be optimized sinuously. Its ability to adjust to various aspect ratios and resolutions is one of its key benefits; it allows for the effective handling of a variety of image sizes without the need for cropping or scaling. Additionally, because EffectiveNetB3 uses compound scaling to obtain high accuracy with fewer parameters, it works well in resource constrained scenarios [26]. The implemented a fine-tuned EffectiveNetB3 model by keeping the convolutional base constant and modifying the classification layer. In our implementation an EfficientNetB3 model using Kera's for two class classification. In the Table 6 the model starts with the EfficientNetB3 base, followed by GlobalAveragePooling2D, two dense layers with ReLU activations, L2 regularization (1e-4), Batch Normalization, and Dropout layers (0.5 and 0.4) for stability and generalization. The final layer uses SoftMax activation for two class output. The model is compiled with the Adam optimizer (learning rate 0.0001), categorical cross-entropy loss, and accuracy as a metric. Training uses Early Stopping to halt if validation performance stops improving (patience of 10) and Reduce LROn Plateau to lower the learning rate if progress stagnates, ensuring efficient and robust training.

Table 6. EfficientNetB3 Training Parameters

Parameter	Value
Model Architecture	EfficientNetB3
Image Size	224×224
Number of Epochs	100
Batch Size	32
Optimizer	Adam
Initial Learning Rate	0.0001
Learning Rate Reduction	Factor: 0.5 (patience: 5 epochs, min: $1e^{-6}$)
Early Stopping	Applied (Patience: 10 epochs, restores best weights)
Regularization	L2 (1e-4)
Augmentations	Rotation (40°), Width/Height Shift (0.2), Shear (0.3), Zoom (0.3), Horizontal Flip, Brightness (0.7 - 1.3)
Normalization	Batch Normalization
Dropout	0.5 (Dense Layer), 0.4 (Additional Layer)
Loss Function	Categorical Crossentropy
Evaluation Metrics	Accuracy

The model was trained for 100 epochs using a batch size of 32, with early stopping applied if no improvement was observed for 10 consecutive epochs. We used the Adam optimizer with an initial learning rate of 0.0001, along with learning rate reduction applied when validation performance plateaued.

4.3 ResNet50

ResNet50, a sophisticated convolutional neural network (CNN) architecture, demonstrates exceptional performance in plant classification by tackling the vanishing gradient issue through the incorporation of skip connections. These pathways enhance gradient propagation during backpropagation, allowing for effective training in extremely deep networks. The 50 layer structure of ResNet50, coupled with its capacity to handle image variations and extract intricate features, positions it as a top choice for plant identification applications. This architecture exhibits superior accuracy, computational efficiency, and adaptability in such tasks. [26]. The implemented a fine tuned ResNet50 model by keeping the convolutional base constant and modifying the classification layer. In the

Table 7 model starts with the base being ResNet50, followed by GlobalAveragePooling2D to reduce the feature dimensions. It has two dense layers with ReLU activations and l1,l2 regularization (l1 = 0.01, l2 = 0.01) to avoid overfitting. Each dense layer is associated with a Batch Normalization layer (momentum set to 0.5) for improving the stability of learning and Dropout layers (0.4 and 0.3) to enhance the generalization by preventing over reliance on certain neurons. The last layer uses SoftMax activation to output probabilities for i class classification. The model is compiled with the SGD optimizer (learning rate 0.001 and momentum 0.9) for effective weight updates, categorical cross entropy loss for i class classification and accuracy as the evaluation metric. This ensures proper training and avoidance of overfitting of the model, while also utilizing ResNet50's pretrained feature extraction abilities.

Table 7. ResNet50 Training Parameters

Parameter	Value
Model Architecture	ResNet50
Image Size	224×224
Number of Epochs	100
Batch Size	32
Optimizer	SGD (Momentum: 0.9)
Initial Learning Rate	0.01
Learning Rate Reduction	Factor: 0.5 (patience: 3 epochs, min: $1e^{-7}$)
Early Stopping	Applied (Patience: 10 epochs, restores best weights)
Regularization	L1-L2 (1e−2, 1e−2)
Augmentations	Rotation (40°), Width/Height Shift (0.2), Shear (0.3), Zoom (0.3), Horizontal Flip, Brightness (0.7 - 1.3)
Normalization	Batch Normalization (Momentum: 0.5)
Dropout	0.4 (Dense Layer), 0.3 (Additional Layer)
Loss Function	Categorical Cross entropy
Evaluation Metrics	Accuracy

In this study, ResNet50, VGG16, and EfficientNetB3 are applied to categorize arecanut leaves as healthy and unhealthy. With the performances compared from all models used, the most suitable one for detection of accurate arecanut leaf diseases is achieved in this study. The comparison highlights the strength of each model in dealing with the intricacies of arecanut leaf classification—deep networks and subtle differences of ResNet50, detailed feature extraction of VGG16, and computational efficiency and scalability of EfficientNetB3.

5 Result and Discussion

The EfficientNetB3 model was quite good due to highly efficient architecture, pre-trained weights on ImageNet, and also a well-designed fine tuning process [26] A data augmentation process helps to increase robustness against variations while l2 penalties, Dropout layers, and Batch Normalization helped curtail the overfitting process [27]. A dynamic learning rate scheduler has optimized the efficiency of training, and early stopping ensured the best model. Saved before overfitting. These strategies, combined with a balanced design of trainable and frozen layers, the model generalized really well to the dataset, with high accuracy and stable validation performance [28].

The Proposed research paper is unique in that, to the best of our knowledge, there has not been any previous deep learning work specifically on arecanut leaf disease analysis. There are many studies employing well-known models like VGG16, ResNet50, and EfficientNetB3, but in this paper, these models have been applied to a new and unique problem. Experimenting with these models on arecanut leaf disease images, we conclude that EfficientNetB3 is the highest in accuracy and hence the ideal model for this particular task. This study not only fills the gap in the studies of agriculture, but also is instructive about which of the deep learning models should ideally be used for the examination of plant leaf images, particularly in the instance of arecanut leaf disease. While we did not present a new model, the application of these existing models in this very particular area makes our contribution pertinent as well as innovative in its own right (Figs. 3, 4, 5, 6, 7 and 8).

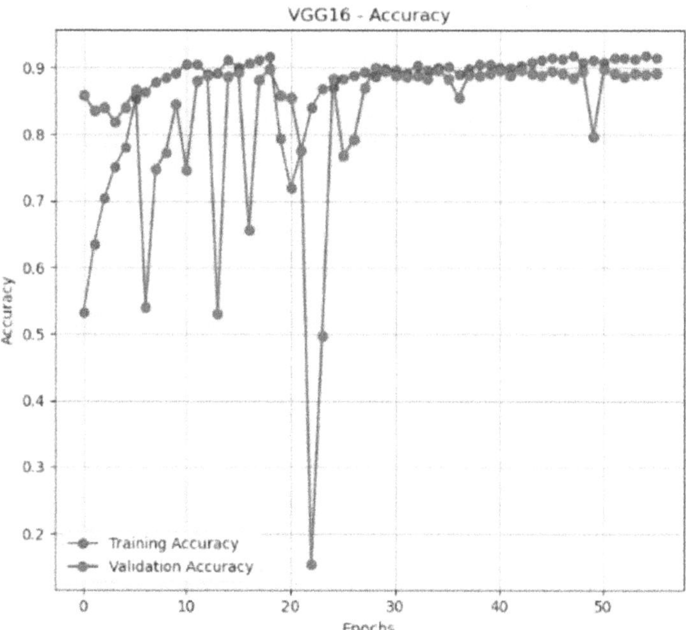

Fig. 3. VGG16 Accuracy Curve for Arecanut Leaf Classification

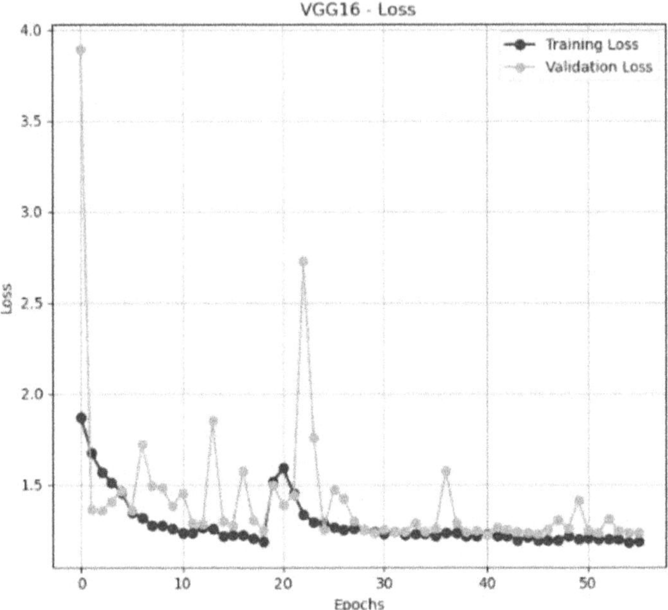

Fig. 4. VGG16 Loss Curve for Arecanut Leaf Classification

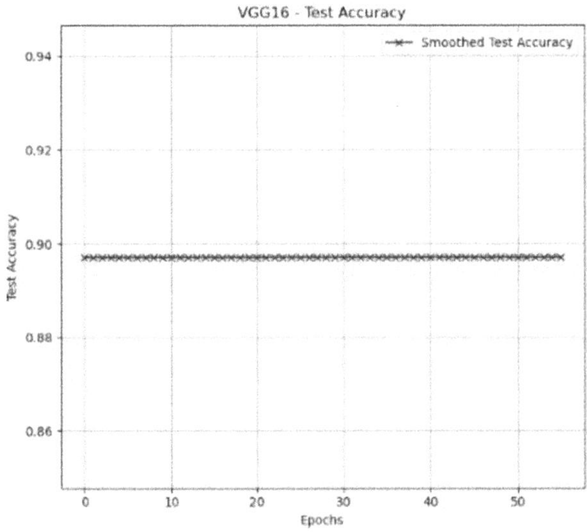

Fig. 5. VGG16 Test Accuracy Curve for Arecanut Leaf Classification

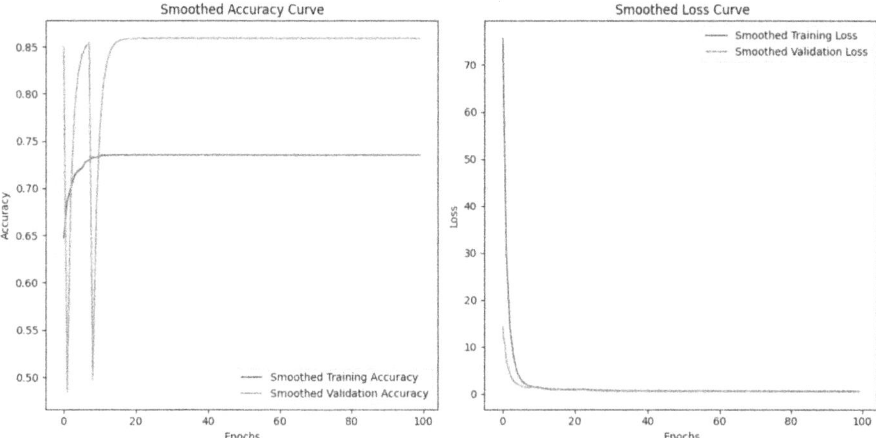

Fig. 6. ResNet50 Accuracy and Loss Curve for Arecanut Leaf Classification

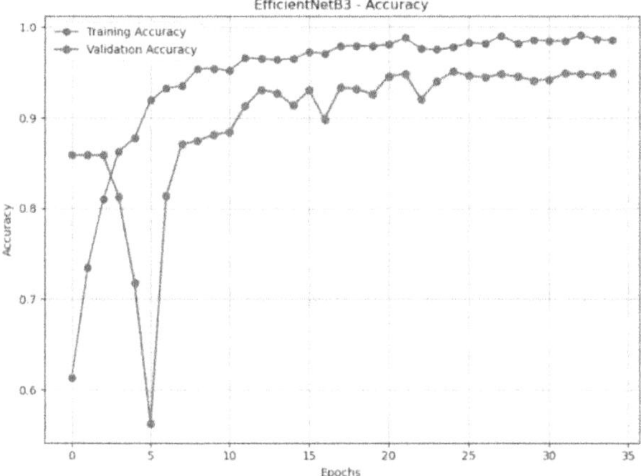

Fig. 7. EfficientNetB3 Accuracy Curve for Arecanut Leaf Classification

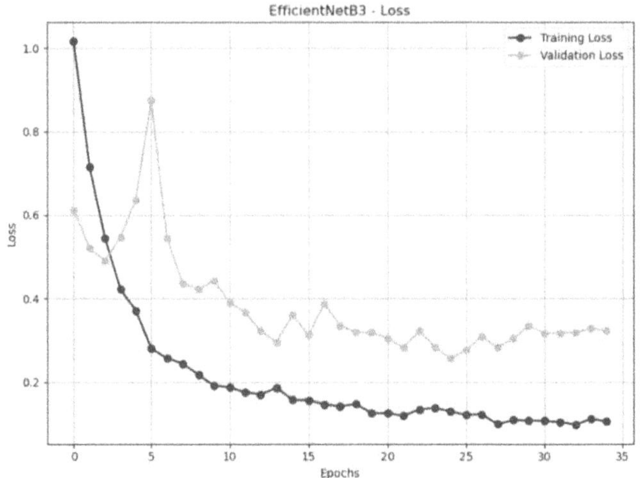

Fig. 8. EfficientNetB3 Loss Curve for Arecanut Leaf Classification

The experimental results validate the effectiveness of various deep learning models in arecanut leaf disease classification. In 3, the accuracy curve of VGG16 exhibits steady improvement in classification performance, while its loss curve 4 reflects steady convergence. The test accuracy curve 5 also confirms the learning efficiency of the model. Similarly, 6 indicates the loss and accuracy curves of ResNet50, with similar performance but weaker generalization compared to EfficientNetB3. Of all models, EfficientNetB3 has the highest accuracy, as illustrated in 7, with minimal overfitting, as indicated from the loss curve in 8. These results confirm that EfficientNetB3 is the optimal model for this task. All the values given in this section are based on our experimental results.

6   Conclusion

The present paper primarily focuses on identifying and classifying arecanut leaf diseases using the deep learning technique. This work is done in steps: first pre-processing, feature extraction, training the model followed by the classification process. The research uses the advanced models such as EfficientNetB3, VGG16, and ResNet50, all of which prove to be highly effective in detecting and classifying the diseases from the leaves. By collecting a diverse set of images from different region of west cost in southIndia.in efffientnetb3 we have got 95.05% of accuracy. This study sheds light on the reality that deep learning can revolutionize farm practices and give better means to manage the diseases that are helpful in promoting growth and sustainability in arecanut production. This may lead to a better practice in farming as well as sustainable prosperity and long run sustainability of the arecanut sector of India. The framework then becomes a good source Realtime disease detection in arecanut leaf. The development of a novel model will be considered in future work to further enhance performance.

References

1. Nagaraja, N.R., Ananda, K.S., Rajesh, M.K.: Genetic variation among varieties, wild species and related genera of Arecanut (Areca catechu L.) as revealed by microsatellite markers. Int. J. Plant Soil Sci. **34**(22), 846–855 (2022)
2. Hebbar, K.B., Ramesh, S.V., Bhat, R., et al.: Predicting current and future climate suitability for Arecanut (Areca catechu L.) in India using ensemble model. Heliyon **10**(4) (2024)
3. Hegde, A., Sadanand, V.S., Hegde, C., Naik, K.K., Shastri, K.: Identification and categorization of diseases in arecanut: a machine learning approach. Indonesian J. Electric. Eng. Comput. Sci. **31**(3), 1803–1810 (2023)
4. Jamanal, H., Murthy, C.: Constraints faced in production and marketing of arecanut in Karnataka. J. Farm Sci. **37**(01), 54–58 (2024)
5. Preeti, Y.H., Thirumal, A., Anil, K.S., Rahul Prasad, R., Priya, Y.H., Arpitha, H.B.: Addressing production and marketing constraints of arecanut growers: Insights from tumcos in Davanagere, Karnataka, India. J. Sci. Res. Reports **30**(6) (2024)
6. Dr. Parveen Kumar. Production and marketing scenarios of arecanut in India. Extension Folder No. 107/2023. All rights reserved. © 2023 ICAR-CCARI, Ela, Old Goa (2023)
7. Karthik, V., Shivaprakash, J., Rajeswari, D.: 2. disease detection in arecanut using convolutional neural network (2024)
8. Chand, G., Akhtar, M.N., Kumar, S.: Diseases of Fruits and Vegetable Crops: Recent Management Approaches. CRC Press (2020)
9. Wei, C., et al.: First report of leaf spot on cucumis melo l. caused by arcopilus aureus in china. Plant Disease (2024)
10. Randy, C.: Ploetz. 8. manejo de enfermedades en cultivos perennes tropicales (2007)
11. Bhat, R., Sujatha, S., Jose, C.T.: 1. role of nutrient imbalance on yellow leaf disease in smallholder arecanut systems on a laterite soil in india. Communications in Soil Science and Plant Analysis (2016)
12. Arecanut yellow leaf disease
13. Raju, J., Jayalakshmi, K., Sonavane, P.S., Raghu, S.: Present scenario of diseases in areca nut or betel nut (areca catechu l.) and their management. In: Diseases of Horticultural Crops: Diagnosis and Management, pp. 3–26. Apple Academic Press (2022)
14. Diannita Harahap. Plant diseases of areca nut. In: Nut Crops-New Insights. IntechOpen (2022)
15. Jose, G., Chowdary, Y.B.K.: Studies on the host range of the endophytic alga cephaleuros kunze in india. Rev. Biol. Trop. **28**(2), 297–304 (1980)
16. Menon, S.P., Yadav, S.S., Hunakunti, M., et al.: Exploring advances in areca nut classification and disease detection: a comprehensive survey. Grenze Int. J. Eng. Technol. **10** (2024)
17. Lei, S., Luo, J., Tao, X., Qiu, Z.: Remote sensing detecting of yellow leaf disease of arecanut based on uav multisource sensors. Remote Sens. **13**(22), 4562 (2021)
18. Jiawei Guo, Y., et al.: Recognition of areca leaf yellow disease based on planetscope satellite imagery. Agronomy **12**(1), 14 (2021)
19. Gupta, A., Rajeshwari, B.S., Ambadas, B., Shreyas, J., Bhuvan, G.: Analysis of areca nut leaf pathology and recommendation system using generative ai. In: 2024 Second International Conference on Networks, Multimedia and Information Technology (NMITCON), pp. 1–8. IEEE (2024)
20. Mahveen, N., Shahneen, U., Hasanuddin, S.M., Fatima, A., Aijaz, S., Qutubuddin, S.: Enhancing areca nut plant wellness: innovative disease detection using deep learning algorithms
21. Dong, X., Yuwei, L., Liang, H., Zhen, L., Lejun, Y., Liu, Q.: Areca yellow leaf disease severity monitoring using uav-based multispectral and thermal infrared imagery. Remote Sens. **15**(12), 3114 (2023)

22. Riza, F., et al.: Android-based areca plant disease detection using convolutional neural network (cnn) algorithm. Instal: Jurnal Komputer **16**(03), 277–288 (2024)
23. Xu, D., Lu, Y., Liang, H., Lu, Z., Yu, L., Liu, Q.: Areca yellow leaf disease severity monitoring using uav-based multispectral and thermal infrared imagery. Remote Sens. **15**(12), 3114 (2023). 24.
24. Karthik, V.A., Shivaprakash, J., Rajeswari, D.; Disease detection in arecanut using convolutional neural network. In: 2024 International Conference on Advances in Computing, Communication and Applied Informatics (ACCAI), pp. 1–6. IEEE (2024)
25. Jiang, D., Li, F., Yang, Y., Yu, S.: A tomato leaf diseases classification method based on deep learning. In: 2020 Chinese Control and Decision Conference (CCDC), pp. 1446–1450. IEEE (2020)
26. Penugonda, G., Singamaneni, R., Kalyani, A.L.: A comparative study for monocot remembrance using vgg16, efficientnet, inceptionv3, and resnet50 on accuracy and response time. Int. J. Plant Soil Sci. (2023)
27. Sharma, J.: 1. efficientnetb3 for high-performance insect identification (2024)
28. Li, L., Spratling, M.: Data augmentation alone can improve adversarial training. arXiv preprint arXiv:2301.09879 (2023)

Explainable AI to Identify Feature Importance in EEG Data for Mild Traumatic Brain Injury

Deepika Nelavagal Sridhara[1] ⓘ, K. S. Hareesha[1][(✉)] ⓘ, Ajay Hegde[2] ⓘ,
and Girish Menon[3] ⓘ

[1] Department of Data Science and Computer Applications, Manipal Institute of Technology,
ManipalAcademyofHigherEducation, Manipal, Karnataka, India
`hareesh.ks@manipal.edu`
[2] Manipal Hospitals, Bengaluru, Karnataka 560102, India
[3] Department of Neurosurgery, Kasturba Medical College, Manipal Academy of Higher
Education, Manipal, Karnataka, India
`girish.menon@manipal.edu`

Abstract. Mild Traumatic Brain Injury (mTBI) is a prevalent yet often underdiagnosed condition, with many patients failing to receive structured follow-up care. While CT scans remain the standard for detecting severe brain injuries, they are often ineffective in mild cases exposing patients to unnecessary radiation. Electroencephalography (EEG) is a promising prediagnostic tool for mTBI detection by capturing brain wave abnormalities. In this study, we utilize an LSTM-based deep learning based model to classify patients as CT positive or CT negative based on EEG data. However, deep learning models, despite their high accuracy, often function as "black boxes," making it difficult to interpret which features contribute most to their predictions. To enhance interpretability, we leverage Explainable AI (XAI) techniques using Captum's FeaturePermutation to identify the most influential electrodes and clinical features. Our findings reveal that central and midline electrodes (Pz, Fz, Cz), along with key clinical factors such as the Glasgow Coma Scale (GCS) and time difference between time of injury to time of recording, significantly contribute to classification performance. The model achieved an impressive accuracy of 99.96%, highlighting the effectiveness of EEG-based assessment. By ranking features based on importance, we demonstrate the potential for reducing feature dimensionality while maintaining accuracy. These insights not only improve model transparency but also pave the way for future research on EEG-based biomarkers for brain injury detection and personalized clinical interventions.

Keywords: Mild Traumatic Brain Injury (mTBI) · Electroencephalography
(EEG) · Machine Learning · explainable AI

1 Introduction

Traumatic Brain Injury (TBI) is a form of brain damage caused by an external force. It can result from a strong impact, such as a bump, blow, or jolt to the head or body, or from a penetrating object. However, not every head injury leads to TBI [5]. The severity of

© The Author(s) 2026
J. Shreyas et al. (Eds.): CODE-AI 2025, CCIS 2689, pp. 223–229, 2026.
https://doi.org/10.1007/978-3-032-19318-6_21

TBI is typically assessed using the Glasgow Coma Scale (GCS), a standardized scoring system that evaluates a person's level of consciousness.

Early diagnosis of mild TBI is critical since it accounts for more than 90% of all TBI cases, and nearly half of the affected patients fail to achieve complete recovery after six months. Despite its prevalence, structured follow-up care is provided to less than 10% of patients, whereas such care is vital in identifying the at-risk individuals for incomplete recovery. Though CT scans can diagnose potentially life-threatening brain injuries, they are often inefficient in mild TBI cases, as 90–95% reveal no intracranial abnormalities while also exposing patients to radiation risks [4].

Electroencephalography (EEG) is a non-invasive, cost-effective neuroimaging method that holds promise for detecting mild TBI by analysing brain wave patterns. By recording the brain's electrical activity, EEG provides valuable insights into abnormalities that may result from injury. Changes in EEG frequency spectra, connectivity between brain regions, and power correlations have all been associated with concussions. These techniques have shown sensitivity to both functional and structural brain damage, thus EEG has proven to be a valuable tool in the classification and diagnosis of individuals with head trauma [2].

Deep learning models, although highly effective in classification tasks, often work as "black boxes," making it challenging to interpret which features contribute most to their decisions. In the context of EEG-based classification, this lack of understanding specific electrodes and clinical variables is, in fact especially challenging for achieving transparency. However, for any research or clinical application utilizing EEG, knowledge of the contribution of specific electrodes is essential. In this research, we will utilize Explainable AI (XAI) in order to elucidate the top-ranked electrodes for classifying if the patient is CT positive or CT Negative for a mild TBI (mTBI) given a model based on an LSTM network. After training the LSTM in order to classify the cases, we use FeaturePermutation from Captum [3] to trace back to find the crucial roles of individual electrodes within the decision-making process of a model on EEG data. In this way, we make the model interpretatable and identify key brain regions required for detecting whether a patient requires a CT scan or not.

2 Related Works

Wickramaratne *et al.* [8] suggested utilization of Long Short-Term Memory (LSTM) units for classifying those suffering from the concussion using an EEG data that is minimally preprocessed. By removing the need of feature engineering suggested utilization of Long Short-Term Memory (LSTM) units towards classifying those suffering from the concussion using an EEG data that is under minimally preprocessed. Removing the need of feature engineering this design automatically extracts key features for the use in the deep model. The model achieves 92.86% with an AUROC of 0.91 performance to distinguish a person with a history of having a concussion to a person without a history.

Thanjavur *et al.* [6] proposed a deep learning-based recurrent neural network that used LSTM to classify concussions from raw resting-state EEG data. The authors considered all 64 EEG channels and ranked them based on their contribution to the LSTM model's classification performance. Their results showed that the model reached 94%

accuracy with only the top six most informative channels, suggesting that concussion classification may be feasible with a minimal number of EEG sensors.

In another study [1], the authors proposed the usage of self-attention-based models for EEG analysis. Using a Self-Explainable Self-Attention Model (SESM), they could obtain better performance on EEG tasks. The intrinsic explainability of the model provides a glimpse into the decision-making process, which makes it a valuable tool for complex neural data.

Furthermore, a survey [10] by authors discussed the importance of interpretability in AI models applied to EEG data. It was noted that understanding the workings of such models is important to allow for transparent and ethical use, especially in a clinical context.

3 Methodology

3.1 Ethical Clearance

Ethical approval for this study was obtained from the Institutional Ethics Committee, Kasturba Medical College, and Kasturba Hospital, Manipal, Karnataka, India (IEC1:41/2024). All participants provided informed consent before their inclusion in the study.

3.2 Dataset Description

The study utilized EEG data and clinical features collected from 100 subjects. The EEG recordings were obtained using a 32-channel EEG system (NicoletOne) with sampling rates of 500 Hz, 2000 Hz, and 512 Hz. Each subject had 23 EEG channels recorded for a duration of 10 minutes. 19 electrodes that were significant were considered. The dataset consisted of three patient categories:

– CT Positive (CT Abnormal Patients): 50 subjects
– CT Negative (CT Normal Patients): 20 subjects
– Normal Patients: 30 subjects

Additionally clinical data namely: Age, Gender, Scalp Wound, GCS, Time difference between date and time of trauma and date and time at which EEG was recorded was collected.

3.3 Data Preprocessing

To standardize the recordings from various EEG devices, the data was resampled to 100 Hz. The preprocessing steps were:

– Resampling: All EEG signals were downsampled to 100 Hz for uniformity.
– Segmentation: EEG signals were divided into segments of 120 timesteps [7]
– Normalization: Standardization was performed using a Standard Scaler, en-Ensuring that the data maintained a constant scale for better model performance.

These preprocessing steps allowed the LSTM model to train effectively while maintaining critical EEG features for classification.

3.4 Classification Process

The model architecture is composed of two LSTM layers, followed by a fully connected (linear) output layer. The LSTM layer has been designed to address sequential data with variable-length input sequences: fed here with sequences of length 120 and 24 features consisting both electrode and clinical features per time step. It consists of two stacked LSTM cells: a two-layer LSTM network that allows the model to catch long-range dependencies and more complex patterns in the data. The LSTM is a batch-first layer where the input is of the following dimension: (1000, 256, 24). Dropout layer of rate 0.03 is applied after the LSTM layer. Final output layer is a fully connected (linear) layer which takes the output from the LSTM and produces two neurons for classification. The model was trained with Adam optimizer which adjusts the learning rate for each parameter, and the CrossEntropyLoss function is used to compute the loss. This design is shown in Fig. 1.

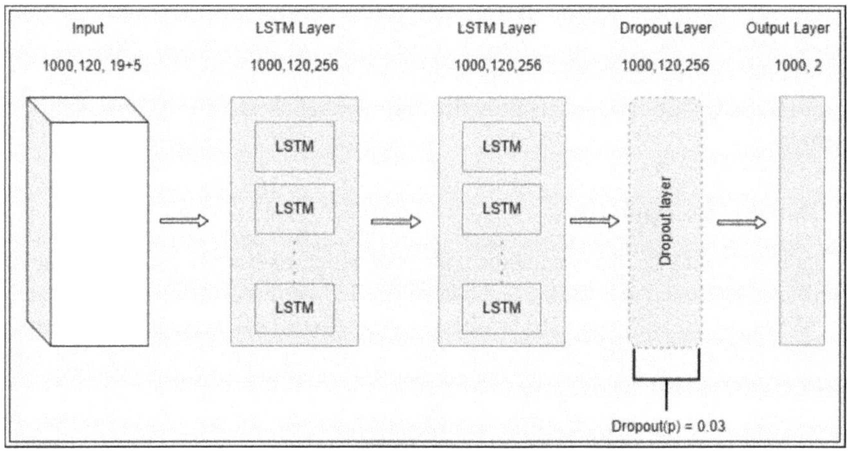

Fig. 1. LSTM Model Design

The LSTM model was trained with the following hyperparameters: 250 epochs, a learning rate of 0.005, weight decay of 0.0001, and a batch size of 1000. The model was trained on an HP EliteDesk 800 G4 workstation equipped with a 48GB NVIDIA GeForce RTX A6000 GPU, 128GB of RAM, and an Intel Core i9 processor running at 3.7 GHz. This ensured proper management of the data and training of the model, which led to the optimal performance by the LSTM model.

Initially, experiments were conducted using only the raw EEG signals, which were directly fed into the LSTM model. The analysis focused on 19 electrodes, allowing the model to learn patterns from the EEG data without handcrafted feature extraction.

Then the same experiments were repeated after adding clinical features namely: Age, Gender, Scalp Wound, GCS, Time difference between date and time of trauma and date and time at which EEG was recorded, totalling to 24 features.

Once the LSTM model was trained, we then assessed feature importance using Captum to understand which features most contribute to predictions of the model. Feature importance scores were computed both for EEG electrode data and clinical features. Features are split into two types: electrodes, that is, first 19 features, and clinical values, which are the remaining 5 features. We sorted those features based upon their importance score in descending order. As a result of this, it helped us ascertain which electrodes/clinical values predominantly played a most important role and contributed to our model's decisive decisions.

3.5 Results and Discussion

The best accuracy obtained by the model was classification of 99.96%, which proved that LSTM performed very well in the differentiation between CT positive and CT Negative patients. To further understand the underlying factors contributing to this high performance, we employed Explainable AI techniques to assess the importance of both EEG electrode features and clinical features in the classification process.

Using Captum, a popular XAI framework, we analyzed the feature importance of the 19 EEG electrodes and 5 clinical features. The feature importance is shown in Fig. 2.

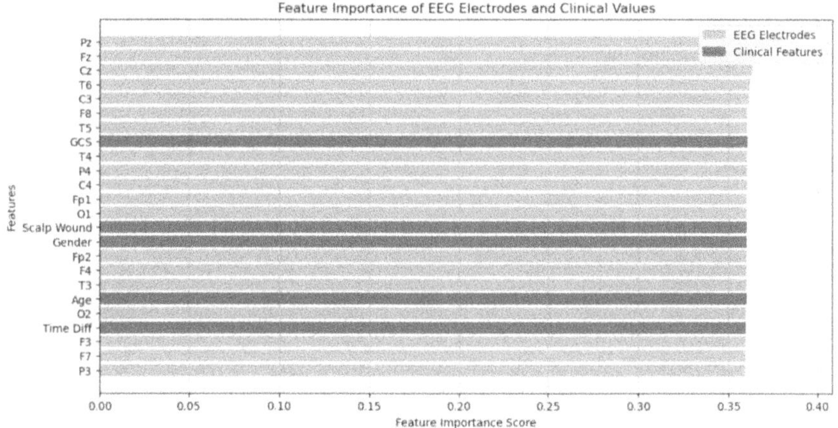

Fig. 2. Electrode Importance

The analysis revealed that electrodes Pz, Fz, and Cz exhibited the highest importance scores, indicating that central and midline electrodes played a crucial role in classification. Additionally, other electrodes such as T6, C3, and F8 also contributed significantly.

Among the clinical features, GCS (Glasgow Coma Scale) and Time Difference were identified as important predictors, further reinforcing their relevance in EEG-based assessment.

3.6 Conclusion

In deep learning models, it is usually very difficult to determine which electrodes contribute most towards classification. Thus, interpretability becomes a major challenge in such models. The application of XAI techniques helped address this challenge as the most relevant features are highlighted to allow further understanding of the decision-making process.

We ranked the features in descending order of importance and concluded that a small subset of electrodes and clinical values were the drivers in the classification task. This insight suggests that reducing the number of features while focusing on the most informative ones could lead to more efficient models without sacrificing accuracy. Integration with Explainable AI helped us learn about the feature contributions as contributing to the reliability and transparency of decisions made by the model.

Our findings may also lay a basis for further investigation, especially regarding the way in which the site of injury could influence brain activity as recorded by EEG. Identification of key features such as particular electrodes and clinical values may thus direct future studies focused on establishing links between electrical brain patterns and sites of injury that would prove invaluable in determining how injury impacts the brain. This may ultimately lead to the development of targeted and individualized treatment approaches in clinical practice.

Acknowledgment. The authors sincerely thank the Department of Neurosurgery at Kasturba Medical College, Manipal, for their invaluable support, guidance, and provision of the data necessary for conducting this study.

References

1. Adey, B., Habib, A., Karmakar, C.: Exploration of an intrinsically explainable selfattention based model for prototype generation on single-channel EEG sleep stage classification. Sci. Rep. **14**(1), 27612 (2024)
2. Hanley, D., et al.: A brain electrical activity electroencephalographic-based biomarker of functional impairment in traumatic brain injury: a multi-site validation trial. J. Neurotrauma **35**(1), 41–47 (2018). https://doi.org/10.1089/neu.2017.5004, pMID: 28599608
3. Kokhlikyan, N., et al.: Captum: a unified library for explanable ai. arXiv preprint arXiv:2009. 07896 (2020)
4. Maas, A.I.R.: https://www.thelancet.com/article/S1474-4422(22)00309-X/fulltext
5. NIH,N.I.o.N.D., Stroke: https://www.ninds.nih.gov/health-information/disorders/traumatic-brain-injury-tbi
6. Thanjavur, K., Hristopulos, D.T., Babul, A., Yi, K.M., Virji-Babul, N.: Deep learning recurrent neural network for concussion classification in adolescents using raw electroencephalography signals: toward a minimal number of sensors. Front. Hum. Neurosci. **15** (2021)
7. Tsiouris, Pezoulas, V.C., Zervakis, M., Konitsiotis, S., Koutsouris, D.D., Fotiadis, D.I.: A long short-term memory deep learning network for the prediction of epileptic seizures using EEG signals. Comput. Biol. Med. **99**, 24–37 (2018). https://doi.org/10.1016/j.compbiomed. 2018.05.019
8. Wickramaratne, S.D., Mahmud, S., Ross, R.S.: Use of brain electrical activity to classify people with concussion: a deep learning approach, pp. 1–6 (2020). https://doi.org/10.1109/ICC40277.2020.9149393

9. Zhang, X., Yao, L.: Deep Learning for EEG-based Brain-Computer Interface: Representations, Algorithms and Applications. World Scientific Publishing (2021)
10. Zhou, X., et al.: Interpretable and robust Ai in EEG systems: a survey. arXiv preprint arXiv: 2304.10755 (2023)

Comparative Analysis of Fine-Tuned CNNs and Segmentation Techniques for Multiclass Skin Cancer Classification

Ananya Gupta, Kashish Bansal, Arti[(✉)], Rishika Anand, Aditi Sabharwal, and S. R. N. Reddy

Department of Computer Science and Engineering, Indira Gandhi Delhi Technical University for Women, New Delhi, Delhi, India
{ananya004btcse21,kashish037btcse21,arti035btcse21, rishika003phd19,aditi001phd21,Srnreddy}@igdtuw.ac.in

Abstract. Skin cancer is one of the major causes of cancer worldwide, affecting approximately 1.5 million individuals annually. While many nonmelanoma skin cancers, such as basal and squamous cell carcinomas, are typically treatable when detected early, melanoma remains particularly dangerous due to its aggressive spread. Although dermoscopy and biopsy are common diagnostic techniques, they are invasive, time-consuming, and subject to considerable variability among practitioners. This situation has created a growing need for automated diagnostic tools that can categorize skin lesions both rapidly and reliably. In this study, skin cancer classification is approached using convolutional neural networks (CNNs) combined with advanced image preprocessing and segmentation techniques. The study evaluates the performance of lightweight architectures against computationally intensive architectures to determine if efficient architectures can achieve comparable classification performance. The method enhances lesion identification by applying specific preprocessing steps, such as hair removal and image sharpening, and by refining lesion edges using segmentation techniques including U-Net, SegNet, COVIDOA and Binary Thresholding. The performance of several advanced CNN architectures—namely DenseNet121, MobileNetV2, ResNet50, and NasNetMobile—is subsequently evaluated. These evaluations utilize several data augmentation techniques aimed at enhancing the representation of underrepresented classes and mitigating the class imbalance in the HAM10000 dataset. The findings could pave the way for integrating efficient preprocessing and segmentation techniques with CNN models in automated dermatological diagnostic systems leading to a more powerful computer-assisted diagnostic tool.

Keywords: Skin Cancer · COVIDOA · Binary Thresholding · SegNet · U-Net · DenseNet121 · MobileNetV2 · ResNet50 · NasNetMobile · HAM10000

1 Introduction

Skin cancer is one of the most common types of cancer globally, with a sharp rise in cases as a result of excessive exposure to sun, genetic predisposition, and environmental influences. Jerant et al. [1] suggests that early diagnosis is essential in reducing mortality rates

© The Author(s) 2026
J. Shreyas et al. (Eds.): CODE-AI 2025, CCIS 2689, pp. 230–242, 2026.
https://doi.org/10.1007/978-3-032-19318-6_22

and improving treatment outcomes, making accurate and efficient diagnostic methods essential. Traditionally, dermatologists rely on clinical examinations and dermoscopic analysis to diagnose skin lesions, but these methods are often subjective and dependent on expertise. Computer-aided diagnosis (CAD) systems that are based on artificial intelligence (AI) have become promising instruments for automated skin cancer diagnosis in order to overcome these constraints.

Deep learning, particularly convolutional neural networks (CNNs) as discussed in [2], has demonstrated remarkable performance in analyzing medical images, enabling automated detection and classification of skin lesions. However, challenges such as class imbalance, low-quality images, and the presence of hair artifacts can hinder model performance. To enhance classification accuracy, various preprocessing techniques like hair removal, image sharpening, and segmentation methods such as Binary Thresholding, COVIDOA, SegNet and U-Net have been explored. Additionally, different CNN architectures, including DenseNet121, MobileNetV2, ResNet50, and NasNetMobile, have been utilized to classify skin lesions effectively.

This study presents a comparative analysis of different CNN architectures for multiclass skin cancer classification and the impact of preprocessing and segmentation techniques on it. Using the publicly available HAM10000 dataset, data imbalance is addressed through augmentation, and various classification approaches are evaluated using standard performance metrics such as accuracy, F1-score, precision, specificity and recall. This research contributes to the ongoing development of efficient, accurate, and scalable AI-based diagnostic systems, ultimately assisting dermatologists in early skin cancer detection and enhancing patient outcomes.

Key contributions of the paper include:

- A comprehensive comparative analysis of lightweight architectures -MobileNetV2 and NasNetMobile against computationally intensive architectures DenseNet121 and ResNet50 for skin cancer classification.
- An evaluation of advanced preprocessing techniques—including hair removal, image sharpening, and segmentation methods (U-Net, SegNet, COVIDOA, Binary Thresholding)—to enhance lesion detection.
- Insights into the optimal combination of preprocessing techniques and CNN architectures for improving the efficacy of AI-based diagnostic systems.

2 Literature Review

The classification of skin cancer into multiple classes by deep learning techniques is a crucial area of research, which aims to improve the accuracy of diagnosis and assist dermatologists in clinical decision-making. Despite significant advancements in computer vision and machine learning, achieving high accuracy in multiclass skin cancer classification remains a challenge because of the visual similarities among multiple types of skin lesions.

A fine-tuned ResNeXt101 model by Chaturvedi et al. [3] which classifies seven types of skin lesions in the HAM10000 dataset achieved a highest accuracy of 93.20% highlighting the effectiveness of transfer learning in improving classification performance.

Similarly, Akter et al. [4] explored various CNN models, including InceptionV3, Xception, MobileNet, DenseNet, ResNet50, and VGG16 all fine-tuned with transfer learning and found that InceptionV3 performed the best, obtaining an accuracy of 90%.

Heenaye-Mamode Khan et al. [5] used 'Deep Generative Adversarial Network (DGAN)' for classifying multiple types of skin diseases. By creating synthetic images and combining them with CNN-based models, their method achieved accuracies of 92.3% on labelled and 91.1% on unlabelled datasets. Kausar et al. [6] used an ensemble model that combined ResNet, InceptionV3, DenseNet, InceptionResNetV2, and VGG-19. By using majority voting and weighted majority voting, their approach improved performance. However, the accuracy of individual models was between 72% and 91.8%. The study finds that ensemble learning could enhance accuracy beyond what individual models could achieve on their own.

Chaturvedi et al. [7] experimented with a pre-trained MobileNet model and transfer learning for classifying seven skin lesion types using the HAM10000 dataset. This model obtained an accuracy of 83.1% as well as the top-2 and top-3 accuracies of 91.36% and 95.34%, respectively. Similarly, Naeem et al. [8] implemented the 'DVFNet' model, which integrates deep feature fusion for multi-class skin cancer classification using dermoscopic images. This model attained an accuracy of 93.10% on the ISIC 2016 dataset, 87.50% on ISIC 2017, and 96.60% on ISIC 2018, indicating strong performance but still facing limitations across different datasets. Arshed et al. [9] used vision transformers with Convolutional Neural Network-Based Pre-Trained Models and found that Vision transformers outperformed CNN based transfer learning models. This ViT based model achieved an accuracy of 92.14%.

After the review of different deep learning algorithms used over a period of time for multi class skin cancer detection, it was inferred that most studies with lightweight models had an accuracy of less than 97% for multi class skin cancer detection.

3   Dataset

The HAM10000 dataset [10] comprises 10,015 dermoscopic images, each categorized into one of seven skin lesion classes: 'melanocytic nevi (nv, 6705 images), melanoma (mel, 1113 images), benign keratosis-like lesions (bkl, 1099 images), basal cell carcinoma (bcc, 514 images), actinic keratoses (akiec, 327 images), vascular lesions (vasc, 142 images), and dermatofibroma (df, 115 images)'. While the dataset contains over 10,000 images, metadata analysis reveals that these correspond to only 7470 unique skin lesions, with the remaining images representing duplicates captured at varying magnifications or angles. To ensure a balanced representation of lesion categories across training, validation, and test sets, stratified sampling was employed during the dataset split to preserve the original class distribution in each subset, mitigating the risk of data imbalance and enhancing model generalization during cross-validation.

4   Methodology

This study follows a systematic approach which includes data preparation, image segmentation, model fine-tuning, and performance evaluation. Using the HAM10000 dataset, this study analyzes how different segmentation techniques affect the accuracy of skin cancer classification, with the goal of enhancing diagnostic capabilities. Figure 1 illustrates the methodology used in this study.

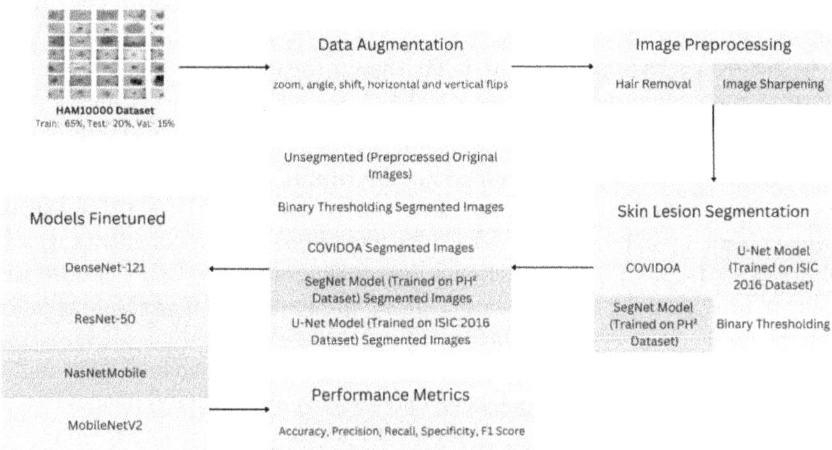

Fig. 1. Methodology Framework for Comparative Analysis

4.1 Data Augmentation

Since the HAM10000 dataset [10] has an uneven distribution of skin cancer types, data augmentation is used to balance the classes by generating additional images for the less common lesion types. This process significantly improves the diversity of the dataset by implementing transformations such as rotations, shifts, zooming, and flipping. By augmenting the representation of underrepresented classes, the study tries to create a balance in sample sizes across all categories, thereby reducing the possibility of the model displaying a bias towards the more frequent classes. This method significantly improves the ability of the model to generalize better, leading to more accurate and reliable diagnosis and classification of skin cancer in real-world applications.

4.2 Image Preprocessing

Image preprocessing is very essential in improving image quality and helps deep learning models to focus on the most important features of skin lesions. In this study, image preprocessing involves hair removal, image sharpening, and normalization. To remove hair artifacts, a black-hat filter is used to detect dark hair structures, followed by inpainting to restore the affected areas. A similar approach was used by Khan et al. [11], that combined Black-hat Morphology with Total Variation for effective hair removal. To improve lesion visibility, an unsharp masking technique is applied to enhance contrast and sharpen edges by subtracting a blurred version of the image from the original. Finally, all images are resized to a standard resolution to obtain consistency across different segmentation methods and to improve model performance.

4.3 Segmentation Techniques

Segmentation is an important step that helps separate skin lesions from the surrounding skin, making it easier for deep learning models to extract meaningful features for classification. To understand how different segmentation techniques impact classification accuracy, this study applies four methods—COVIDOA, Binary Thresholding, U-Net, and SegNet to images from the HAM10000 dataset. The models are then tested on both segmented and unsegmented (pre-processed original images), allowing for a thorough comparison of how segmentation influences performance.

- **COVIDOA (COVID-19 Optimization Algorithm)**, is a metaheuristic optimization technique inspired by the transmission patterns of COVID-19 [12]. It enhances segmentation by improving edge detection and refining lesion boundaries based on intensity variations. Unlike traditional thresholding methods, COVIDOA dynamically selects optimal threshold values to maximize lesion visibility, resulting in more precise and accurate segmentation.
- **Binary Thresholding**, is a traditional image segmentation technique that simplifies grayscale images by converting them into black-and-white (binary) images based on a set intensity threshold [13]. Pixels brighter than the threshold appear white, while darker ones turn black. This method offers a fast and efficient way to separate lesions from the background, allowing classification models to focus on the lesion's features without distractions from surrounding skin.
- **U-Net**, is a deep learning-based segmentation model designed to generate highly accurate lesion masks and its architecture is described in [14]. Miradwal et al. [15] and Behera et al. [16] shows how UNet architecture can be used for lesion segmentation in skin cancer detection. For this analysis, this study trained U-Net on the ISIC 2016 dataset using an encoder-decoder architecture with skip connections, to preserve important spatial details which refined the segmentation process. U-Net improves classification accuracy by combining high-resolution details with contextual information which creates precise lesion masks. This ensures that the model focuses specifically on the lesion area that leads to more precise and reliable results.
- **SegNet**, is a deep learning-based model designed for skin lesion segmentation [17]. This study trained it on the PH^2 dataset. Unlike U-Net, SegNet uses a pooling indexbased upsampling technique which makes it more memory-efficient while still delivering high segmentation accuracy. Its encoder-decoder architecture reduces computational demands, while enabling precise lesion segmentation which makes it a better choice for real time applications.

4.4 Pretrained CNN Models for Classification

To classify skin lesions from both segmented and unsegmented (pre-processed original images), this study fine-tunes four pre-trained convolutional neural networks: DenseNet121, ResNet50, MobileNetV2, and NasNetMobile. These models were selected for their strong track record in medical image classification.

- **DenseNet121 (Densely Connected Convolutional Network)** as described in [18], is designed to minimize computational redundancy and enhance feature propagation through dense layer connections. By establishing direct links between layers, the model efficiently reuses features, improving generalization and boosting classification accuracy. In this study, DenseNet121 is fine-tuned on images processed with different segmentation techniques to assess its adaptability to various lesion representations.

- **ResNet50 (Residual Network)** as proposed in [19], employs a deep residual learning network to prevent vanishing gradients. Its skip connections enable efficient feature extraction across multiple layers that captures both fine details and broader patterns that are important for accurate skin lesion classification. To determine its efficiency, ResNet50 is tested on images processed with different segmentation techniques, evaluating its performance across various lesion representations.

- **MobileNetV2** was first proposed in [20] and is chosen for its lightweight architecture, which makes it suitable for deployment on mobile devices. By utilizing depth wise separable convolutions and inverted residuals, it reduces computational complexity while ensuring strong classification accuracy. In this study, MobileNetV2 is tested to evaluate its effectiveness in resource-limited environments.

- **NasNetMobile**, a model designed using neural architecture search (NAS) [21], is fine-tuned for skin lesion classification. It is optimized for mobile applications and dynamically adjusts its structure through reinforcement learning to enhance feature extraction. In this study, NasNetMobile is tested on images processed with different segmentation techniques to analyze how well automated architecture search adapts to variations in input preprocessing.

4.5 Performance Evaluation

To assess each classification mode's performance, this study uses various assessment metrics, including specificity, accuracy, recall, F1-score and precision. Accuracy estimates the overall correctness of the classifications and precision shows how well the model avoids false positives. Recall or sensitivity, evaluates the capacity of the model to accurately determine cases that are positive. Furthermore, specificity measures the ability of the model to correctly identify negative cases, minimizing false positives. The F1-score provides a balanced measurement of precision and recall. Additionally, AUC score (Area Under the ROC Curve) is obtained to determine the model's ability in differentiating between the multiple types of skin lesions.

For a fair evaluation, a stratified validation strategy is employed to divide the dataset into training, validation, and test sets. The models are trained on one portion of the data, while a separate validation set is used to evaluate model during training and prevent overfitting. Finally, an independent test set assesses real-world classification accuracy. To further optimize training and improve model generalization, techniques such as learning rate scheduling, early stopping, and model checkpointing are applied.

The study also compares classification performance across different segmentation techniques to understand their impact on multiclass skin cancer detection. By determining whether segmentation improves accuracy and identifying the most effective method, this analysis helps improve the reliability of deep learning models for automated skin cancer diagnosis, which could support early detection and lead to better patient outcomes.

5 Implementation

To evaluate the balance between computational efficiency and classification performance, two lightweight models—MobileNetV2 and NasNetMobile—and two more computationally intensive models—DenseNet121 and ResNet50 were selected. This comparison aimed to determine whether lightweight models, designed for efficiency and deployment on resource-constrained devices, could achieve comparable performance to heavier architectures in medical image classification.

To analyze the effectiveness of various segmentation techniques in skin lesion classification, four methods—U-Net, SegNet, COVIDOA, and Binary Thresholding—were implemented to segment dermoscopic images. These techniques help enhance lesion visibility, eliminate background noise, and refine feature extraction, ultimately improving classification accuracy. The segmented images and the unsegmented images were fed into four pretrained models—DenseNet121, MobileNetV2, NasNetMobile, and ResNet50—each then carefully fine-tuned for multi-class skin cancer classification.

To establish a balanced evaluation, the dataset was split into 65% for training, 15% for validation, and 20% for testing. Since class imbalance is a major challenge in medical imaging, data augmentation techniques were applied to generate diverse samples for underrepresented classes. The augmentation included random rotations (up to $180°$), width and height shifts (10% of the image size), zooming (10%), and horizontal and vertical flipping to enhance variability in the dataset. All the images in the dataset were resized to (224×224) pixels, and model-specific preprocessing functions were used to normalize inputs before training, ensuring optimal performance for each model.

The selected models pre-trained on ImageNet, were fine-tuned by modifying their fully connected layers. This included adding a Global Average Pooling layer, a dropout layer to prevent overfitting, and a softmax classifier for classifying skin lesions into seven categories. Training was accomplished using the Adam optimizer with a learning rate of 0.0001 and categorical cross-entropy loss. To enhance efficiency, Early Stopping was applied to halt training if the validation loss stopped improving, while ReduceLROnPlateau dynamically adjusted the learning rate for better optimization. Additionally, Model Checkpoint was used to automatically save the best-performing model during training. Each model was trained for 30 epochs, and classification performance was evaluated using many performance metrics, including precision, F1-score, accuracy, recall, and top-2 and top-3 accuracy, ensuring a well-rounded assessment. To further analyze the impact of segmentation, confusion matrices, Receiver Operating Characteristic (ROC) curves, and Area Under the Curve (AUC) scores were generated, providing deeper insights into how well the models distinguished between different lesion types.

6 Results and Analysis

The performance of fine-tuned models was evaluated on different segmentation techniques, with each approach yielding distinct classification outcomes. Tables 1, 2, 3 and 4 showcase the performance of DenseNet121, MobileNetV2, NasNetMobile and ResNet50 respectively. The results reveal that training on the unsegmented (preprocessed original images) produced the highest accuracy across all models, followed closely by U-Net and COVIDOA, which effectively segmented lesions while preserving critical features. SegNet also performed well, but Binary Thresholding yielded the lowest accuracy due to its inability to retain fine lesion details. Among the deep learning models, DenseNet121 consistently outperformed the rest, achieving 97.64% accuracy on original images and 97.22% accuracy on U-Net-segmented images. Despite being a lightweight model, NasNetMobile performed comparably to more computationally intensive architectures, proving that efficient models can achieve results on par with complex networks. The ability of MobileNetV2 and NasNetMobile to deliver high classification accuracy while using significantly fewer computational resources makes them promising candidates for real-world clinical applications, especially in portable and embedded medical imaging systems.

Table 1. Performance of Fine Tuned DenseNet121 with Different Segmentation Techniques

Metrics (Micro-average in %)	Accuracy	F1 Score	Recall	Specificity	Precision
Unsegmented Images	97.64	91.74	91.74	98.62	91.74
UNet	97.22	90.29	90.29	98.38	90.29
COVIDOA	96.81	88.84	88.84	98.14	88.84
SegNet	96.45	87.57	87.57	97.92	87.57
Binary Thresholding	96.21	86.76	86.76	97.79	86.76

Table 2. Performance of Fine Tuned MobileNetV2 with Different Segmentation Techniques

Metrics (Micro-average in %)	Accuracy	F1 Score	Recall	Specificity	Precision
Unsegmented Images	97.35	90.75	90.75	98.45	90.75
UNet	96.89	89.12	89.12	98.18	89.12
COVIDOA	96.37	87.30	87.30	97.88	87.30
SegNet	96.60	88.12	88.12	98.02	88.12
Binary Thresholding	96.08	86.31	86.31	97.71	86.31

Table 5 compares NasNetMobile (unsegmented images) with different state of the art models used in multi class skin cancer classification.

Table 3. Performance of Fine Tuned NasNetMobile with Different Segmentation Techniques

Metrics (Micro-average in %)	Accuracy	F1 Score	Recall	Specificity	Precision
Unsegmented Images	97.56	91.47	91.47	98.57	91.47
UNet	96.71	88.48	88.48	98.08	88.48
COVIDOA	96.42	87.48	87.48	97.91	87.48
SegNet	96.55	87.94	87.94	97.99	87.94
Binary Thresholding	96.50	87.76	87.76	97.96	87.76

Table 4. Performance of Fine Tuned ResNet50 with Different Segmentation Techniques

Metrics (Micro-average in %)	Accuracy	F1 Score	Recall	Specificity	Precision
Unsegmented Images	97.59	91.56	91.56	98.59	91.56
UNet	96.78	88.75	88.75	98.12	88.75
COVIDOA	96.81	88.84	88.84	98.14	88.84
SegNet	96.76	88.66	88.66	98.11	88.66
Binary Thresholding	96.47	87.67	87.67	97.94	87.67

Table 5. Comparison of Fine Tuned NasNetMobile with different state of the art models

References	Dataset Used	Model Used	Accuracy
[3]	HAM10000	Fine-tuned ResNeXt101	93.2%
[4]	ISIC 2019	InceptionV3 (Best among tested models)	90.0%
[6]	HAM10000	Ensemble of, InceptionV3, ResNet, DenseNet, InceptionResNetV2, VGG-19	Varied (72% -91.8%)
[8]	ISIC2016,ISIC 2017, ISIC 2018	DVFNet (Deep Feature Fusion)	ISIC 2016: 93.1% ISIC 2017: 87.5% ISIC 2018: 96.6%
[9]	ISIC 2019	Vision Transformers + CNN Models	92.14%
Proposed Model	**HAM10000 (unsegmented images)**	**Fine Tuned NasNetMobile**	**97.56%**

7 Conclusion

Our evaluation of fine-tuned DenseNet121, MobileNetV2, NasNetMobile, and Res-Net50 with different image preprocessing techniques reveals a key insight: unsegmented (pre-processed original images) consistently deliver the best performance across all models. Metrics like accuracy, specificity, F1 score, recall and precision are highest when images remain unsegmented, highlighting the capability of these deep learning models to effectively extract features without additional modifications. The consistently superior performance observed with unsegmented (pre-processed original images) can be credited to their ability to preserve all diagnostic details, including fine lesion structures and subtle patterns in the surrounding skin. While segmentation techniques are effective at enhancing certain features, they can sometimes unintentionally remove valuable contextual information or alter lesion boundaries, especially in cases involving complex or irregular lesions. These results highlight that, although segmentation is important in specific situations, unsegmented images provide an ideal balance of lesion features and surrounding context, making them highly effective for medical imaging classification tasks.

This study also observed a significant boost in model performance when the training data was augmented. This improvement can be attributed to several factors. Data augmentation increases the diversity of training samples which allows the model to learn more generalizable features while reducing class bias. By exposing the model to more diverse patterns, particularly from underrepresented classes, it becomes better equipped to handle real-world variations. Unlike class weighting, which merely adjusts the loss function, augmentation actively introduces new samples, making the model more resilient and adaptable. In summary, deep learning models perform best with unsegmented (pre-processed original images), but when segmentation is necessary, UNet is the most effective method. Furthermore, augmenting the training data—without altering the validation and test sets—improves model robustness, enabling more reliable and accurate classification in real-world (imbalanced) scenarios. The results highlight that light-weight models (MobileNetV2, NasNetMobile) give comparable performance to computationally intensive models (DenseNet121, ResNet50).

8 Future Scope

Building on these findings, the next step is to develop an intuitive, web and mobile based platform that allows clinicians to upload images of skin lesions for real-time analysis and classification. To make AI-driven diagnostics more trustworthy and interpretable, the future plan is to incorporate Grad-CAM and similar other visualization techniques, enabling clinicians and medical professionals to see which parts of an image influenced the model's decision. This transparency will help build confidence in the system and support clinical validation.

One of the major constraints in detecting skin cancer is the absence of a diverse dataset. The future scope could be to collect a diverse dataset of skin lesion dermoscopic images incorporating images from different regions, genders and skin colour to improve dataset diversity. Additionally, ensemble learning with lightweight models could be used to enhance reliability, particularly for multi-class classification. Hybrid and new segmentation techniques can be applied to skin lesion images to compare their performance as well. Wu et al. [22] proposed an 'MHorUNet enhanced U-shaped' model design that intelligently incorporates a recursive gated convolutional (gnConv) mechanism to better understand complex spatial relationships. Narayanan et al. [23] discusses the use of hybrid SegNet namely 'IARS SegNet' for melanoma segmentation. Zhong et al. [24] used 'Dual-Stage U-Net' which is based on CNN and Transformer for skin lesion segmentation. Hao et al. [25] used arithmetic optimization algorithm for multithreshold image segmentation in dermoscopic skin images. By integrating these advancements, the future scope is to ensure accurate multi class skin cancer detection using lightweight models that can easily be integrated into mobile devices.

Acknowledgments. The authors express their appreciation to 'Indira Gandhi Delhi Technical University for Women' for fostering a supportive research environment. They also appreciate the valuable feedback from reviewers that contributed to improving this work.

Author Contributions. Ananya Gupta designed and conducted the experiments which involved data augmentation, image preprocessing, segmentation, finetuning pretrained models used in this analysis, and analyzing the data and results. Kashish Bansal contributed to data analysis and preprocessing, carried out a thorough literature review, and worked on writing the manuscript. Arti supported the project by collecting data and contributing to the manuscript preparation. Throughout the process, they worked closely with S.R.N. Reddy, Rishika Anand and Aditi Sabharwal who provided valuable guidance and constructive feedback to improve the manuscript.

Funding Sources. This analysis was carried out independently without any financial support from commercial, public, or non-profit organizations.

Disclosure of Interests. The authors confirm that they have no financial or personal conflicts of interest to declare.

References

1. Jerant, A.F., Johnson, J.T., Sheridan, C.D., Caffrey, T.J.: Early detection and treatment of skin cancer. Am. Fam. Phys. **62**(2), 357–368 (2000)
2. Malo, D.C., Rahman, M.M., Mahbub, J., Khan, M.M.: Skin cancer detection using convolutional neural network. In: 2022 IEEE 12th Annual Computing and Communication Workshop and Conference (CCWC), pp. 0169–0176. IEEE (2022)
3. Chaturvedi, S.S., Tembhurne, J.V., Diwan, T.: A multi-class skin Cancer classification using deep convolutional neural networks. Multimedia Tools Appl. **79**(39), 28477–28498 (2020)
4. Akter, M.S., Shahriar, H., Sneha, S., Cuzzocrea, A.: Multi-class skin cancer classification architecture based on deep convolutional neural network. In: 2022 IEEE International Conference on Big Data (Big Data), pp. 5404–5413. IEEE (2022)
5. Heenaye-Mamode Khan, M., et al.: Multi-class skin problem classification using deep generative adversarial network (DGAN). Comput. Intell. Neurosci. **2022**(1), 1797471 (2022)
6. Kausar, N., et al.: Multiclass skin cancer classification using ensemble of fine-tuned deep learning models. Appl. Sci. **11**(22), 10593 (2021)

7. Chaturvedi, S.S., Gupta, K., Prasad, P.S.:. Skin lesion analyser: an efficient sevenway multi-class skin cancer classification using MobileNet. In: Hassanien, A., Bhatnagar, R., Darwish, A. (eds.) Advanced Machine Learning Technologies and Applications. AMLTA 2020. AISC, vol. 1141. Springer, Singapore (2021). https://doi.org/10.1007/978-981-15-3383-9_15

8. Naeem, A., Anees, T.: DVFNet: a deep feature fusion-based model for the multiclassification of skin cancer utilizing dermoscopy images. PLoS ONE **19**(3), e0297667 (2024)

9. Arshed, M.A., Mumtaz, S., Ibrahim, M., Ahmed, S., Tahir, M., Shafi, M.: Multi-class skin cancer classification using vision transformer networks and convolutional neural network-based pre-trained models. Information **14**(7), 415 (2023)

10. Tschandl, P., Rosendahl, C., Kittler, H.: The HAM10000 dataset, a large collection of multi-source dermatoscopic images of common pigmented skin lesions. Sci. Data **5**(1), 1–9 (2018)

11. Khan, A.H., Iskandar, D.A., Al-Asad, J.F., El-Nakla, S.: Classification of skin lesion with hair and artifacts removal using black-hat morphology and total variation. Int. J. Comput. Dig. Syst. **10**(1) (2021)

12. Alsahafi, Y.S., Elshora, D.S., Mohamed, E.R., Hosny, K.M.: Multilevel threshold segmentation of skin lesions in color images using coronavirus optimization algorithm. Diagnostics **13**(18), 2958 (2023)

13. Sahoo, P.K., Soltani, S.A.K.C., Wong, A.K.: A survey of thresholding techniques. Comput. Vis. Graph. Image Process. **41**(2), 233–260 (1988)

14. Ronneberger, O., Fischer, P., Brox, T.:. U-net: convolutional networks for biomedical image segmentation. In: Navab, N., Hornegger, J., Wells, W., Frangi, A. (eds.) Medical Image Computing and Computer-Assisted Intervention – MICCAI 2015. MICCAI 2015. LNCS, vol. 9351. Springer, Cham (2015). https://doi.org/10.1007/978-3-319-24574-4_28

15. Miradwal, S., Mohammad, W., Jain, A., Khilji, F.: Lesion segmentation in skin cancer detection using UNet architecture. In: Buyya, R., Hernandez, S.M., Kovvur, R.M.R., Sarma, T.H. (eds.) Computational Intelligence and Data Analytics. LNDECT, vol. 142. Springer, Singapore (2022). https://doi.org/10.1007/978-981-19-3391-2_25

16. Behera, N., Singh, A.P., Rout, J.K., Balabantaray, B.K.: Melanoma skin cancer detection using deep learning-based lesion segmentation. Int. J. Inform. Technol. 1–16 (2024)

17. Badrinarayanan, V., Kendall, A., Cipolla, R.: Segnet: a deep convolutional encoder-decoder architecture for image segmentation. IEEE Trans. Pattern Anal. Mach. Intell. **39**(12), 2481–2495 (2017)

18. Huang, G., Liu, Z., Van Der Maaten, L., Weinberger, K.Q.: Densely connected convolutional networks. In: Proceedings of the IEEE Conference on Computer Vision and Pattern Recognition, pp. 4700–4708 (2017)

19. He, K., Zhang, X., Ren, S., Sun, J.: Deep residual learning for image recognition. In: Proceedings of the IEEE Conference on Computer Vision and Pattern Recognition, pp. 770–778 (2016)

20. Sandler, M., Howard, A., Zhu, M., Zhmoginov, A., Chen, L.C.: Mobilenetv2: Inverted residuals and linear bottlenecks. In: Proceedings of the IEEE Conference on Computer Vision and Pattern Recognition, pp. 4510–4520 (2018)

21. Zoph, B., Vasudevan, V., Shlens, J., Le, Q.V.: Learning transferable architectures for scalable image recognition. In: Proceedings of the IEEE Conference on Computer Vision and Pattern Recognition, pp. 8697–8710 (2018)

22. Wu, R., et al.: MHorUNet: high-order spatial interaction UNet for skin lesion segmentation. Biomed. Signal Process. Control **88**, 105517 (2024)

23. Narayanan, V.S., Sikha, O.K., Benitez, R.: IARS SegNet: interpretable attention residual skip connection SegNet for melanoma segmentation. IEEE Access (2024)

24. Zhong, L., Li, T., Cui, M., Cui, S., Wang, H., Yu, L.: DSU-Net: dual-stage U-Net based on CNN and Transformer for skin lesion segmentation. Biomed. Signal Process. Control **100**, 107090 (2025)

25. Hao, S., et al.: A multi-threshold image segmentation method based on arithmetic optimization algorithm: a real case with skin cancer dermoscopic images. J. Comput. Des. Eng. qwaf006 (2025)

Enhanced Deep Learning Framework with Vision Transformer and Slime Mould Optimization for Lung Cancer Detection in CT Images

K. Muthulakshmi and T. Gopalakrishnan[✉]

Manipal Institute of Technology Bengaluru, Manipal Academy of Higher Education, Manipal, India
`muthulakshmi.mitblr2024@learner.manipal.edu, gopalakrishnan.t@manipal.edu`

Abstract. Lung cancer is the world's second biggest cause of cancer-related fatalities, and it poses a serious health risk due to uncontrolled cell proliferation in the lungs. Detecting lung cancer at an early stage remains difficult, especially when discriminating between different types of nodules on CT imaging. This research is intended to improve the accuracy and efficiency of lung cancer detection with deep learning techniques. A dataset of 1,000 CT scan pictures with cases of large cell carcinoma, squamous cell carcinoma, adenocarcinoma, and normal lungs were used. An enhanced deep learning hybrid model ViT-SMA Xpert was generated by combining a Vision Transformer (ViT) with an Enhanced Xception backbone, which was optimized using the Slime Mould Algorithm. The suggested model attained a test accuracy of 92% across 25 epochs, proving the efficacy of merging Vision Transformers with CNNs for lung cancer diagnosis. The findings verify the use of hybrid deep learning architectures for accurate and scalable lung cancer diagnosis, opening the way for future clinical applications.

Keywords: Deep Learning · Computed Tomography (CT) · Optimization

1 Introduction

According to World Health Organization (WHO) projections, lung cancer–which is characterized by unchecked lung cell growth–will rank as the second most common disease worldwide by 2020. Because cancer cells are aggressive, prevention and early detection are very difficult. Since the size and spread of the tumor indicate the cancer's stage, early diagnosis is essential for successful therapy [1]. Every year, millions of people globally, irrespective of gender or region, are affected by lung cancer, which continues to be one of the main causes of cancer. The improvement of survival rates depends on early identification. Computer-aided design (CAD) solutions have been developed to meet this requirement by

© The Author(s) 2026
J. Shreyas et al. (Eds.): CODE-AI 2025, CCIS 2689, pp. 243–255, 2026.
https://doi.org/10.1007/978-3-032-19318-6_23

lowering physician workloads and improving diagnostic efficiency and accuracy. The primary approaches for detecting malignant nodules are computed tomography (CT) scans and X-rays, however low-dose CT (LDCT) scans offer a helpful screening method for at-risk patients. However, occasionally these scans can yield false-positive results because benign and malignant nodules often appear similar in their early stages. Non-small cell lung cancer (NSCLC) can be detected more effectively using CT imaging than with LDCT because of its higher resolution and better tumor visualization. Automatic methods for making a diagnosis. The most common kind of lung cancer, non-small cell lung cancer (NSCLC), is divided into three subtypes: large cell carcinoma (LCC), squamous cell carcinoma (SCC), and adenocarcinoma (ACA) shown in Fig. 1 This study uses 1,000 histological lung cancer CT images to identify the different subtypes of NSCLC. On the other hand, although less frequent, small cell lung cancer (SCLC) has a greater death rate. Rapid diagnosis is made more difficult by symptoms such as chronic cough, hoarseness, breathing difficulties, weight loss, chest tightness, and respiratory problems that frequently appear in advanced stages. Before being underlying lung diseases. Cigarette smoking, exposure to radon gas, air pollution, and hazardous chemical exposure at work are important risk factors. Often, early detection of lung cancer is not possible using conventional diagnostic techniques. Deep learning algorithms are proposed as a more accurate and efficient way to diagnose lung cancer early on in order to get around these challenges.

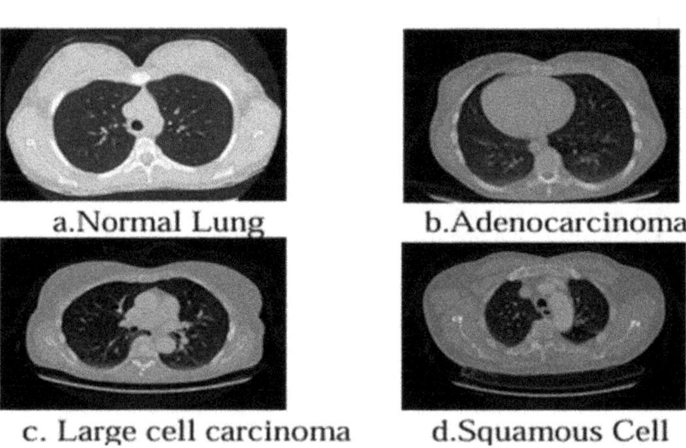

Fig. 1. Diseases of Lung Cancer

Based on artificial neural networks, deep learning is a subfield of machine learning that takes cues from the structure and functions of the human brain. Models developed with deep learning are excellent at analyzing large amounts of data, identifying complex patterns, and making decisions based on those patterns. These features make deep learning ideal for jobs like picture recognition, medical image analysis, and natural language processing. Picture recognition issues benefit greatly from deep learning. These properties make deep learning very useful for image recognition, natural language processing, and most importantly, medical image analysis challenges. In medical imaging, the DL method has been proved to work when combined with CT scan pictures. Large volumes of CT data can be used with deep learning, a potent technique that consistently and accurately extracts, detects, and classifies cancer cells. DL algorithms can identify and evaluate patterns that indicate the presence of lung cancers as well as predict their presence by training on a significant number of CT image data.

2   Literature Review

Models for detecting lung cancer that use transfer learning have shown increased robustness and performance [2]. The best-performing transfer learning models were VGG16, which produced good classification results. On the original dataset, it achieved an accuracy of 81.42%, and on the updated dataset, 91.64%. [3]. To increase efficiency and accuracy in early lung cancer diagnosis, deep learning approaches are recommended. Computer recognition algorithms based on transfer learning have shown promise in enhancing radiologists' expertise [4]. For both classification and segmentation problems, pre-training can be a very useful tactic for improving the Swin Transformer model's accuracy [5]. Ensemble approaches provide a powerful means of improving Writing with AI accuracy of submission by combining the advantages of several classifiers to reduce their shortcomings, leading to improved performance overall [6]. To help radiologists make decisions, this work proposes an automated computer-aided diagnosis (CAD) method for lung nodule detection that uses the Vision Transformer architecture supplemented with Bayesian Optimization [7]. The framework, with its course based on morphology and attentiveness, uses lung cancer histopathology images to precisely and successfully capture the morphological variations of lung cancer subgroups [8].

To differentiate between benign and cancerous lung tumors, deep learning is employed to address problems with small data in medical picture classification [9]. Use cutting-edge deep learning techniques to increase the accuracy and efficacy of lung nodule categorization [11]. Pictures from the non-small cell lung cancer CT scan dataset to forecast the occurrence of lung cancer [12]. To better comprehend CAD technology for pulmonary cancer identification, categorization, classification, and retrieving, radiologists and researchers are working together [13] to introduce a condensed convolutional transformer designed to detect cancer to use a small number of factors and get excellent classification accuracy [14] collecting datasets, honing photos, using segmentation strategies, selecting the best feature extraction and selection strategies, evaluating metrics, and putting classifiers into practice [15]. This method reduces variance and avoids overfitting to ensure a precise diagnosis of lung cancer. The Deep CNN with Dual State Transfer Learning (DSTL) utilizing ResNet50 helps with autonomous data classification by achieving high accuracy [16] (Table 1).

Table 1. Overview of related works and Research gaps in Lung Cancer Detection

Author	Objective	Methodology	Research Gap
Kumar, Vinod, et al. 2024	Investigates how well the Fusion Model diagnoses Lung cancer.	ResNet-50, ResNet-101, and EfficientNet-B3	One or two indicators, which is insufficient to assess efficacy and accuracy.
Ezhilraja, K., et al. 2024	Aims to diagnose NSCLC using deep transfer learning-based pre-trained models	Pretrained Models, Multi Level Dualistic Sub Image Histogram Equalization	Generalizability is impacted by its limited testing on various datasets.
Sun, Ruina, et al. 2023	An accurate segmentation technique for classifying lung cancer.	Swin transformer	The scalability of the experiment was limited because models were trained on a single GPU.
Quasar, Syeda Reeha, et al. 2023	Accurately diagnosing lung cancer, by promoting early detection and lowering false positives.	BEiT, DenseNet, and Sequential CNN.	Possible overfitting and restricted generalization to complicated variations.
Mkindu, Hassan, et al. 2023	Optimizing hyperparameters to gain better lung nodule detection performance.	Vision Transformer, Bayesian Optimization	Restricted generalization to complicated variations
Halder, Amitava, et al. 2023	The study attempts to create a deep learning framework for classifying lung cancer subtypes.	Morphology-based Attention Network	In medical image segmentation, identifying suitable techniques and pertinent features is a difficult issue.
Wankhade, S., et al. 2023	Proposes a novel method for early and accurate lung cancer diagnosis	Cancer Cell Detection using Hybrid Neural Network (CCDC-HNN)	Possible overfitting and restricted generalization to complicated variations.
Xie, Y., et al. 2019	The SSAC model avoids parameter sharing and uses both labeled and unlabeled data to enhance training.	SSAC model	Due to the scarcity of data and the difficulty of annotation, deep learning in medical imaging encounters difficulties
Xie, H., et 2018	Precisely identify the pulmonary nodules that are latent.	2D Convolutional neural network	Need to include additional nodule background information.
Atiya, S. U., et al. 2022	The research aims to classify and detect lung cancer.	Dual-state transfer learning	The suggested method's generalizability is impacted by its limited testing on various datasets.
Proposed Model	Developed a hybrid model to improve classification accuracy of lung cancer	ViT-SMA Xpert	Limited testing on various datasets impacted generalizability.

3 Proposed Methodology

The development of vision transformers (ViTs) and convolutional neural networks (CNNs) is crucial to the identification of lung cancer. CNNs are highly proficient in lung due to their robust spatial feature extraction capabilities, which stem from their architecture of fully connected layers, pooling, and convolution, from CT scans. The functional block diagram of the proposed model is described in Fig. 2.   Architecture for Hybrid Model ViT-SMA Xpert

A. Data Preprocessing To prepare clear and well-contrasted CT scan pictures, preprocessing is essential. The given Eq. 1 will Eliminates the noise using Wavelet Transformers analyse lung CT scans for noise and break them down into different frequency components. Utilize thresholding strategies to modify wavelet coefficients and eliminate noise. After removing the noise, reconstruct the image. Dynamic Histogram Equalization improves visual contrast in CT images by adjusting the distribution of pixel intensity is given in the Eq. 2. Vision Transformer (ViT) with CNN to capture global content across CT scans

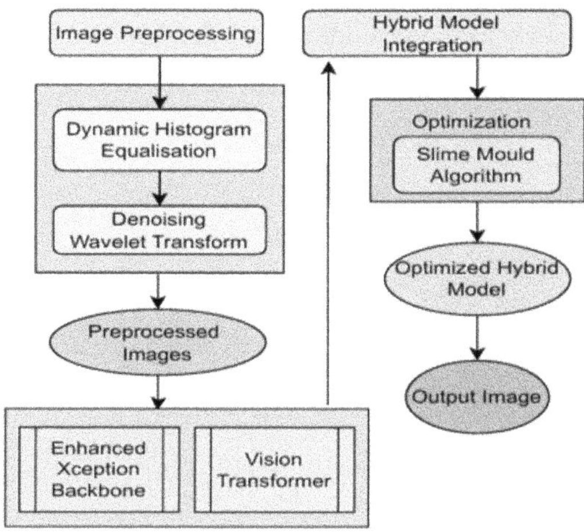

Fig. 2. Functional Block Diagram of proposed model

and to preserve local feature details, integrate it using an Enhanced Xception backbone.

$$X^{(a,b)} = \begin{cases} X(a,b), & \text{if } |X(a,b)| > T \\ 0, & \text{otherwise} \end{cases} \tag{1}$$

where, $X^{(a,b)}$: Denoised pixel at position (a,b)
$X(a,b)$: Original wavelet coefficient
T : Threshold determined

$$I_{\text{new}}(a,b) = \text{CDF}(I(a,b)) \cdot (L-1) \tag{2}$$

where, $I(a,b)$: Actual pixel intensity
$I_{\text{new}}(a,b)$: Enhanced pixel intensity
CDF : Image's cumulative histogram
L : Number of intensity levels

B. Feature Extraction This method takes CT scans and uses Enhanced Xception to extract spatial characteristics. To collect local features, feed input CT images through the Enhanced Xception backbone. These attributes are fed into the Vision Transformer to extract context based on global attention.
C. Vision transformer for global Context Patches are created from the feature maps and to process each patch, the Vision Transformer is utilized for every patch, linear embeddings are made have shown in the Eq. 3 and Eq. 4. Relationships between patches are captured by multi-head self-attention. After combining the outputs, a global representation is created. Visual Representation can be done by input feature map broken down patch by patch. Attention-layered transformer blocks that given in Eq. 5 show how patches are connected. By combining the global context modelling capabilities of ViTs with the local extraction of fea-

tures of CNNs, the hybrid models offer a dependable and all-inclusive approach that significantly increases the efficacy and reliability of lung cancer prediction.

$$\text{Patch}_a = \text{flatten}(X_{[l,b]}) \tag{3}$$

Where,

$-\ X_{[l,b]}$: Patch of size $l \times b$ extracted from the image.
• Patches in fixed size:

$$F = W_x \cdot \text{Patch}_a + b_x \tag{4}$$

where W_x: Weight matrix- b_x: Bias term
F: Patch vector embeded

• Multiheaded Special Attention:

$$\text{Attention}(X, Y, Z) = \text{softmax}\left(\frac{XY^T}{\sqrt{d_k}}\right) Z \tag{5}$$

Where, X,Y,Z: Query, Key and value matrix P D. Parameter Tuning The Eq. 6 utilize the Slime Mould Algorithm to optimize the following hyperparameters: learning rate, batch size, and optimizer parameters. The Optimizer is a quicker convergence and fast training dynamics that are more stable than those of more conventional optimizers like AdamW and RMSprop. Slime Mould Optimization

$$A_I^{(t+1)} = \begin{cases} \text{Best}(A^t) + P \cdot S, & \text{if } r < p \\ A_i^t + \text{Rand}(-1, 1) \times (BU - BL), & \text{otherwise} \end{cases} \tag{6}$$

where $A_I^{(t+1)}$, iteration t+1.
$Best(At)$: Best solution
P:Spiral operator

Developing precise and effective deep learning models for proposed work requires adjusting hyperparameters. The brief description of the hyperparameters given in
Convolution Layers: This method improves feature extraction while lowering processing overhead by employing 36 depth wise separable convolutional layers.
Pooling layers: which are followed by a global average pooling layer, aid in reducing dimensionality and capturing global characteristics.
Training sample sets: 80% of the dataset, or 800 images, are used to train the model, giving it a significant amount of data for learning.
Testing Sample Sets: To evaluate the model's capacity for interpreting new, unidentified data, 200 images (20% of the dataset) are used. Total Images (1000 CT scan images): This dataset provides a varied collection for testing and training because it includes 1000 CT scan images.
Vision Transformer Depth (12): By capturing intricate relationships in the data,

---

Algorithm 1. ViT-SMA Xpert Algorithm

Require: Lung cancer dataset D, Image size $(128, 128)$, hyperparameters: d_{model}, num_{heads}, num_{layers}, epochs, $batch_{size}$, lr

Ensure: Trained Model with Optimized Parameters

1: **Step 1: Load and Preprocess Data**
2: Load images from directory D
3: **for** each image in D **do**
4:    Resize to $(128, 128)$
5:    Normalize pixel values to $[0, 1]$
6: **end for**
7: Encode class labels using one-hot encoding
8: **Step 2: Split Data**
9: Split data into training and testing sets
10: **Step 3: Define Vision Transformer with Xception Backbone**
11: Initialize Xception as a feature extractor (pre-trained on ImageNet)
12: Extract feature maps
13: Reshape extracted feature maps into patch sequences
14: Apply Transformer layers:
15:    Multi-head Attention
16:    Feedforward Network
17: Perform Global Average Pooling
18: Apply Fully Connected Softmax layer
19: **Step 4: Hyperparameter Optimization using Slime Mould Algorithm (SMA)**
20: **for** each iteration in num_{iter} **do**
21:    Randomly select $batch_{size} \in \{16, 32, 64\}$
22:    Randomly select learning rate $lr \in \{0.0001, 0.001, 0.01\}$
23:    Compile model with Adam optimizer (lr)
24:    Train model for $epochs$ using training data
25:    Evaluate model accuracy on test data
26:    **if** accuracy is best **then**
27:      Update best hyperparameters
28:    **end if**
29: **end for**
30: **Step 5: Train Final Model with Optimized Hyperparameters**
31: Retrain model using best $batch_{size}$ and lr
32: Apply ImageDataGenerator for data augmentation
33: Train model with:
34:    Early stopping
35:    Learning rate scheduling
36: **Step 6: Evaluate Performance**
37: Compute final test accuracy
38: Generate Confusion Matrix
39: Generate Classification Report
40: Plot Accuracy and Loss Curves

---

a 12-layer transformer model improves feature representation (Table 2).

Table 2. HYPERPARAMETER VALUE

Hyperparameter	Value
Number of Epochs	25
Convolution Layers	36 [Depthwise separable layers]
Pooling Layers	3 + 1 Global Average Pooling (GAP)
Batch SizeKernel Size	[(3, 3), (5, 5)]
Filter Size	[64, 128, 256]
Stride Size	(1, 1)
Padding	Zero-padding
Training Sample Sets	800 images (80% of the dataset)
Testing Sample Sets	200 images (20% of the dataset)
Total Images	1000 CT scan images
Loss Function	Cross-Entropy Loss
Patch Size	16 × 16
Number of Attention Heads	12
Xception Backbone Filters	64, 128, 256
Slime Mould Iterations	100

E. Evaluation The model that was learned is evaluated on the test set to determine the accuracy, loss, confusion matrix, and performance parameters. Performance indicators are shown, and the model is evaluated using these metrics shown in the Eq. 7. Cross Loss Entropy:

$$L = -\frac{1}{R}\sum_{i=1}^{R}\sum_{c=1}^{D} y_a^c \log(z_a^c) \tag{7}$$

R: No of Samples
D: No of classes
y_a^c:Truth label ground
z_a^c: Probability of Predicted classes.

4   Experimental Results and Discussion

The development of identification and categorization models requires the collecting of data. This part discusses the NSLC dataset that is now being utilized in research on deep learning techniques for lung cancer diagnosis. Four classes were

Table 3. Performance Metrics Equations

Performance Metrics	Formula
Accuracy	$\frac{TP+TN}{TP+TN+FP+FN}$
Precision	$\frac{TP}{TP+FP}$
Recall	$\frac{TP}{TP+FN}$
F1 Score	$\frac{2 \times Precision \cdot Recall}{Precision+Recall}$

identified from the 1,000 CT scan images: normal cells, big cell carcinoma, adenocarcinoma, and squamous cell cancer. The dataset was split into three subsets for training, testing, and validation: 80% for training and 20% for testing. The proposed model's test accuracy and test loss over 25 epochs are visualized as follows: The left graph Fig. 3 illustrates how test accuracy has steadily increased over time. The right graph Fig. 4 shows how test loss decreases as the model gains knowledge during training.

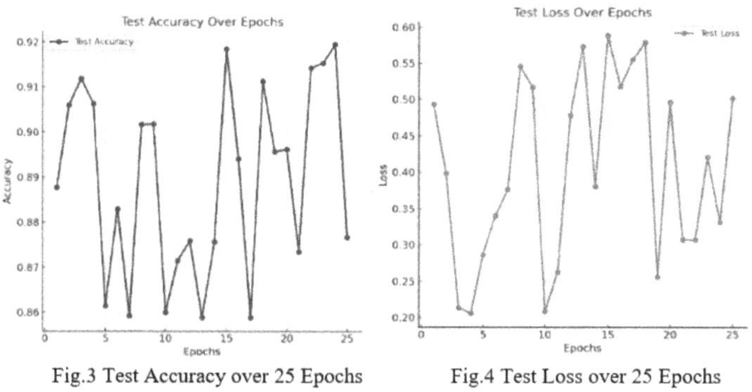

Fig.3 Test Accuracy over 25 Epochs   Fig.4 Test Loss over 25 Epochs

Fig. 3. Test Accuracy and Loss of 25 Epochs

The performance metrics F1-score, recall, accuracy, and precision are measured for the four classes. The proposed approach successfully detects lung cancer ViT-SMA Xpert, attains the 92% test accuracy and balanced performance across healthy and malignant lung types shown in Table 3. Important discoveries include strong parameter tweaking through SMA and effective feature extraction with an Enhanced Xception backbone, which guarantee quicker convergence and better outcomes. The creative hybrid design is sophisticated preprocessing (dynamic histogram equalization, denoising wavelet transform) and excellent interpretability and generalizability through attention maps. One approach for assessing how well your classification model is working is the confusion matrix shown in Fig. 5.

The dataset's actual classes are represented by the True Labels (Y-axis). The model's predictions are represented by the predicted labels (X-axis). Classes are

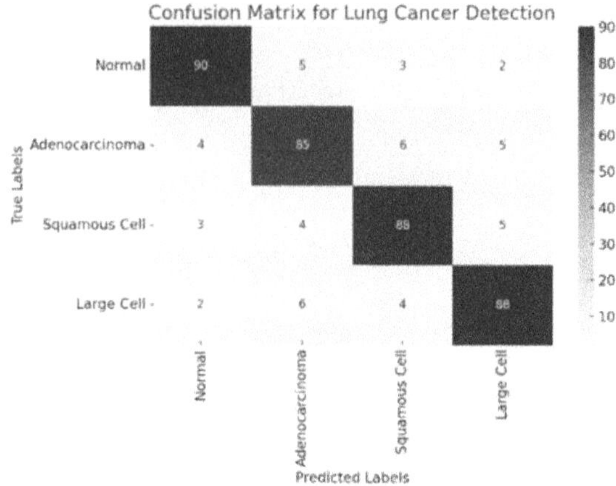

Fig. 4. Confusion Matrix

Table 4. Preformance Metrics

Classes	Precision	Recall	F1 Score	Support
Normal	0.92	0.95	0.93	190
Adenocarcinoma	0.93	0.91	0.92	190
Squamous Cell	0.91	0.88	0.89	185
Large Cell	0.92	0.94	0.93	195
Macro Avg	0.92	0.92	0.92	0.760
Weighted Avg	0.92	0.91	0.92	0.760

used as Typical Lung Function, Adenocarcinoma, Squamous Cell Cancer, Large Cell Carcinoma, Matrix Dissection. The number of images in each category is represented by the matrix entries: Correct predictions (e.g., Normal lungs identified as Normal) are examples of diagonal elements. Issues with dataset size and actual variability. The sensitivity and specificity in Fig. 6 show the high-performance metrics. Whereas the ROC curve in Fig. 6 shows how well the proposed methodology performs well. For clinical application, explainability needs to be improved. Prioritize using tools like Grad-CAM to improve interpretability, condense the model for effective deployment, integrate it into clinical workflows, and validate the model on a variety of datasets. Tables 3 and 4 shows the performance metrics and the comparison of the proposed model with the other pretrained models that are used for detection of Lung cancer (Table 5).

5   Conclusion

With its high accuracy and reliable performance, this work demonstrates the promise of ViT-SMA Xpert for lung cancer detection. The features are extracted

Table 5. Comparison of proposed model with other Deep Learning Methods used in LIDC-IRDA CT Lung Cancer Datasets

Authors	Methodology	Accuracy
Ezhilraja, K., et al., 2024	VGG16 ML-DSIHE	91.64%
Ramesh, J. V. N., et al.	VGG16	84.76%
Atiya, S. U., et al.	INCEPTION VGG16	91.02% 88.09%
Zhao, X., et al.	DCNN	91.12%
Proposed Model	ViT-SMA Xpert	92.00%

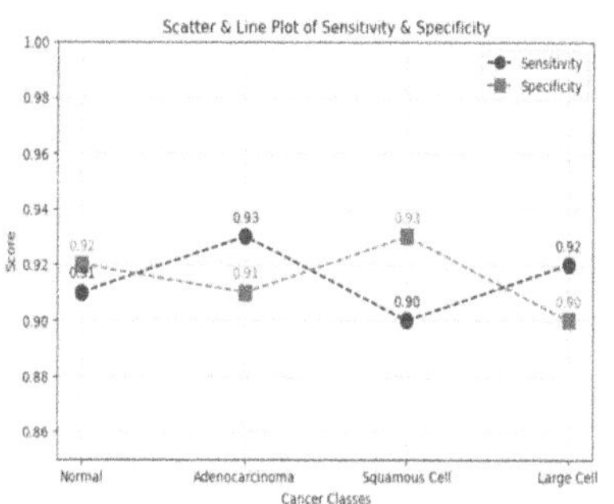

Fig. 5. Performance Metrics

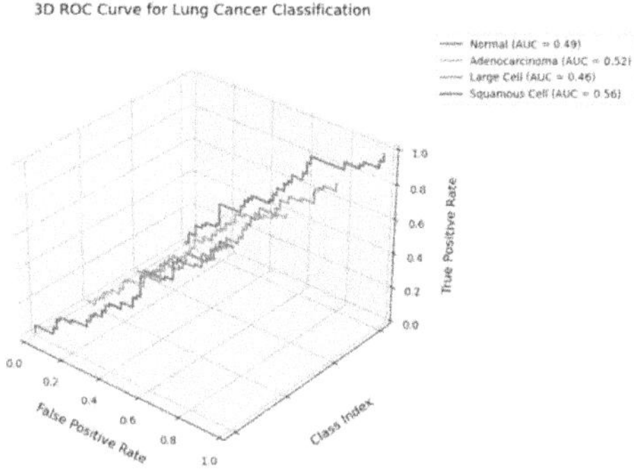

Fig. 6. ROC Curve

using attention mechanism to show higher classification accuracy. The hyperparameters are fine-tuned by slime mould algorithm to improve accuracy. The generalisation and computability are enhanced using data augmentation and early stopping. Reviewed the state-of-the-art deep learning CAD (computer-aided design) tools for diagnosing and detecting lung nodules from CT data. Comprehensive analysis that explores the use of proposed methods in relation to applications involving LC and comparison of the technologies using the LIDC-IDRI. This strategy could be an imperative tool in helping doctors diagnose lung cancer early and accurately if the drawbacks are fixed and potential improvements are investigated. Further improvements can be made by fine tuning and ensemble on larger datasets for enhancing real world applications.

References

1. Sung, H., et al.: Global Cancer Statistics 2020: GLOBOCAN estimates of incidence and mortality worldwide for 36 cancers in 185 countries. Cancer J. Clin. **71**(3), 209–249 (2020). https://doi.org/10.3322/caac.21660
2. Kumar, V., et al.: Unified deep learning models for enhanced lung cancer prediction with ResNet-50–101 and EfficientNet-B3 using DICOM images. BMC Med. Imag-

ing **24**(1), 2024 (2024). https://doi.org/10.1186/s12880-024-01241-4. Accessed 19 Jun 2024

3. Ezhilraja, K., Shanmugavadivu, P.: Performance assessment of deep learning models on non-small cell lung cancer type classification. EAI Endorsed Trans. Pervasive Health Technol. **10** (2024b). https://doi.org/10.4108/eetpht.10.6423

4. Ramesh, J.V.N., Agarwal, R., Deekshita, P., Elahi, S.A., Bindu, S.H.S., Pavani, J.S.: Application of several transfer learning approaches for early classification of lung cancer. EAI Endorsed Trans. Pervasive Health Technol. **10** (2024b). https://doi.org/10.4108/eetpht.10.5434

5. Sun, R., et al.: Efficient lung cancer image classification and segmentation algorithm based on an improved swin transformer. Electronics **12**(4), 1024 (2023). www.mdpi.com/2079-9292/12/4/1024, https://doi.org/10.3390/electronics12041024. Accessed 8 Aug 2023

6. Quasar, S.R., et al.: Ensemble methods for computed tomography scan images to improve lung cancer detection and classification. Multimedia Tools Appl. **83**(17), 52867–52897 (2023). https://doi.org/10.1007/s11042-023-17616-8. Accessed 22 Sept 2024

7. Mkindu, H., et al.: Lung nodule detection in chest CT images based on vision transformer network with bayesian optimization. Biomed. Signal Process. Control **85**, 104866–104866 (2023). https://doi.org/10.1016/j.bspc.2023.104866. Accessed 23 Nov 2023

8. Halder, A., Dey, D.: MorphAttnNet: an attention-based morphology framework for lung cancer subtype classification. Biomed. Signal Process. Control **86**(105149–105149), 2024 (2023). https://doi.org/10.1016/j.bspc.2023.105149. Accessed 22 Sept 2023

9. Wankhade, S., Vigneshwari, S.: A novel hybrid deep learning method for early detection of lung cancer using neural networks. Healthc. Anal. **3**, 100195 (2023). https://doi.org/10.1016/j.health.2023.100195

10. Xie, Y., Zhang, J., Xia, Y.: Semi-supervised adversarial model for benign–malignant lung nodule classification on chest CT. Med. Image Anal. **57**, 237–248 (2019). https://doi.org/10.1016/j.media.2019.07.004

11. Xie, H., Yang, D., Sun, N., Chen, Z., Zhang, Y.: Automated pulmonary nodule detection in CT images using deep convolutional neural networks. Pattern Recogn. **85**, 109–119 (2018). https://doi.org/10.1016/j.patcog.2018.07.031

12. Naik, A., Edla, D.R.: Lung nodule classification on computed tomography images using deep learning. Wireless Pers. Commun. **116**(1), 655–690 (2020). https://doi.org/10.1007/s11277-020-07732-1

13. Shakeel, P.M., Burhanuddin, M.A., Desa, M.I.: Automatic lung cancer detection from CT image using improved deep neural network and ensemble classifier. Neural Comput. Appl. **34**, 9579–9592 (2022). https://doi.org/10.1007/s00521-020-04842-6

14. Gu, Y., et al.: A survey of computer-aided diagnosis of lung nodules from CT scans using deep learning. Comput. Biol. Med. **137**, 104806 (2021). https://doi.org/10.1016/j.compbiomed.2021.104806

15. Atiya, S.U., Ramesh, N.V.K., Reddy, B.N.K.: Classification of non-small cell lung cancers using deep convolutional neural networks. Multimedia Tools Appl. **83**(5), 13261–13290 (2023). https://doi.org/10.1007/s11042-023-16119-w

16. Mridha, M.F., et al.: A comprehensive survey on the progress, process, and challenges of lung cancer detection and classification. J. Healthc. Eng. **2022**, 1–43 (2022). https://doi.org/10.1155/2022/5905230

Cellular Automata Based Brain Tumor Segmentation from Magnetic Resonance Images

K. Indrakumar[(✉)] and M. Ravikumar

Department of Computer Science, Kuvempu University, Shimoga, Karnataka, India
indk214@gmail.com

Abstract. This paper presents a novel methodology for brain tumor segmentation in MR images, integrating K-Means clustering with Cellular Automata (CA) to improve segmentation accuracy and spatial uniformity. The proposed approach begins with MR image acquisition, grayscale conversion and clustering using the K-Means algorithm to segment tumor regions based on intensity similarities. To overcome the limitation of fragmented boundaries in K-Means, Cellular Automata is applied to refine segmentation by including spatial relationships. The segmentation process iteratively adjusts the number of clusters and applies morphological post-processing techniques, including erosion and dilation, for boundary enhancement. The proposed method is evaluated using quality metrics such as MSE, RMSE, PSNR, SNR, SSIM and CNR, providing insights into segmentation accuracy, structural integrity and contrast clarity. Radar charts visually demonstrate performance across multiple test cases. The results exhibit that fusion of K-Means and Cellular Automata improves tumor boundary detection and segmentation quality, making this approach valuable for accurate tumor diagnosis in clinical settings.

Keywords: K-Means clustering · Cellular Automata · MR images · Erosion · Dilation

1 Introduction

Brain tumors are one of the most severe neurological disorders, formed by abnormal cell growth in the brain that can be either benign or malignant [2, 3]. Early detection and accurate segmentation of brain tumors using Magnetic Resonance Imaging (MRI) play a crucial role in diagnosis and treatment planning [4, 5]. MRI is widely used due to its non-invasive nature, high soft tissue contrast and ability to provide detailed anatomical information about the brain [6, 7]. However, detecting and segmenting tumors from MR images remain challenging due to variations in tumor size, shape, intensity and location. The complexity increases further due to the presence of noise and intensity variations in MRI scans [9].

Brain tumor incidence and mortality rates have been rising globally, with malignant tumors such as glioblastomas exhibiting particularly high mortality rates [10]. The challenges associated with brain tumor segmentation are primarily due to the heterogeneous nature of tumors, which exhibit irregular growth patterns and varying intensity

J. Shreyas et al. (Eds.): CODE-AI 2025, CCIS 2689, pp. 256–269, 2026.
https://doi.org/10.1007/978-3-032-19318-6_24

distributions across different MR imaging modalities. Manual segmentation by radiologists is time-consuming, prone to human error and highly dependent on expertise [11, 12]. Subsequently, automated and semi-automated segmentation techniques have gained importance in medical image analysis, significantly helping in exact tumor localization and volume estimation [13, 14].

K-means clustering, a popular unsupervised machine learning algorithm, has contributed significantly to brain tumor segmentation by effectively classifying pixels based on intensity similarities [3]. The algorithm works by iteratively partitioning the image into clusters and assigning pixels to the nearest cluster centroid, making it useful for segmenting tumors from surrounding tissues [3]. Although K-means is computationally efficient and straightforward, its major limitations include sensitivity to initialization, difficulty in handling overlapping intensities and dependency on the pre-defined number of clusters. Even though these drawbacks, K-means clustering has been widely used in hybrid approaches, often combined with other techniques such as morphological processing and deep learning to enhance segmentation accuracy.

Cellular Automata (CA) is another robust computational technique that has been successfully applied in brain tumor segmentation [15]. CA consists of a grid of cells, each following predefined transition rules based on the states of its neighboring cells. The fundamental rules of cellular automata in tumor segmentation include:

Initialization Rule:
Assign initial states to cells based on pixel intensity thresholds.

$$\text{If } I\ (x, y) > T \ \rightarrow\ \text{Cell} = \text{Tumor} \tag{1}$$

$$\textit{Else} \rightarrow\ \text{Cell} = \text{Background} \tag{2}$$

where $I\ (x, y)$ is the intensity of the pixel and T is a predefined threshold.

Neighborhood Rule (Local Interaction):
Update cell states based on neighboring pixels (Moore neighborhood).
If most neighbors belong to the tumor class, the current cell may transition to a tumor state, ensuring smooth region growth.

$$\textit{If} \text{If} \ \geq 5 \text{ of } 8 \text{ neighbors are Tumor} \rightarrow\ \text{Cell} = \text{Tumor} \tag{3}$$

$$\textit{Else} \rightarrow\ \text{Cell} = \text{Background} \tag{4}$$

Growth Rule:
If neighboring cells have high intensity or texture similarity, expand the tumor region. This helps in capturing tumor boundaries accurately.

Shrinkage Rule:
If a tumor cell is surrounded largely by background cells or shows intensity variations inconsistent with tumor texture, it reverts to background.

$$\textit{If} \leq 2 \text{ neighbors are Tumor} \rightarrow\ \text{Cell} = \text{Background} \tag{5}$$

Convergence Rule (Stopping Condition):
Stop updating when changes between successive iterations fall below a predefined threshold.

If the percentage of state changes in the current iteration is less than 1%, stop the segmentation.

The efficiency of Cellular Automata in brain tumor segmentation lies in its ability to model complex spatial patterns and dynamically evolve segmentation boundaries [15]. Unlike traditional methods that rely on static clustering, CA-based approaches effectively capture spatial dependencies and ensure smoother segmentation results. Moreover, CA algorithms can be integrated with machine learning and deep learning techniques to improve segmentation accuracy and strength [15]. The method's ability to adapt to different tumor shapes and sizes makes it highly effective in segmenting both well-defined and distributive tumor regions. Still, CA models require careful parameter tuning and rule selection to achieve optimal segmentation performance, while methods like K-means clustering offer simplicity and efficiency in tumor segmentation, Cellular Automata provide a more dynamic and adaptable approach. Combining these techniques can further enhance the accuracy and reliability of automated brain tumor segmentation systems, eventually contributing to improved diagnosis and patient outcomes.

1.1   Challenges of K-Means in Brain Tumor Segmentation

1. **Choosing the Optimal k Value:** K-Means requires specifying the number of clusters (k) early. An inappropriate k value can result in under-segmentation (merging tumor with healthy tissue) or over-segmentation (fragmented tumor regions).
2. **Sensitivity to Initialization:** The algorithm's results depend on the initial placement of centroids. Poor initialization can lead to local optima, resulting in inconsistent segmentation outcomes.
3. **Inability to Capture Spatial Relationships:** K-Means clusters pixels based only on their intensity values, ignoring spatial information. As a result, it may produce noisy or fragmented boundaries, especially in complex tumor regions.
4. **Difficulty with Overlapping Intensity Values:** Tumor and healthy tissues can have similar intensity levels in MRI images. K-Means may fail to differentiate between them, leading to inaccurate segmentation.

To overcome these challenges, this study combines K-Means with Cellular Automata (CA) to improve segmentation accuracy and boundary clarity. For validation of our proposed methodology, we use the publicly available Br35H dataset from Kaggle [16].

Please note that the first paragraph of a section or subsection is not indented. The first paragraphs that follows a table, figure, equation etc. does not have an indent, either. Subsequent paragraphs, however, are indented.

1.2   Contributions of This Study

1. **Integration of K-Means and Cellular Automata (CA):** The study combines K-Means clustering with Cellular Automata to improve segmentation accuracy and maintain spatial consistency.

2. **Automated Tumor Segmentation:** Provides an efficient and automated methodology for detecting and segmenting brain tumors from MRI scans.
3. **Adaptive Cluster Refinement:** The proposed approach iteratively adjusts the number of clusters (k) to achieve optimal segmentation.
4. **Spatial Consistency Enhancement:** CA ensures smoother tumor boundaries and reduces misclassified pixel regions.
5. **Post-processing for Improved Accuracy:** Morphological operations such as erosion and dilation refine the segmentation and enhance tumor contour clarity.
6. **Quantitative Evaluation:** The method is validated using quality metrics to ensure segmentation reliability.

This paper is organized into five sections. Section 2 presents a literature review of existing research on brain tumor detection, highlighting deep learning techniques integrated with cellular automata and optimization methods. It highlights previous approaches, their strengths and limitations, providing a strong foundation for the proposed methodology. In Sect. 3, the methodology for brain tumor segmentation is detailed, using a hybrid approach that combines K-means clustering with cellular automata to improve segmentation accuracy. Section 4 presents the results and discussion, analyzing the performance of the proposed method with quantitative metrics and visual comparisons. Finally, Sect. 5 offers the conclusion, summarizing key outcomes and contributions of the study.

2 Literature Review

In this section we review existing research on brain tumor detection, focusing on deep learning techniques with cellular automata and optimization methods.

The study [1] presents a hybrid approach for breast tumor segmentation by integrating M3D-Neural Cellular Automata (M3D-NCA) with Shape-Guided Segmentation (SGS) to enhance segmentation accuracy. The M3D-NCA framework models tumor growth and segmentation using a biologically inspired cellular automata approach, while SGS refines segmentation boundaries by joining shape priors. This hybrid model effectively balances exploration and exploitation in segmentation, reducing the dependence on extensive training data while maintaining high accuracy. Additionally, by eliminating unnecessary convolutional layers, the proposed method optimizes computational efficiency, making it suitable for large-scale medical imaging applications. The proposed M3D-NCA+SGS model achieved the highest Dice Score (62.71%) and mIoU (54.10%) on INbreast dataset, while also securing 82.13% accuracy on DBT dataset and 83.76% accuracy on CBIS-DDSM dataset.

The proposed methodology in the study [2] integrates Bacteria Foraging Optimization Algorithm (BFOA) with Learning Automata (LA) to enhance the classification performance of a Convolutional Neural Network (CNN) for brain tumor detection. The segmentation is performed using Otsu thresholding, which effectively separates tumor regions from MRI images. The BFOA optimizes hyperparameters, while LA automates learning to refine classification accuracy. This hybrid approach addresses challenges such as computational efficiency and convergence speed, ensuring detailed segmentation and classification. Experimental results on the BRATS dataset demonstrate the superiority of the proposed BFOA-LA-CNN model, achieving an accuracy of 99.41%, outperforming existing methods like DCNN-G-HHO (97%) and IIB-based Deep Residual Network (91.9%). This proposed method also improves precision (99.18%), recall (98.80%) and F1-score (97.65%), highlighting its effectiveness in tumor classification. The integration of BFOA and LA significantly enhances CNN's learning efficiency, making it a robust and efficient model.

In study [3], authors proposed a notable approach which uses Firefly Optimization (FFO) for segmentation, using intra-cluster distance, inter-cluster distance and entropy to refine image partitioning, with K-means clustering providing initial segmentation points. Additionally, Local Binary Patterns (LBP) are extracted to capture texture-based features, further helping classification. A parallel CNN architecture combining 1D and 2D CNNs processes these extracted features, with SoftMax classification determining tumor type. These advancements have proved improved accuracy, with studies reporting classification rates above 98%, highlighting the effectiveness of combining optimization and deep learning techniques in medical image analysis.

The study [4] introduces a hybrid approach, the Support Vector Machine optimized with Seagull Optimization Algorithm (SVM-SOA), aims to enhance classification accuracy and computational efficiency of brain tumor MR images. This methodology involves preprocessing MRI images using the Savitzky Golay denoising method, feature extraction with Residual Exemplars Local Binary Pattern (RELBP) and classification via SVM with weight parameters optimized by SOA. The results demonstrate that the proposed SVM-SOA model outperforms existing methods, achieving 33.78% higher accuracy and a 30.62% improvement in F-score compared to Deep-CNN-DSCA, CNN-WHHO and AFDNN-FLA-based classifiers. These findings indicate that integrating optimization algorithms with classification techniques significantly enhances brain tumor detection accuracy and efficiency.

GANs have emerged as powerful tools, with architectures like Vox2Vox, Parasitic GAN and Segmentor GAN significantly improving segmentation accuracy. These models enhance MRI image quality, address data scarcity and refine tumor outlining. For classification, authors in study [5] have implemented Multi-Scale Gradients GAN (MSG-GAN) and Progressive Growth GAN (PGGAN) to generate realistic synthetic images, achieving high classification accuracy. Results indicate that GAN-based segmentation models achieve Dice scores above 85%, while classification methods report accuracy improvements exceeding 98%. Despite these advancements, challenges include class imbalance, annotation quality and computational complexity remains. Similar studies can be found in [16–20].

3   Proposed Methodology

In this section, we discuss our proposed methodology for brain tumor segmentation using K-Means with Cellular Automata.

In Fig. 1, the proposed methodology for brain tumor MR image segmentation integrates K-Means clustering with Cellular Automata (CA) to enhance segmentation accuracy and spatial coherence. The process begins with input image acquisition, where an MRI scan is obtained for segmentation. Since MRI images often contain multiple color

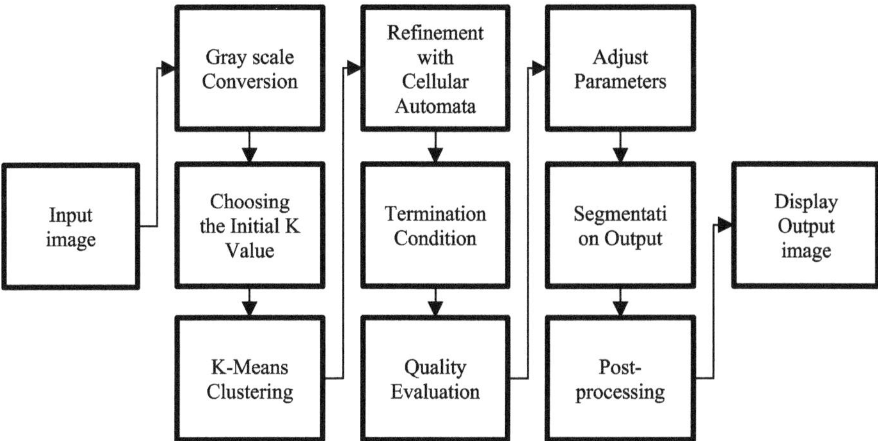

Fig. 1. Block diagram of proposed Methodology for Brain Tumor Segmentation from MR Image Using K-Means and Cellular Automata.

channels, the image is first converted into grayscale to simplify processing and reduce computational complexity. This transformation ensures that segmentation relies purely on intensity differences rather than color variations. Next, an initial value of k is chosen for the K-Means algorithm, determining the number of clusters. Selecting the right k value is crucial, as a smaller k may result in under-segmentation, while a larger k can lead to over-segmentation, splitting the tumor into multiple smaller clusters.

The K-Means clustering algorithm is then applied to group pixels based on intensity similarities. This involves initializing k random cluster centroids, assigning each pixel to the nearest centroid based on Euclidean distance, updating the centroids and iterating until cluster assignments stabilize. However, K-Means clustering does not consider spatial relationships between pixels, leading to fragmented segmentations with rough edges. To refine the segmentation, Cellular Automata (CA) rules are applied iteratively. CA enhances spatial coherence by updating each pixel's cluster assignment based on its neighbors, ensuring smoother boundaries and reducing small misclassifications. The refinement continues until the segmentation reaches convergence, meaning no significant changes occur between successive iterations.

If the segmentation is still unstable, the k value is adjusted and the clustering-refinement process is repeated. Once the segmentation stabilizes, the quality of the segmentation is evaluated using both visual inspection and quantitative metrics. Visual inspection involves checking whether the segmented tumor region corresponds to expected anatomical structures, while quantitative evaluation uses metrics to measure segmentation accuracy. If the results are unsatisfactory, the k value is adjusted for better segmentation performance. Once an optimal segmentation is achieved, the final segmented image is generated.

To further refine the output, morphological post-processing techniques, including erosion and dilation are applied. Erosion helps eliminate small misclassified regions, while dilation fills gaps and enhances boundary clarity. Finally, a contour graph is covered on the segmented image to highlight the detected tumor boundaries visually, helping in accurate diagnosis and interpretation. This integrated approach of K-Means and Cellular Automata ensures robust and accurate segmentation of brain tumor MR images.

3.1 Proposed Algorithm for Brain Tumor Segmentation

Step 1: Input Image Acquisition

Given an MRI image $I(x, y)$, where x,y represent spatial coordinates, define the image function:

$$I : R^2 \to R \tag{6}$$

where I (x, y) gives the intensity at each pixel (x,y)

Step 2: Grayscale Conversion

Convert the image I (x, y) to grayscale using a weighted sum:

$$I_{g(x,y)} = 0.2989 * I_{R(x,y)} + 0.5870 * I_{G(x,y)} + 0.1140 * I_{B(x,y)} \tag{7}$$

where I_R, I_G, I_B are the red, green and blue channel intensities.

Step 3: Choosing Initial k (Clustering Parameter)

Define an initial number of clusters k, where:

$$k = \text{argmin}_{c_1,\ldots\ldots,c_k} \sum_{i=1}^{k} \int_{\Omega} \left(\left(\sum_{i=1}^{k} ||I_{g(x,y)} - C_i||^2 \right) \right) dA \tag{8}$$

where, C_i is the centroid of the ith cluster and dA represents the differential area element over the image domain Ω.

Step 4: K-Means Clustering.

1. Initialize k centroids C_i randomly:

$$C_i = \left(x_i, y_i, I_{g(x_i,y_i)} \right), \quad i \in \{1, 2, \ldots, k\} \tag{9}$$

2. Assign each pixel to the nearest centroid using euclidean distance:

$$S_i = \{(x, y) | ||I_{g(x,y)} - C_i||^2 \le ||I_{g(x,y)} - C_j||^2, \text{ for all } j! = i\} \tag{10}$$

3. Update centroids:

$$C_i = \left(\frac{1}{|S_i|} \right) * \sum_{((x,y) \in S_i)} I_{g(x,y)} \tag{11}$$

4. Repeat until convergence:

$$\lim_{n \to \infty} \sum_{i=1}^{k} \int_{\Omega} \left(||C_i^n - C_i^{(n-1)}||^2 \right) dA = 0 \tag{12}$$

where C_i^n is the centroid at iteration n.

Step 5: Refinement using Cellular Automata (CA)
Apply CA rules iteratively to refine segmentation:

1. Define the neighborhood function N (x, y):

$$N(x, y) = \{I_g(x', y') | (x', y') \in N(x, y)\} \tag{13}$$

where N (x, y) is the neighborhood of (x, y).
2. Update each pixel state based on neighboring states:

$$I_{g(t+1)(x,y)} = f\left(I_{g(t)(x,y)}, N(x, y)\right) \tag{14}$$

where f is the transition function that smooths boundaries and corrects misclassified pixels.
3. Iterate until segmentation stabilizes:

$$\lim_{t \to \infty} \sum_{i=1}^{k} \int_{\Omega} \left(\left\|I_g^{(t+1)}(x, y) - I_g^t(x, y)\right\|^2\right) dA = 0 \tag{15}$$

Step 6: Convergence Check.
Define a stability condition where no significant changes occur:

$$\max_{((x,y) \in Omega)} \left\|I_g^{(t+1)}(x, y) - I_g^t(x, y)\right\| < \epsilon \tag{16}$$

for a small threshold ϵ. If the condition is not met, return to Step 3 and adjust k.
Step 7: Segmentation Quality Evaluation
Use metrics to quantitatively evaluate the segmentation performance.
Step 8: Parameter Adjustment
If the segmentation quality is unsatisfactory, update k and repeat from Step 3.
Step 9: Final Segmentation Output
The final segmented tumor region S_f is obtained as:

$$S_f = \bigcup_{i=1}^{k} S_i \tag{17}$$

Step 10: Post-processing.
Apply morphological operations such as erosion and dilation:

1. Erosion to remove noise:

$$S_{e(x,y)} = \min_{((x',y') \in B)} S_f(x + x', y + y') \tag{18}$$

where B is the structuring element.
2. Dilation to restore the tumor shape:

$$S_{d(x,y)} = \max_{((x',y') \in B)} S_e(x + x', y + y') \tag{19}$$

3. Contour Extraction using gradient:

$$\nabla S_d = \left| \frac{\partial S_d}{\partial x} \right| + \left| \frac{\partial S_d}{\partial y} \right| \tag{20}$$

The final output is S_d, the refined segmented image with contours overlaid.

In this methodology, the k value, centroids and CA implementation play crucial roles in achieving accurate results. The k value determines how many regions (clusters) the image is divided into and it is chosen by minimizing the sum of squared distances between pixels and their cluster centroids (Eq. 3). If the initial segmentation produces unclear or fragmented tumor boundaries, we adjust the k value and rerun the clustering, trying different values until a stable and clear tumor region is obtained. During K-Means clustering, centroids are initialized randomly and iteratively updated based on the average intensity of pixels in each cluster (Eq. 6), continuing until union, when centroid values no longer change significantly. Once K-Means segmentation is complete, we apply Cellular Automata (CA) as a post-processing step to refine boundaries and correct misclassifications. In CA, each pixel's cluster is updated based on the intensity patterns of its neighbors, using a neighborhood function (8-neighbors) and a transition rule that assigns the most frequent cluster from the neighborhood (Eq. 9). This iterative process continues until no significant changes occur, ensuring smooth and spatially clear tumor boundaries (Eq. 10). Together, K-Means provides an initial tumor region and CA enhances the segmentation by smoothing fragmented edges and improving boundary accuracy.

3.2 Challenges of the Proposed Work

- **Selection of Optimal k:** The accuracy of segmentation highly depends on choosing an appropriate number of clusters.
- **Computational Complexity:** Cellular Automata introduces additional computational overhead, making the method slower for large MRI datasets.
- **Sensitive to Initial Conditions:** The segmentation results can vary based on the initial centroid selection in K-Means clustering.
- **Difficulty in Handling Irregular Tumor Shapes:** Complex tumor structures with highly irregular boundaries may still pose challenges.

4 Results and Discussions

In this section, we will discuss the results of our proposed methodology for brain tumor MR image segmentation and their quality evaluation.

Figure 2 illustrates the segmentation process of brain tumor MRI images using the K-Means algorithm combined with Cellular Automata (CA). The Fig. 2 presents different stages of the segmentation pipeline for five test cases. The original MRI image represents the raw input data, while the grayscale image converts the input into a single intensity channel to simplify processing. The segmentation using K-Means classifies tumor regions based on pixel intensity and spatial features. The segmentation using CA and

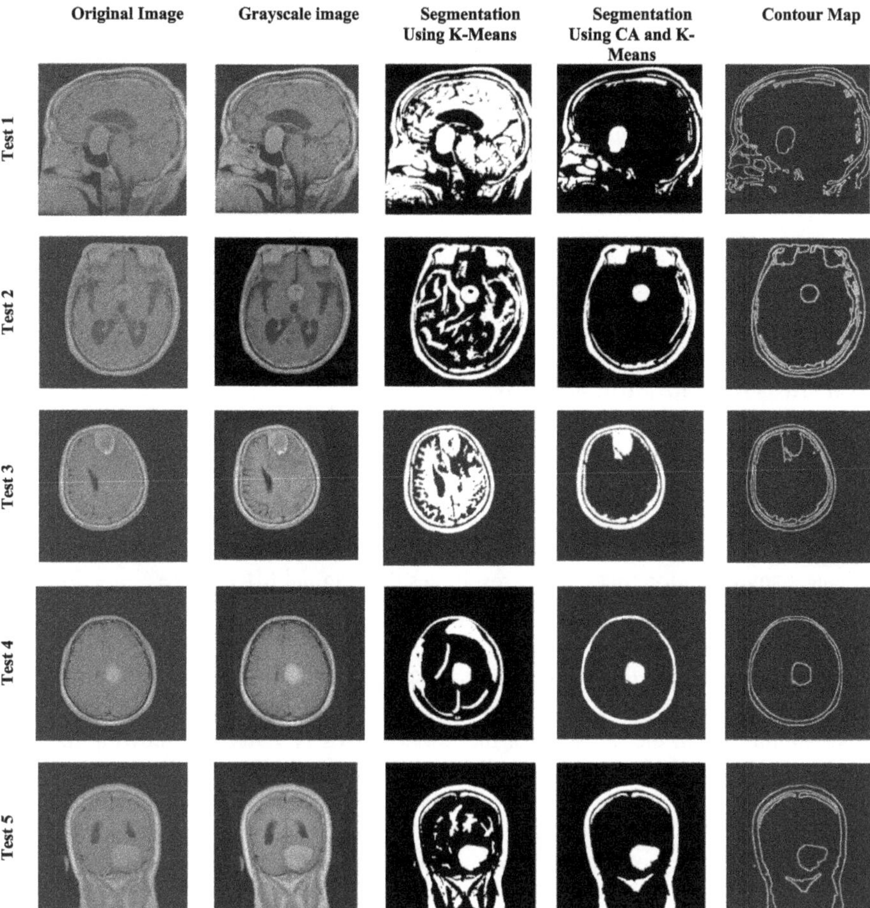

Fig. 2. Segmentation of Brain tumor from MR Images using K-Means algorithm with Cellular Automata.

K-Means enhances this process by integrating Cellular Automata, which refines boundary detection and improves segmentation accuracy. Finally, the contour map highlights the segmented tumor regions, providing a clear visualization of tumor boundaries. This Fig. 2 demonstrates the effectiveness of combining K-Means with Cellular Automata for detailed tumor segmentation in MRI images.

In Table 1, the segmentation quality of brain tumor MR images is measured using multiple metrics, each offering a different view on accuracy and structural preservation. These include Mean Squared Error (MSE), Root Mean Squared Error (RMSE), Peak Signal-to-Noise Ratio (PSNR), Signal-to-Noise Ratio (SNR), Structural Similarity Index Measure (SSIM) and Contrast-to-Noise Ratio (CNR).

Table 1. Evaluation of Segmented Brain Tumor from MRI Images Using Quality Metrics

	MSE	RMSE	PSNR	SNR	SSIM	CNR
Test 1	6556.9	80.9	9.964	−1.005	0.065	10.941
Test 2	5766.1	75.9	10.522	−1.327	0.0567	12.972
Test 3	4288.6	65.4	11.808	−0.701	0.0479	67.732
Test 4	3496.1	59.1	12.695	−0.472	0.0345	11.757
Test 5	4979.8	70.5	11.159	−0.456	0.0545	56.460

Error Measurement: MSE and RMSE

The highest MSE observed in the tests is 6556.9, indicating the most significant deviation between the segmented and input images. The lowest MSE recorded is 3496.1, representing the least amount of error. Similarly, RMSE, which provides an interpretable error magnitude, has its highest value at 80.9 and its lowest at 59.1, further reflecting differences in segmentation accuracy across tests. Lower values of MSE and RMSE indicate a segmentation that closely resembles the original MRI image.

Image Quality: PSNR and SNR

PSNR, which measures the quality of the segmented image in terms of signal preservation, reaches a highest value of 12.695 dB, indicating the best segmentation in terms of visual similarity. On the other hand, the lowest PSNR is 9.964 dB, showing the presence of higher distortion in that particular segmented image. SNR, which represents the ratio of meaningful information to noise, has a highest observed value of -0.456 and a lowest value of -1.327, suggesting varying degrees of noise interference in different segmentations.

Structural Similarity: SSIM

SSIM quantifies how well the segmented image maintains the structural integrity of the original. The highest SSIM recorded is 0.065, while the lowest is 0.0345. Since SSIM values range from 0 to 1, these values suggest that structural retention is relatively low in all tests, with some segmentations preserving slightly more spatial consistency than others.

Contrast and Feature Distinguishability: CNR

Contrast-to-Noise Ratio (CNR) helps evaluate how well the tumor region stands out from its surroundings. The highest CNR observed is 67.732, indicating excellent contrast between the tumor and background, while the lowest CNR recorded is 10.941, showing a segmentation where the tumor is less distinguishable from surrounding tissues (Fig. 3).

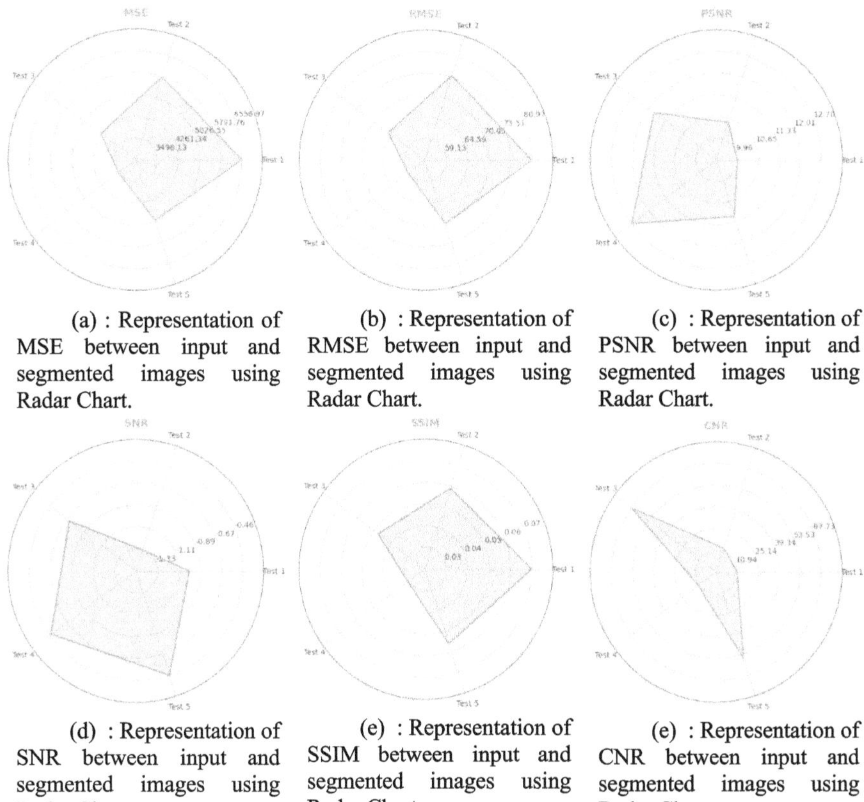

(a) : Representation of MSE between input and segmented images using Radar Chart.

(b) : Representation of RMSE between input and segmented images using Radar Chart.

(c) : Representation of PSNR between input and segmented images using Radar Chart.

(d) : Representation of SNR between input and segmented images using Radar Chart.

(e) : Representation of SSIM between input and segmented images using Radar Chart.

(e) : Representation of CNR between input and segmented images using Radar Chart.

Fig. 3. (a–e) illustrate the performance evaluation of the proposed segmentation method using various quality measures like MSE, RMSE, PSNR, SNR, SSIM, and CNR through radar charts. Each chart presents a comparative analysis of these metrics across different segmented images to measure the accuracy and visual quality of the results.

5 Conclusion

In this study, a hybrid brain tumor segmentation approach combining K-Means clustering and Cellular Automata (CA) was proposed and evaluated on MR images. The methodology successfully uses the clustering capabilities of K-Means and the spatial refinement properties of Cellular Automata, producing more clear and accurate segmentations. Performance was measured using quality metrics such as MSE, RMSE, PSNR, SNR, SSIM and CNR, with results highlighting the method's effectiveness in preserving tumor structure and enhancing boundary clarity. Future work will explore advanced post-processing techniques, adaptive clustering methods and deep learning-based enhancements to further improve segmentation accuracy and clinical applicability. This integrated approach marks a step forward in automated brain tumor detection, offering a valuable tool for medical imaging analysis and computer-aided diagnosis.

Acknowledgments. This study was funded by Indrakumar K and Ravikumar M.

Disclosure of Interests.   The authors have no competing interests to declare that are relevant to the content of this article.

References

1. Ali, M., Wu, T., Hu, H., MahmoodI, T.: Breast tumor segmentation using neural cellular automata and shape guided segmentation in mammography images. PLOS ONE **19**(10), e0309421 (2024)
2. Keerthi, D.S., Maithri, C., Dayanidhi, M., Gowda, M.V., Babu, K.J.A.: Segmentation and classification of medical big data on brain tumor using bacteria foraging optimization algorithm along with learning automata. Internet Technol. Let. **7**, e468 (2024)
3. Li, C., Zhang, F., Du, Y., Huachao, L.: Classification of brain tumor types through MRIs using parallel CNNs and firefly optimization. Sci. Reports **14**, p. 15057 (2024)
4. Lavanya, S., Annlin Jeba, S.V., Bhuvaneswari, P., Shajin, F.H.: Support vectormachine classifier optimized with seagull optimization algorithm for brain tumor classification. Concurr. Computat Pract Exper **35**, e7396 (2023)
5. Kaur, J., Singh, A.K., Jindal, N.: Systematic survey on generative adversarial networks for brain tumor segmentation and classification. Concurr. Computat Pract Exper **35**, e7850 (2023)
6. Yamuna, R., Rajalingam, R., Usha Rani, M.: Performance analysis of task offloading in mobile edge cloud computing for brain tumor classification using deep learning **4**(2), 1038–1056 (2023)
7. Shekari, M., Rostamian, M.: Brain tumor segmentation from MRI using FCM clustering, morphological reconstruction, and active contour. Multimedia Tools Appl. **83**, 42973–42998 (2024)
8. Syga, S., Jain, H.P., Krellner, M., HatzikirouID, H., Deutsch, A.: Evolution of phenotypic plasticity leads to tumor heterogeneity with implications for therapy. PLOS Comput. Biol. **20**(8), e1012003 (2024)
9. Dehghani, F., Karimian, A., Arabi, H.: Joint brain tumor segmentation from multi-magnetic resonance sequences through a deep convolutional neural network. J. Med. Sign. Sens. **14**, 9 (2024)
10. Zaitoon, R., Syed, H.: Enhanced brain tumor detection and classification in MRI scans using convolutional neural networks. Int. J. Adv. Comput. Sci. Appl. **14**(9), 266–275 (2023)
11. Qadir, I., Devendran, V., Qadir, F.: Performance comparison of decision-based median filtering for brain MR images with high multiplicative noise. Int. J. Comput. Appl. **46**, 9, 687–701 (2024)
12. Merdin Shamal Salih, Adnan Mohsin Abdulazeez.: A Fusion-Based Deep Approach for Enhanced Brain Tumor Classification. Journal of Soft Computing and Data Mining 5(1), pp. 183–193 (2024)
13. Gottipati, S.B., Thumbur, G.: A comprehensive approach to brain tumor classification in MRI: unifying classical local binary patterns and convolution neural networks. J. Electrical Syst. **20**(3), 519–531 (2024)
14. Ramtekkar, P.K., Pandey, A., Pawar, M.K.: Accurate detection of brain tumor using optimized feature selection based on deep learning techniques. Multimedia Tools Appl. **82**, 44623–44653 (2023)
15. Messina, L.: Hybrid cellular automata modeling reveals the effects of glucose gradients on tumour spheroid growth. Cancers **15**, 5660 (2023)

16. Shivaprasad, B.J., Ravikumar, M.: Enhancement of brain magnetic resonance images using cascade of notch filter and linear transformation methods. Pattern Recognit Image Anal. **33**(1), 66–79 (2023)
17. Ravikumar, M., Shivaprasad, B.J., Guru, D.S.: Enhancement of MRI brain images using notch filter based on discrete wavelet transform. Int. J. Image Graph. **22**(01), 2250010 (2022)
18. Ravikumar, M., Rachana, P.G., Shivaprasad, B.J.: Segmentation of tumor region from mammogram images using deep learning approach. In: International Conference on Advanced Informatics for Computing Research, pp. 30–42 (2021)
19. Ravikumar, M., Shivaprasad, B.J.: Bidirectional ConvLSTMXNet for brain tumor segmentation of MR images Tehnički glasnik **15**(1), 37–42 (2021)
20. Ravikumar, M., Shivaprasad, B.J., Guru, D.S.: Enhancement of MRI brain images using fuzzy logic approach. Recent Trends in Image Processing and Pattern Recognition Conference, RTIP2R, pp. 131–137 (2021)
21. https://www.kaggle.com/datasets/ahmedhamada0/brain-tumor-detection

Role of Artificial Intelligence in Cyber Security: Systematic Study on Attacks, Techniques and Tools

Ayush Shanmukha[1]([✉]), C. N. Sowmyarani[2], and Shanmukha Nagaraj[3]

[1] Computer Science and Engineering (Cyber Security), RNS Institute of Technology, Uttarahalli Road, Bangalore 560098, Karnataka, India
ayushshan2005@gmail.com
[2] Computer Science and Engineering, RV College of Engineering, Mysuru Road, Bangalore 560059, Karnataka, India
sowmyaranicn@rvce.edu.in
[3] Mechanical Engineering, RV College of Engineering, Mysuru Road, Bangalore 560059, Karnataka, India
shanmukhan@rvce.edu.in

Abstract. There is a rapid growth in terms of complexity, volume, and sophistication of cyber threats in the field of cyber security. The conventional cyber security methods often struggle to detect or prevent various types of existing attacks. Currently, in the field of cyber security, there is a drastic development in the field of Artificial Intelligence (AI) and Machine Learning. This growth has created a huge scope for improvement in detection and prevention mechanisms against continuously evolving attacks.

There are AI-enabled mechanisms and AI powered tools are used for detection and preventing threats which lead to attack besides analysing huge amounts of data. AI mechanisms help to identify the patterns, anomalies, irregular activities in the system at a larger scale. This dynamic and adaptive nature of these approaches significantly improves the accuracy, efficiency, and speed of threat detection and prevention. Due to threats and attacks, huge losses have incurred significantly and damaged the financial status and reputational quality of many firms. Embracing AI to combat threats is not just a strategic advantage-it's a powerful way to protect what matters most.

Its ability to detect attacks early, faster response time, and continuous adaptation can drastically minimize the impact of harmful actions before they escalate. The resilience of AI lies in its constant vigilance and rapid adjustment to developing risks. The current study focuses on various AI-based approaches to detect and defend the system against cyber-attacks. This work presents a vast and state of the art study on various AI-based methods and tools used in detection and prevention mechanisms against most common types of attacks and threats to systems.

Keywords: Artificial Intelligence · Cyber Security · Threats · Attacks

C. N. Sowmyarani and S. Nagaraj—These authors contributed equally to this work.

J. Shreyas et al. (Eds.): CODE-AI 2025, CCIS 2689, pp. 270–279, 2026.
https://doi.org/10.1007/978-3-032-19318-6_25

1 Introduction

Artificial Intelligence (AI) has grown widely since a few years and has the power to significantly create impact [1] in the cybersecurity [2]. There are different approaches in AI where, it is highly advanced and efficient in terms of detecting the threats and preventing attacks.

In the current era of Cybersecurity there is a great requirement for security mechanisms which protects data [3] belonging to persons, firms and agencies from attacks and threats, [4]. As the routine tasks are performed online, there is a need for robust cybersecurity measures to defend systems has become greatly important.

AI-based systems are capable of analyzing larger volumes of data, recognize patterns [4], behaviours. This leads to identification of attacks and malicious activities. Additionally, it can be used to monitor and safeguard data, devices, and networks while assisting in the identification and remediation of vulnerabilities. Some of the latest cyber threats and attacks [5] which are making severe damage to the sensitive information and undermining user trust are–Malwares, Phishing attacks, Spoofing attempts, Backdoor Trojan threat, Ransomware attacks, password attacks, Cryptojacking, Drive by download and Denial of Service attacks.

An extensive study has been carried out related to the attacks listed above. Various attack detection and prevention mechanisms using AI approach is presented by organising the contents of this article into several sections mentioned as: Sect. 1 being introduction, the next is Sect. 2-AI-based Prevention and Detection techniques and tools for attacks in cyber security which presents various AI-based mechanisms used or can be adapted in the tools. The Sect. 3 presents existing AI-enabled open-source tools to investigate cybersecurity attacks and depicts the tools employed for having specific types of attack. In Sect. 4 the implications of the study are discussed followed by Sect. 5 as a conclusion.

2 AI-Based Prevention and Detection Techniques and Tools for Attacks in Cyber Security

In cyber security, there is a need for efficient techniques to handle numerous types of attacks. Artificial Intelligence has been integrated with tools to investigate the threats or attacks in the current era of technology. Most popular attacks against the prevention and detection techniques have been discussed as follows. Malware is any malicious code [6] that can compromise a system further spreading to other systems too. Phishing is a social engineering attack where the victims are manipulated to reveal personal information [7]. Spoofing refers to impersonating or faking a legitimate service or website and deceive the victim to gain unauthorised access particularly in the context of digital security.

Backdoor Trojan [8] is a code which appears to be legitimate but acts as a spy for the attacker and compromises the system when instructed. In ransomware the attacker encrypts the victim's sensitive information and demands a ransom to be paid for decryption [9]. Through password cracking, cybercriminals are gaining unauthorised access

[10] to accounts having weaker and vulnerable passwords. Cryptojacking is an attack to mine the cryptocurrency [11] by gaining unauthorised access to computer or device.

There are various approaches [1, 12] employed for prevention and detection of these attacks in AI technology. Each subsection under Sect. 2 depicts the AI-powered tools which are built based on several approaches. In each table, each row shows the various techniques employed by each tool with the cell value as "✓" mark. The cross mark "*" indicates the technique in the respective column is not employed by the tool (Fig. 1).

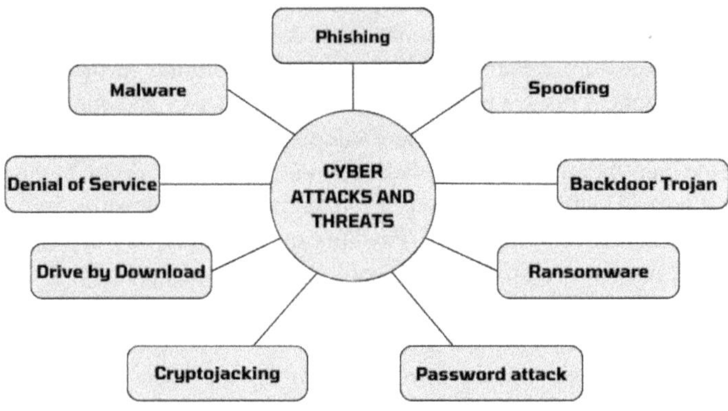

Fig. 1. Cyber Attacks and Threats

2.1 Malware

Table 1. Malware open-source tools and AI-based techniques

Existing Tools	Machine Learning	Deep Learning	Neural Networks	Pattern Recognition	Anomaly Detection
DeepInstinct	*	✓	✓	*	*
TensorFlow Malware Detection	✓	✓	✓	*	*
YARA with ML Integration	✓	*	*	✓	*
Snort with ML Integration [13]	✓	*	*	✓	✓

The analysis of various open-source tools for malware detection [14] and prevention is mentioned in Table 1. DeepInstint analyzes intricate patterns and forecasts possible dangers by implementing deep learning. While TensorFlow Malware Detection combines machine learning [15] and deep learning to identify malware patterns and behaviors. By utilizing machine learning to improve its rules, YARA with ML Integration enables adaptive threat detection and anomaly detection to identify malicious activity. In order to categorize and identify network threats, Snort with ML Integration combines machine learning and supervised learning. It also uses pattern recognition and anomaly detection to spot deviations from typical traffic, which aids in the identification of sophisticated assaults [16].

2.2 Phishing

The tools in the table enhance the identification of fraudulent behavior and phishing by utilizing a variety of AI techniques. Machine learning [17] and anomaly detection are combined in OpenDXL to enhance phishing prevention and identify odd activity patterns. Suricata with AI-based Detection analyzes traffic to find possible attacks and uses machine learning techniques to detect network-based threats (Table 2).

Table 2. Phishing Tools vs AI Techniques

Existing Tools	Machine Learning	Supervised Learning	Anomaly Detection
OpenDXL with Phishing Prevention Integration	✓	*	✓
Suricata with AI-based Detection	✓	*	✓
SpamAssassin with AI rules	✓	✓	✓

By learning from labeled data and adjusting to new spam kinds, SpamAssassin with AI rules improves accuracy in identifying and filtering spam messages using supervised learning. By identifying and stopping phishing and harmful activity, these solutions combine machine learning, supervised learning and anomaly detection [18] to provide sophisticated techniques for enhancing cybersecurity.

2.3 Spoofing

The products in the table use various AI techniques to improve cybersecurity, especially when it comes to implementing intelligent rules and identifying threats. SpamAssassin applies predetermined intelligence to detect dangerous material and uses AI-based rules to identify and filter spam messages. In order to detect network-based assaults, Snort (with custom AI Integration) [19] combines anomaly detection and machine learning. It does this by examining traffic for odd patterns and using historical data to enhance detection. By using intelligent detection rules and analyzing traffic for anomalies, Suricata (with AI-based Detection) enhances its capacity to identify sophisticated network threats by fusing machine learning with anomaly detection. By detecting anomalous activity and reacting with clever, adaptive rules, Modsecurity improves web application firewall capabilities through the use of both anomaly detection and AI-based rules (Table 3).

Table 3. Spoofing tools vs AI Techniques

Existing Tools	Machine Learning	Anomaly Detection	AI-based Rules
SpamAssassin	✓	✓	✓
Snort (with custom AI Integration)	✓	✓	*
Suricata (with AI-based Detection)	✓	✓	*
Modsecurity	*	✓	✓

2.4 Backdoor Trojan

The tools in the table use machine learning, anomaly detection, and other cutting-edge methods to identify and assess risks and improve cybersecurity. OSSEC uses different techniques to identify and address security events, but it does not mandate the use of anomaly detection or machine learning. By combining anomaly detection, deep learning for network traffic analysis, and machine learning for dynamic analysis, Zeek [19] assists in identifying possible threats and malicious activity based on anomalous behavior. Suricata employs machine learning algorithms and anomaly detection to spot possible intrusions and questionable traffic patterns. In order to help find patterns and signatures suggestive of malware or questionable activities, YARA incorporates machine learning for heuristic analysis (Table 4).

Table 4. Backdoor Trojan Tools vs AI Techniques

Existing Tools	Anomaly Detection	ML for Dynamic Analysis	ML for Heuristic Analysis	Deep Learning	Behaviour Based Analysis	Supervised Learning/ML
OSSEC	✓	*	*	*	✓	✓
Zeek	✓	*	✓	✓	*	✓
Suricata	✓	*	*	*	*	*
YARA	*	✓	*	*	*	*
Cuckoo Sandbox	*	✓	*	✓	✓	*
ClamAV	*	*	✓	*	*	*

Through the use of deep learning for behavior-based analysis and machine learning for dynamic analysis, Cuckoo Sandbox is able to study malware, trojan and identify malicious activity in a controlled setting with behaviour based analysis. Based on recognized patterns and behavior, ClamAV use machine learning for heuristic analysis to identify viruses and other harmful threats. To build a thorough defense mechanism against a broad spectrum of cyberthreats, these technologies combine a number of techniques, including anomaly detection, dynamic analysis, heuristic analysis, deep learning, and supervised learning.

2.5 Ransomware

To improve cybersecurity, the technologies in the table employ a variety of detection strategies and cutting-edge approaches [20]. Based on established patterns and signatures, YARA uses signature-based detection to find malware and other threats. It also incorporates anomaly detection to discover departures from typical system behavior. ClamAV use anomaly detection to spot new threats or odd activity in addition to signature-based detection to find viruses and malicious software. In order to monitor and examine system logs and assist in identifying security incidents and intrusions, OSSEC uses anomaly detection and signature-based detection (Table 5).

Suricata monitors network traffic using behavioral analysis, spotting possible threats by looking at actions and behaviors rather than signs. Rita employs behavioral analysis to look for trends in traffic and behavior in order to identify network-based risks. To provide a multi-layered approach to security, these technologies combine behavioral analysis [21], anomaly detection, signature-based detection.

Table 5. Ransomware Tools vs AI Techniques

Existing Tools	Signature-based Detection	Anomaly Detection	Behavioural Analysis
YARA	✓	✓	*
ClamAV	✓	✓	*
OSSEC	✓	✓	*
Suricata	*	*	✓
Rita	*	*	✓

2.6 Well Known Cyber Attacks

2.6.1 Password Attacks

Password attacks are techniques performed by the cybercriminals to gain unauthorized access to sensitive information or systems by cracking weak or vulnerable passwords. Password attacks also known as password cracking has various methods [22] like brute force, phishing, dictionary, rainbow, MITM used by malicious actors to breach the systems or account by compromising user credentials. However various machine learning solutions [23] are establishing for efficient detection and mitigations.

2.6.2 Cryptojacking

Cryptojacking is a kind of threat where it is embedded in a device and then uses its resources to mine cryptocurrencies [24]. It is a form of cybercrime where hackers covertly get into the victim's computing environment, also browser-based [25] access their resources to mine cryptocurrencies. This malicious crypto-mining essentially rewards the criminals with cryptocurrency like Bitcoin or Monero, which can be traded for the other types of cryptocurrencies which are coined as fiat currency. Users must employ strong cybersecurity approaches such as up-to-date antivirus software, firewalls, frequent software patches, even AI-based methods [26] caution while clicking any suspicious links or visiting unfamiliar websites.

2.6.3 Drive By Download

Drive-by download is one of the well known attacks where malicious software is downloaded and installed via the victim's browser [24] without their explicit consent. This attack occurs when the user visits malicious or compromised websites, where the malicious code exploits vulnerabilities taking control of the browser so that criminals can perform illegal activities. The downloaded malware may range from viruses and spyware to ransomware or trojans, often leading to unauthorized access to the highly-sensitive information, data theft, or system damage.

2.6.4 Denial of Service

Denial of Service is a dangerous ongoing cyber attack where the hackers try to deny the service to the legitimate users from accessing the information through servers, therefore bringing heavy traffic in the network with unwanted packets to the resource server.

Distributed Denial of service (DDOS) is the one where, multiple systems attack a single system, the traffic of the arriving messages makes the target system crash, therefore denying service to genuine users. However, DDOS attack are detected and mitigated through various AI-based techniques [27,28].

3   AI-Powered Tools and Technology for Existing Threats in Cybersecurity

Table 6 depicts the attacks on one side and tools on another side. A tabular analysis of all the attacks discussed and the open-source tools used for the same and we can also see which all tools are applicable for multiple attacks detection and prevention.

Figure 2 shows the different AI techniques used - Machine Learning, Anomaly Detection, Deep Learning, Behavior based Analysis, and Neural Networks. The different open-source tools that integrate these techniques for efficiently detecting and preventing the evolving cybersecurity threats.

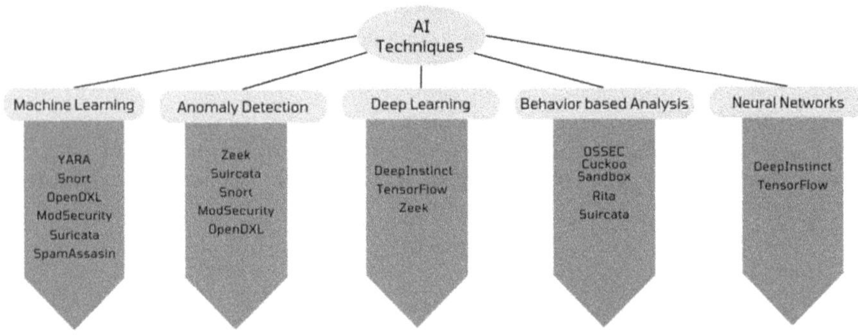

Fig. 2. AI-based Techniques and tools

Table 6. Different Attacks vs Tools used for detection and prevention

Attacks	OSSEC	YARA with ML Integration	Zeek	Snort with ML Integration	ClamAV	Suricata	SpamAssassin	ModSecurity	Fail2Ban	Falco
Malware	*	✓	*	✓	*	*	✓	*	*	*
Phishing	*	*	*	*	*	✓	✓	✓	*	*
Spoofing	*	*	*	✓	*	✓	✓	*	*	*
Backdoor Trojan	✓	✓	✓	*	✓	✓	*	*	*	*
Ransomware	✓	✓	*	*	✓	✓	*	*	*	*
Password attack	✓	*	✓	✓	*	✓	*	*	✓	*
Cryptojacking	✓	*	✓	*	*	✓	*	*	*	✓
Drive by Download	✓	*	✓	✓	✓	✓	*	✓	*	*
Denial of Service	*	*	✓	✓	*	✓	*	✓	*	*

4   Implications

Our study expands the theoretical understanding of cybersecurity by emphasizing the potential of various open-source AI-integrated tools to enhance real-time threat detection and prevention. It challenges the conventional defence models that rely on proprietary software by providing information on how AI-based techniques can be adapted/extended for these tools (Table 6).

From a practical perspective, our findings suggest that open-source AI tools could significantly reduce the costs associated with cybersecurity for businesses, especially small and medium enterprises (SMEs) that may not have the budget for proprietary solutions. Moreover, these open-source tools have greater allowance for customization, public-driven progress and transparency. In Future, research must be exploring the long-term effectiveness of these open-source tools and its reliability.

5   Conclusion

Artificial Intelligence is critical to cybersecurity, asset protection, and defence against evolving cyber threats and attacks. To conclude, our study aimed to show trending threats and attacks in the current era of technology – what would be the detection and prevention techniques and tools for it. This would be helpful for further research on this field which is the rising scope in cybersecurity.

References

1. Khder, M.A., Shorman, S., Showaiter, D.A., Zowayed, A.S., Zowayed, S.I.: Review study of the impact of artificial intelligence on cyber security. In: 2023 International Conference on IT Innovation and Knowledge Discovery (ITIKD), pp. 1–6 (2023). https://doi.org/10.1109/ITIKD56332.2023.10099788
2. Shanthi, R.R., Sasi, N.K., Gouthaman, P.: A new era of cybersecurity: The influence of artificial intelligence. In: 2023 International Conference on Networking and Communications (ICNWC), pp. 1–4 (2023). https://doi.org/10.1109/ICNWC57852.2023.10127453
3. Nautiyal, R., Jha, R.S., Kathuria, S., Singh, R., Kathuria, A., Pandey, V.: Artificial intelligence indulgence in protection of cybercrime. In: 2023 3rd International Conference on Pervasive Computing and Social Networking (ICPCSN), pp. 518–522 (2023). https://doi.org/10.1109/ICPCSN58827.2023.00090
4. Mehra, A., Badotra, S.: Artificial intelligence enabled cyber security. In: 2021 6th International Conference on Signal Processing, Computing and Control (ISPCC), pp. 572–575 (2021). https://doi.org/10.1109/ISPCC53510.2021.9609376
5. Murali, R.J., Binu Siva Singh, S.K., Onyx Nathanael Nirmal Raj, S.S.: An extensive examination of cyber attacks and cyber security, encompassing recent advancements and emerging trends. In: 2024 Ninth International Conference on Science Technology Engineering and Mathematics (ICONSTEM), pp. 1–5 (2024). https://doi.org/10.1109/ICONSTEM60960.2024.10568588
6. Kim, M.: Research on malware detection system using artificial intelligence. In: 2022 IEEE/ACIS 7th International Conference on Big Data, Cloud Computing, and Data Science (BCD), pp. 211–213 (2022). https://doi.org/10.1109/BCD54882.2022.9900792

7. Basit, A., Zafar, M., Liu, X., Javed, A.R., Jalil, Z., Kifayat, K.: A comprehensive survey of AI-enabled phishing attacks detection techniques. Telecommun. Syst. **76**, 139–154 (2021)
8. Shen, K.J., Selvarajah, V.: The impact of attacking windows using a backdoor trojan. In: 2023 International Conference on Evolutionary Algorithms and Soft Computing Techniques (EASCT), pp. 1–5 (2023). https://doi.org/10.1109/EASCT59475.2023.10393460
9. Çalışkan, B., Gülataş, Kilinc, H.H., Zaim, A.H.: The recent trends in ransomware detection and behaviour analysis. In: 2024 17th International Conference on Security of Information and Networks (SIN), pp. 1–8 (2024). https://doi.org/10.1109/SIN63213.2024.10871663
10. Chakraborty, A., Biswas, A., Khan, A.K.: Artificial intelligence for cybersecurity: threats, attacks and mitigation. In: Artificial Intelligence for Societal Issues, pp. 3–25. Springer (2023)
11. Tekiner, E., Acar, A., Uluagac, A.S., Kirda, E., Selcuk, A.A.: SoK: Cryptojacking malware. In: 2021 IEEE European Symposium on Security and Privacy (EuroS&P), pp. 120–139 (2021). https://doi.org/10.1109/EuroSP51992.2021.00019
12. Shaukat, K., Luo, S., Varadharajan, V., Hameed, I.A., Xu, M.: A survey on machine learning techniques for cyber security in the last decade. IEEE Access **8**, 222310–222354 (2020). https://doi.org/10.1109/ACCESS.2020.3041951
13. Aeraj, O.E., Leghris, C.: Study of the snort intrusion detection system based on machine learning. In: 2023 7th IEEE Congress on Information Science and Technology (CiSt), pp. 33–37 (2023). https://doi.org/10.1109/CiSt56084.2023.10409876
14. Mohapatra, N., Satapathy, B., Mohapatra, B., Mohanta, B.K.: Malware detection using artificial intelligence. In: 2022 13th International Conference on Computing Communication and Networking Technologies (ICCCNT), pp. 1–6 (2022). https://doi.org/10.1109/ICCCNT54827.2022.9984218
15. Ucci, D., Aniello, L., Baldoni, R.: Survey of machine learning techniques for malware analysis. Comput. Secur. **81**, 123–147 (2019). https://doi.org/10.1016/j.cose.2018.11.001
16. Maniriho, P., Mahmood, A.N., Chowdhury, M.J.M.: A study on malicious software behaviour analysis and detection techniques: taxonomy, current trends and challenges. Futur. Gener. Comput. Syst. **130**, 1–18 (2022). https://doi.org/10.1016/j.future.2021.11.030
17. Rathee, D., Mann, S.: Detection of e-mail phishing attacks-using machine learning and deep learning. Int. J. Comput. Appl. **183**(1), 7 (2022)
18. Beaman, C., Isah, H.: Anomaly detection in emails using machine learning and header information. arXiv preprint arXiv:2203.10408 (2022)
19. AbdulRaheem, M., et al.: Machine learning assisted snort and zeek in detecting DDoS attacks in software-defined networking. Int. J. Inf. Technol. **16**(3), 1627–1643 (2024)
20. Ferdous, J., Islam, R., Mahboubi, A., Zahidul Islam, M.: AI-based ransomware detection: a comprehensive review. IEEE Access **12**, 136666–136695 (2024). https://doi.org/10.1109/ACCESS.2024.3461965
21. Arabo, A., Dijoux, R., Poulain, T., Chevalier, G.: Detecting ransomware using process behavior analysis. Procedia Comput. Sci. **168**, 289–296 (2020). https://doi.org/10.1016/j.procs.2020.02.249. Complex Adaptive Systems Malvern, PennsylvaniaNovember 13–15 2019
22. Prozorova, E.A.: Thermal residue-based password attacks and the ways to counteract the same. In: 2022 International Siberian Conference on Control and Communications (SIBCON), pp. 1–5 (2022). https://doi.org/10.1109/SIBCON56144.2022.10002981
23. Sharma, A., Babbar, H., Vats, A.K.: Empowering security: machine learning solutions for detecting brute force attacks. In: 2024 4th Asian Conference on Innovation in Technology (ASIANCON), pp. 1–5 (2024). https://doi.org/10.1109/ASIANCON62057.2024.10838096
24. Gajmal, Y.M., More, P., Kale, K.D., Jagtap, A.: A survey on a novel cryptojacking covert attack. In: 2024 2nd International Conference on Intelligent Data Communication Technologies and Internet of Things (IDCIoT), pp. 359–364 (2024). https://doi.org/10.1109/IDCIoT59759.2024.10467823

25. Sachan, R.K., Agarwal, R., Shukla, S.K.: DNS based in-browser cryptojacking detection. In: 2022 Fourth International Conference on Blockchain Computing and Applications (BCCA), pp. 259–266. IEEE (2022)
26. Caprolu, M., Raponi, S., Oligeri, G., Di Pietro, R.: Cryptomining makes noise: detecting cryptojacking via machine learning. Comput. Commun. **171**, 126–139 (2021)
27. Musa, N.S., Mirza, N.M., Rafique, S.H., Abdallah, A.M., Murugan, T.: Machine learning and deep learning techniques for distributed denial of service anomaly detection in software defined networks—current research solutions. IEEE Access **12**, 17982–18011 (2024). https://doi.org/10.1109/ACCESS.2024.3360868
28. Gadallah, W.G., Ibrahim, H.M., Omar, N.M.: A deep learning technique to detect distributed denial of service attacks in software-defined networks. Comput. Secur. **137**, 103588 (2024). https://doi.org/10.1016/j.cose.2023.103588

Content-Based Smart E-Mail Dispatcher Using Large Language Models

K. Paramesha$^{(\boxtimes)}$ ⓘ, K R Sriram, Sujan Shetty, Shamanth Kishore,
and R. Tejaswini

Vidyavardhaka College of Engineering, Mysuru, India
`paramesha.k@vvce.ac.in`
`https://vvce.ac.in/departments/department-of-computer-science-engineering/`

Abstract. Email communication has become an integral part of personal and professional life, but handling its vast volume is still a significant issue for large organizations. Manual perusal of emails and forwarding their contents and attachments to intended recipients using other instant messaging platforms has proven to be error-prone and time-consuming leading to losses in terms of productivity and creating undue stress. The main objective of this paper is to explore an alternative mechanism that is to automate the task of dispatching emails based on their contents to the respective WhatsApp groups of students of various semesters of programs in an engineering college, facilitating a smooth flow of information from one end to another end in an organization. The dispatcher system is built using agents querying large language models (LLMs) to enable it to analyze the contents of emails and route them to the relevant groups of students for their information and consumption. The system harnesses the capabilities of LLMs in analyzing the textual contents for decision-making. With a well-structured agent framework prompting that includes email content as input with instructions and context, the system figures out the relevant groups to which the email message is dispatched thus providing the required information on time. The proposed system does not rely on labeled datasets and provides several benefits, including enhanced productivity and a reduction in the cognitive load associated with reading emails.

Keywords: LLMs · Prompt · AI agents

1 Introduction

Emails have gone from obscure to ubiquitous in half a century. Each day, on average, a person can deal with as many as 21 to 26 emails in a short span of time, creating an overwhelming cognitive load. Studies highlight that each email interruption costs 64 s to recover focus, significantly affecting productivity [1]. With over 4.4 billion email accounts globally and more than 320 billion emails exchanged daily, manual peruse remains inefficient and stressful for employees, as

© The Author(s) 2026
J. Shreyas et al. (Eds.): CODE-AI 2025, CCIS 2689, pp. 280–286, 2026.
https://doi.org/10.1007/978-3-032-19318-6_26

evidenced by increased stress levels and reduced productivity when emails are checked continuously [7]. Therefore, decreasing the usage of emails might not be the right solution. Although continuous interruption of work due to emails can have negative impacts, through proper techniques. These findings highlight the need for an automated solution to reduce the cognitive burden, and time consumption to effectively manage emails [11] caused by manual email perusal.

In this paper, we study the problems associated with manual reading and forwarding email messages received in the department mailbox to the concerned students for their information and follow-up actions. From the students' point of view, these academic notifications and circulars are important to keep in sync with the academic calendar of events. Figure 1 shows the structure of the information flow from higher authorities to students through the respective department. The departments receive emails containing different types of information from higher-ups. These email messages and attachments are routed to the students registered to the respective batch's Whatsapp group. The assigned staff in the department is responsible for manually going through the emails and forwarding them to the respective WhatsApp group for student consumption. This process is burdening the staff, affecting their productivity. Operating manually is bound to face a lot of issues. The solution to these types of manual operations is automation. With the latest advancements in natural language processing (NLP) techniques, checking emails and forwarding them to the intended recipients could be automated. The level of automation demands analyzing their contents to identify the target WhatsApp groups to which the email message is relevant rather than just channeling incoming emails to groups. As email messages are from diverse sources, understanding the sense of the content is key to providing an efficient solution. With efficient development and deployment of this type of system, two challenges are addressed. 1) Manual email checking and forwarding operations can be eliminated significantly thereby productivity of the staff is enhanced. 2) Since students frequently use instant messaging apps such as WhatsApp and Telegram, posting the latest information will enable them to stay updated with the latest happenings.

2   Related Work

The potential of LLMs is enormous and has demonstrated a variety of use cases on multi-modal data [3]. LLMs are becoming more relevant and useful when annotated datasets are unavailable as preparing them requires extensive manual effort. They can classify a URL link that is commonly seen embedded in email as malicious or not by one-shot prompting [10]. LLMs trained on large data have demonstrated the ability to understand input text and generate suitable responses as per the requirements of the applications such as generating documentation [2], machine translation [5], medical transcription [12], malware detection [9], etc. Many automated machine learning applications have been developed to analyze email content for classification tasks [6] and optimize workload management online [14]. There are some automated services online [15]

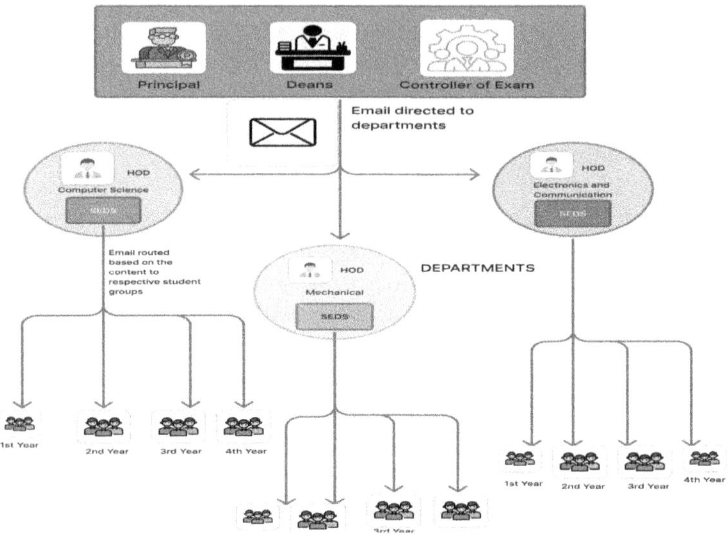

Fig. 1. Information flow from higher authorities to students.

wherein the system is configured to forward the received emails sent by a particular account as WhatsApp messages without content analysis. Keywords in the subject line can drive the email forwarding to WhatApp [8]. Artificial Intelligence(AI) is being integrated into platforms that use LLMS to perform a variety of automation tasks specifically in question and answering mode in chat and email applications. If the task is complex, it is decomposed into subtasks which solved through agents making the model adaptable and scalable [13].

3   Methodology

The proposed solution to the challenges is to build and validate an automated system before deployment. The system functionality is divided into three steps. Step 1: Data capturing, in this step reading emails is performed. In the development phase, the data in the textual format is obtained by reading email messages and their attachments offline, whereas, in the deployment phase, email messages and their attachments are read from the mail inbox. Step 2: Content Analysis, in which the content is analyzed by agents using LLM through prompts. Step 3: Dispatching emails, in which the email message is posted to WhatsApp groups specified by LLM. The important and core computation is the content processing step which depends on the LLM and design of the prompt.

3.1   Data Collection

The system is a data-driven model that analyzes email content to make decisions for dispatching them to appropriate WhatsApp groups. Therefore data collection and analysis to know its sources and categories are of primary importance to design prompt for LLMs. The dataset used for model validation comprises over 1000 carefully selected email messages related to students and faculty members.

3.2   AI Agent Prompt Engineering

Prompt Engineering refers to the techniques for designing prompts for LLMs to generate accurate, relevant, and contextually appropriate responses. We employ the Langchain framework to facilitate agents to perform the task step by step. Figure 2 shows a smart email dispatcher system(SEDS) to route the email to WhatsApp groups using custom tools used by the agents. Initially, the category of email content is identified by an agent and then the content is passed to the respective agent for the follow-up action. The agent for circular notifications and announcements needs to identify which of the target WhatsApp groups to which the message needs to be transferred. Having a WhatsApp group for each batch of students by their joined academic year and one faculty member group, the agents use the custom-built dispatcher to post email attachments and messages recommended by the LLM.

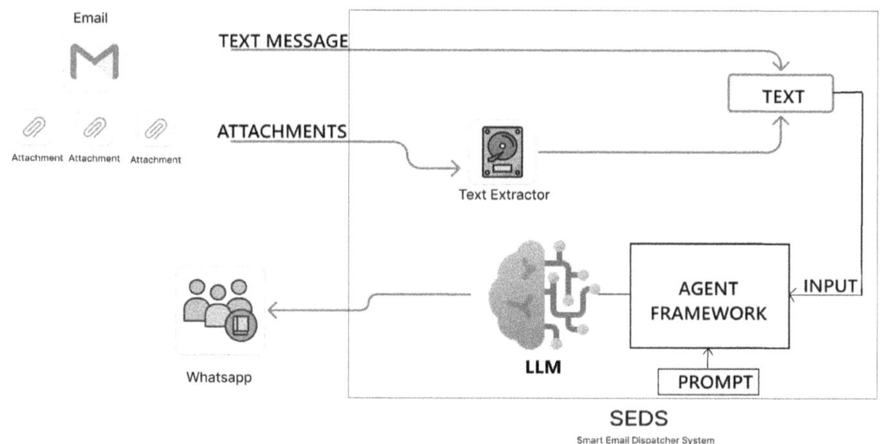

Fig. 2. Email dispatcher system using agent-based framework.

4   Results and Discussion

The results of the proposed model using various LLMs are furnished in Table 1. The agents query the LLM using Apis. We consider the perfect classification if the model identifies all the target WhatsApp groups. Even a single extra or missing target group will be considered a misclassification. ChatGPT and Gemini models perform much better than Deepseek and Perplexity AI models. When we closely observed misclassification results, we found that the model would recommend one or more groups other than targeted ones, especially in the case of circulars regarding valuation and revaluation, which are intended only for faculty members and are recommended to be shared with the student groups as well. Circulars mentioning only the batch series without year were able to map to the correct target groups.

Table 1. Accuracies of models on the dataset.

MODELS/PLATFORMS	ACCURACY (in %)
Gemini	93.5
DeepSeek	80
Perplexity	82.5
ChatGPT	92

4.1 Ablation of LLM

The replacement of LLM with a machine learning/deep learning model poses a lot of challenges for yielding the desired results. A labeled dataset is needed for training and validating the machine learning model to identify the target WhatsApp groups. Moreover, multiple selection of groups requires altogether a different technique in obtaining the expected results as machine learning prediction favours the best label. Clustering or any unsupervised technique will not lead to the solutions because the objectives are different. The WhatsApp group batches are not fixed as a new batch is added every year. However, if the system is implemented by overcoming all the challenges of the machine learning/deep learning models, the model's performance in the content analysis will not match that of LLMs as benchmarks have proven LLMs' supremacy in NLP tasks [4].

5 Conclusion and Future Work

The smart email dispatcher work is a giant leap forward in the automation and streamlining of email management for institutions and organizations. It simplifies workload management with the integration of advanced technologies such as transformer-based models and unsupervised learning, the system addresses challenges like semantic understanding, unstructured data processing, and dynamic email categorization. This results in a robust solution that can extract, classify, and route emails with remarkable accuracy and efficiency. Through automation, the system not only reduces the time and effort involved in handling manual emails but also minimizes errors and ensures consistency in performance. Its adaptability to changing communication patterns and the ability to process diverse formats like PDFs, images, and Word documents make it a versatile tool for modern workflows. This project's successful implementation showcases AI's potential to change the nature of operations. This is because it allows users to focus on high-priority tasks and strategic decisions, thereby improving productivity and efficiency in operations. This project is a stepping stone for future innovations in email management, which would eventually lead to smarter, adaptive communication systems across domains. In future work, we plan to provide email messages in a concise manner where the email messages will be summarized and pumped to WhatsApp channels for quick and clear notifications of events and information to the students' community.

References

1. Analysis of email management strategies and their effects on email management performance. Omega **124** (2024). https://doi.org/10.1016/j.omega.2023.103002
2. Automating software documentation: Employing LLMs for precise use case description. Procedia Comput. Sci. **246**, 1346–1354 (2024b). https://doi.org/10.1016/j.procs.2024.09.568
3. Chiarello, F., Giordano, V., Spada, I., Barandoni, S., Fantoni, G.: Future applications of generative large language models: a data-driven case study on ChatGPT. Technovation **133**, 103002 (2024). https://doi.org/10.1016/j.technovation.2024.103002
4. Gaitan, N.C.: Performance and accuracy research of the large language models. Int. J. Adv. Comput. Sci. Appl. **15**(8) (2024)
5. Heakl, A., Zaghloul, Y., Ali, M., Hossam, R., Gomaa, W.: Arzen-LLM: code-switched Egyptian Arabic-English translation and speech recognition using LLMs. Procedia Comput. Sci. **244**, 113–120 (2024). https://doi.org/10.1016/j.procs.2024.10.184
6. Jáñez-Martino, F., Barrón-Cedeño, A., Alaiz-Rodríguez, R., González-Castro, V., Muti, A.: On persuasion in spam email: a multi-granularity text analysis. Expert Syst. Appl. **265**, 125767 (2025). https://www.sciencedirect.com/science/article/pii/S0957417424026344
7. Mark, G., Iqbal, S.T., Czerwinski, M., Johns, P., Sano, A., Lutchyn, Y.: Email duration, batching and self-interruption: patterns of email use on productivity and stress. In: Proceedings of the 2016 CHI Conference on Human Factors in Computing Systems, pp. 1717–1728. Association for Computing Machinery (2016). https://doi.org/10.1145/2858036.2858262
8. pabbly: Easily Connect Multiple Applications — Automate your Tasks! (2025). https://www.pabbly.com/connect/inr/
9. Patsakis, C., Casino, F., Lykousas, N.: Assessing LLMs in malicious code deobfuscation of real-world malware campaigns. Expert Syst. Appl. **256**, 124912 (2024). https://doi.org/10.1016/j.eswa.2024.124912
10. Rashid, F., Ranaweera, N., Doyle, B., Seneviratne, S.: LLMs are one-shot URL classifiers and explainers. Comput. Netw. **258**, 111004 (2025). https://doi.org/10.1016/j.comnet.2024.111004
11. Russell, E.: Strategies for effectively managing email at work. In: ACAS Research Papers, vol. 6, p. 17 (2017)
12. Tan, Y., et al.: Medchatzh: a tuning LLM for traditional Chinese medicine consultations. Comput. Biol. Med. **172**, 108290 (2024). https://doi.org/10.1016/j.compbiomed.2024.108290

13. Wang, Y., Wu, Z., Yao, J., Su, J.: TDAG: a multi-agent framework based on dynamic task decomposition and agent generation. Neural Netw. **185**, 107200 (2025). https://doi.org/10.1016/j.neunet.2025.107200, https://www.sciencedirect.com/science/article/pii/S0893608025000796
14. Wassenger: WhatsApp communication and automation for all businesses made easy (2025). https://wassenger.com/
15. Zapier: Send new Gmail emails as WhatsApp messages in Tellephant (2025). https://zapier.com/apps/gmail/integrations/tellephant/809733/send-new-gmail-emails-as-whatsapp-messages-in-tellephant

Dynamic Delay-Based Rapid Serial Visual Presentation Using Linguistic Features

M. Shanthi[(✉)], Vaishali Gajendra Shende, and G. Rakshith

Department of Information Science and Engineering, East Point College of Engineering and Technology, Bengaluru 560049, Karnataka, India
`shanthimd.sm@gmail.com`

Abstract. Rapid Serial Visual Presentation (RSVP) is a technique traditionally used to increase reading speed by displaying words sequentially at fixed time intervals. However, a uniform delay does not account for the varying cognitive demands of different words. In this paper, we introduce a Dynamic RSVP system that leverages spaCy's natural language processing capabilities, specifically its part-of-speech tagging and named entity recognition, to tailor word display durations based on linguistic significance. By assigning longer delays to words that carry greater semantic or syntactic importance and shorter delays to less complex words, our approach aims to improve comprehension without sacrificing reading speed. We describe the system architecture, discuss the methodology for extracting linguistic features with spaCy, and present preliminary evaluations demonstrating the potential benefits of adaptive pacing in RSVP.

Keywords: RSVP · Natural Language Processing · spaCy · Named Entity Recognition · Part-of-Speech Tagging · Reading Comprehension · Adaptive Systems

1 Introduction

Reading is a fundamental human skill that has evolved over millennia from the days of handwritten manuscripts to printed books and now digital media. Traditional reading involves a complex interplay of eye movements (saccades and fixations) that naturally adapt to the complexity and structure of text [1]. Over time, researchers have recognized that the pace at which information is delivered can significantly influence comprehension and retention.

Rapid Serial Visual Presentation (RSVP) emerged as an innovative technique to enhance reading speed by rapidly presenting words at a fixed location on a screen [2,3]. Although effective for boosting speed, conventional RSVP systems apply a uniform delay across all words. This static approach neglects the intrinsic variability in language; some words require more cognitive processing than others due to their semantic or syntactic roles. Indeed, studies have shown that fixed-delay RSVP can negatively impact comprehension [4].

J. Shreyas et al. (Eds.): CODE-AI 2025, CCIS 2689, pp. 287–297, 2026.
https://doi.org/10.1007/978-3-032-19318-6_27

SpeedReed is our implementation of a Dynamic RSVP system that addresses this limitation by harnessing the power of spaCy, a state-of-the-art natural language processing library [5]. In SpeedReed, each word's display duration is adapted based on a detailed linguistic analysis. Our methodology classifies words into four distinct categories:

1. **Named Entity:** Words identified as named entities (such as names, organizations, or locations) are considered to have high contextual importance and are assigned longer delays.
2. **Content:** This category includes key carriers of meaning–nouns, proper nouns, verbs, adjectives, and interjections–ensuring that critical information is given sufficient processing time.
3. **Function:** Words that serve a structural role (e.g., determiners, pronouns, adpositions, auxiliary verbs, and conjunctions) receive moderate delays reflecting their role in maintaining sentence structure.
4. **Modifiers:** Symbols and other modifying elements, which adjust or nuance the meaning of adjacent words, are tuned with specific delays to preserve their subtle contributions to overall comprehension.

Beyond these adaptive delays, SpeedReed incorporates additional features aimed at enhancing the reading experience. For example, a burst mode provides extra delay at sentence boundaries to facilitate natural information integration, while a bionic reading mode helps guide visual attention by highlighting parts of a word with different colors. A contextual reference panel displays a window of the preceding and following words, offering readers immediate context without disrupting the flow of text.

By integrating these components, our Dynamic RSVP system seeks to bridge the gap between rapid reading and deep comprehension. In the following sections, we detail the system architecture, the process for extracting and categorizing linguistic features with spaCy, and our initial findings on the effectiveness of adaptive RSVP pacing. This work represents a step toward more personalized, cognitively attuned reading interfaces in digital environments.

2   Related Work

2.1   Traditional RSVP Implementations

Rapid Serial Visual Presentation (RSVP) has been investigated for decades as a method to accelerate reading by displaying words or images sequentially at a fixed location [2]. Early studies demonstrated the feasibility of RSVP for both experimental and commercial applications, as seen in classic work by Potter [2] and more recent analyses by Spence [3]. Commercial systems such as Spritz have capitalized on this concept by employing a fixed delay between words to enhance reading speed. However, these systems generally do not account for the varying cognitive demands of individual words, a limitation that our work seeks to address. Indeed, recent studies like [4] have highlighted the limitations of fixed-delay RSVP in maintaining comprehension.

2.2   Studies on Reading Speed and Comprehension

Numerous studies have examined the trade-off between increased reading speed and comprehension in RSVP-based interfaces. While RSVP can significantly boost the rate at which text is processed, fixed timing has been shown to negatively impact comprehension; particularly when complex or contextually rich words are presented too briefly [2]. [6] further explores the relationship between reading speed and comprehension in dynamic text presentation. These findings highlight the need for adaptive pacing mechanisms that can dynamically modify word exposure times based on text complexity, thereby supporting both speed and comprehension.

2.3   NLP in Reading Systems

The integration of Natural Language Processing (NLP) techniques into reading systems has recently gained attention as a means of enhancing user experience. State-of-the-art NLP libraries like spaCy provide robust part-of-speech (POS) tagging and Named Entity Recognition (NER) capabilities, which can be used to assess the linguistic importance of each word [5]. The editorial by Atanassova et al. [7] discusses the growing role of NLP in analyzing scientific literature, underscoring the broader trend of leveraging NLP for text understanding. Recent work has begun exploring how such linguistic features can inform the design of adaptive text presentation systems. By leveraging NLP, systems can tailor the reading experience on a per-word basis, which is a key motivation for our Dynamic RSVP approach.

2.4   Adaptive Timing in Text Presentation

Adaptive timing approaches seek to overcome the limitations of fixed-delay RSVP systems by dynamically adjusting the presentation time for each word. Early research, such as Öquist's study on Adaptive RSVP [8], explored methods for altering display durations based on text characteristics like readability. More recent systems, including the Thoth model [9], have utilized NLP parsing to refine this process further. Zambarbieri et al. [10] evaluated adaptive RSVP on mobile devices, showing its potential for improved readability. In our implementation SpeedReed we build upon these ideas by categorizing words into four distinct groups:

- **Named Entity:** Words recognized as named entities (e.g., names, locations) are given increased delays to emphasize their contextual importance.
- **Content:** This category includes nouns, proper nouns, verbs, adjectives, and interjections–elements that carry the core meaning of the text.
- **Function:** Words that serve grammatical roles (e.g., determiners, pronouns, ad-positions, auxiliary verbs, conjunctions) are assigned moderate delays.
- **Modifiers:** Symbols and other modifying elements are tuned with specific delays to capture their subtle contribution.

This adaptive approach aims to better align word exposure with cognitive processing needs, thereby improving comprehension without sacrificing speed.

3   System Design

In this section, we describe the architecture and implementation details of our Dynamic RSVP system. Our approach leverages spaCy's NLP capabilities on a Flask-based backend to extract linguistic features and compute adaptive delays. The system is complemented by a React-based user interface that enables users to customize delay parameters and control the RSVP reading experience.

3.1   Overall Architecture

The system is composed of two primary components:

- **Backend Server:** A Flask server, written in Python, loads the spaCy model (using en_core_web_lg) to perform tokenization, part-of-speech tagging, and named entity recognition (NER). The server exposes a RESTful API (e.g., the /pos-tag endpoint) that accepts text input and returns a JSON object containing a list of tokens along with their assigned delay groups.
- **Frontend Interface:** A React-based application (implemented using Next.js) provides four main pages: the Home page for users to enter their corpus, the settings page, the RSVP reader page called Reed and the profile page for Google sign ups. All of which are made available through a bottom navigation bar. The settings page allows users to customize reading parameters (such as words-per-minute, speech speed, and advanced delay multipliers for different token categories) with values stored in Firebase. The RSVP reader page retrieves user settings and fetches POS tagging results from the Flask server to dynamically adjust the display timing for each word.

3.2   NLP Pipeline

The NLP pipeline, implemented on the Flask server, is responsible for processing the input text and categorizing each token. The pipeline performs the following steps:

1. **Text Processing:** Upon receiving a POST request at the /pos-tag endpoint, the server processes the input text using spaCy's en_core_web_lg model. This step includes tokenization, POS tagging, and NER.
2. **Token Grouping:** Tokens are then assigned to one of four delay groups based on their linguistic features:
    - **Named Entity:** Tokens recognized as named entities.
    - **Content:** Tokens tagged as NOUN, PROPN, VERB, ADJ, or INTJ.
    - **Function:** Tokens with POS tags such as DET, PRON, ADP, AUX, CCONJ, or SCONJ.
    - **Modifiers:** Tokens tagged as ADV or NUM and those considered as punctuation or symbols.
3. **Token Combination:** A helper function combines tokens with their following punctuation to maintain natural groupings.
4. **Output:** The server returns a JSON response containing the combined tokens and their corresponding group labels.

3.3   Delay Calculation Mechanism

The RSVP reader uses the linguistic data from the NLP pipeline to dynamically calculate the display duration for each word. The mechanism operates as follows:

1. A base delay is computed using the formula:

$$\text{Base Delay} = \frac{60000}{\text{Words Per Minute}}$$

2. For each token, the system retrieves its delay group (e.g., named_entity, content, function, or modifiers) and applies a corresponding multiplier. For example, tokens in the Named Entity category, which are considered highly significant, are given a higher multiplier.
3. If the token contains punctuation (e.g., at sentence boundaries) and burst mode is enabled, an additional extra time is added to provide a brief pause.
4. The final delay is then calculated as:

$$\text{Delay} = \text{Base Delay} \times \text{Multiplier} + \text{Extra Time}$$

This adaptive mechanism ensures that words requiring more processing time are displayed longer, thereby enhancing comprehension without compromising overall reading speed.

3.4   User Interface for Delay Customization

The user interface is designed to give readers fine-grained control over the RSVP experience. Key features include:

- **Default Settings:** Users can adjust the base reading speed (words per minute) and speech speed, which influence the overall timing.
- **Mode Selection:** Options such as Bionic mode (for enhanced visual presentation) and burst mode (for extra delay at sentence boundaries) are available.
- **Advanced Options:** Sliders allow users to fine-tune the delay multipliers for each token category:
  - **Named Entity:** Emphasizes words with higher contextual importance.
  - **Content:** Adjusts delays for key content words (nouns, proper nouns, verbs, adjectives, interjections).
  - **Function:** Sets delays for grammatical elements (determiners, pronouns, ad positions, auxiliary verbs, conjunctions).
  - **Modifiers:** Controls the emphasis on symbols and other modifying elements.
- **Persistent Customization:** User preferences are stored in Firebase, ensuring that personalized settings are maintained across sessions.

The settings page and RSVP reader page are integrated seamlessly; the settings page feeds into the delay calculation on the RSVP page, allowing users to immediately observe the effects of their customizations.

Overall, our system design combines robust NLP processing with a dynamic timing mechanism and a user-friendly interface, resulting in a highly adaptive RSVP system that can improve the reading experience by aligning word display times with cognitive processing demands.

4   Evaluation

This section presents the evaluation of our Dynamic RSVP system. We describe the testing methodology, report the experimental results, and summarize user feedback.

4.1   Testing Methodology

We recruited 20 volunteers, all university students aged 18–22. The study was structured into three reading conditions:

1. **Normal Reading (Baseline)** - Participants read passages in a conventional, non-RSVP manner.
2. **RSVP (Fixed Delay)** - Participants used a standard RSVP reader with a uniform delay for each word.
3. **Dynamic RSVP (Proposed)** - Participants used our Dynamic RSVP system, which adjusts display durations based on linguistic features.

All participants were first given a brief tutorial on how to use the RSVP system and were told to practice for a week. The following week they completed the following tasks under each reading condition:

- **Reading Speed Test:** Participants read text passages while the system recorded the effective words per minute (WPM).
- **Comprehension Test:** After reading, participants answered multiple-choice questions related to the passages.
- **User Feedback Survey:** Participants provided subjective ratings regarding ease of use, cognitive load, and overall satisfaction.

4.2   Results

4.2.1   Performance Metrics Figure 1 shows the performance metrics for reading speed (WPM) and comprehension score (%) under each condition. Notably:

- **Normal Reading (Baseline):** 301.14 WPM, 88% comprehension
- **RSVP (Fixed Delay):** 350 WPM, 65.6% comprehension

- **Dynamic RSVP (Proposed):** 400 WPM, 81.33% comprehension

Compared to normal reading, both RSVP methods increased reading speed. However, the standard RSVP condition experienced a significant drop in comprehension (65.6%). Our Dynamic RSVP approach improved speed and preserved comprehension at a much higher level (81.33%) than the fixed-delay RSVP.

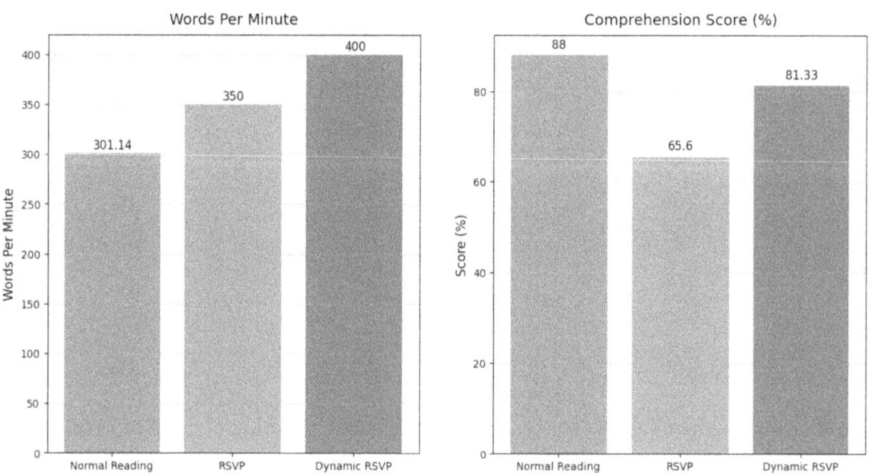

Fig. 1. Performance Metrics Comparison. Left: Words Per Minute. Right: Comprehension Score.

4.2.2   User Feedback Figure 2 presents the average user feedback (on a 1–5 scale) across three dimensions:

- **Ease of Use:**
  - Normal Reading: 4.8
  - RSVP: 3.8
  - Dynamic RSVP: 4.2
- **Cognitive Load (lower is better):**
  - Normal Reading: 2.5
  - RSVP: 4.2
  - Dynamic RSVP: 3.8
- **Overall Satisfaction:**
  - Normal Reading: 4.5
  - RSVP: 3.5
  - Dynamic RSVP: 4.5

Users found normal reading the most intuitive (Ease of Use = 4.8), but Dynamic RSVP (4.2) still outperformed the standard RSVP mode (3.8). For cognitive load, normal reading was rated lowest (2.5), while Dynamic RSVP (3.8) proved significantly less demanding than standard RSVP (4.2). Overall satisfaction was tied at 4.5 for normal reading and Dynamic RSVP, with standard RSVP scoring 3.5.

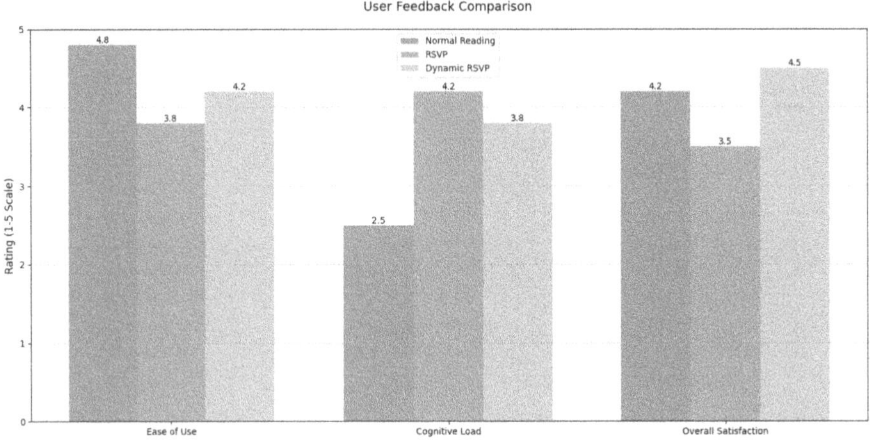

Fig. 2. User Feedback Comparison. Left: Ease of Use. Middle: Cognitive Load. Right: Overall Satisfaction.

4.3   Discussion of Findings

4.3.1   Reading Speed Dynamic RSVP achieved the fastest reading speed (400 WPM), surpassing both normal reading (301.14 WPM) and fixed-delay RSVP (350 WPM). This suggests that our adaptive approach can push reading speeds to higher levels without severely compromising comprehension.

4.3.2   Comprehension While normal reading yielded the highest comprehension (88%), Dynamic RSVP still retained a strong average of 81.33%. In contrast, fixed-delay RSVP dropped to 65.6%, indicating that uniform pacing may cause readers to miss critical information or struggle with complex words.

4.3.3   User Feedback Participants reported that Dynamic RSVP reduced cognitive load relative to fixed-delay RSVP, and provided an overall experience comparable to normal reading. The adaptive pacing helped users feel less rushed when encountering semantically or syntactically important words.

4.4   Summary

Overall, these findings demonstrate that a linguistically informed, adaptive approach to RSVP can significantly enhance both reading speed and comprehension compared to a standard, fixed-delay RSVP system. User feedback underscores the importance of adaptive pacing for maintaining ease of use and reducing cognitive strain.

5   Future Work

There are several promising directions for extending the Dynamic RSVP system:

1. **Automated Delay Adjustments Using Deep Learning:** We plan to integrate recurrent neural network (RNN) architectures to automatically adjust word display delays based on features extracted from spaCy, sentiment analysis tools, and transformer-based models (e.g., BERT or GPT) applied to the user-submitted corpus. By incorporating user feedback mechanisms, this approach aims to offer more personalized and adaptive timing that continuously learns from individual reading patterns and text complexity.
2. **Incorporating Human Speech Rhythm:** Future work will explore modeling word delays based on human speech rhythm. By leveraging forced alignment tools (such as the Montreal Forced Aligner or Gentle), we intend to extract word-level timing from human speech and test various speech rhythm styles with RSVP. Mimicking the natural cadence of speech could further enhance user engagement and comprehension.
3. **Enhanced Text Preprocessing and UI Improvements:** We plan to improve the SpeedReed user interface by developing a specialized summarizer and parser tool. This tool will transform user-submitted text into a version more suitable for RSVP by using techniques that shorten words, simplify grammar, and condense the content. Additionally, further UI enhancements including real-time feedback and interactive controls will be investigated to provide a more intuitive and efficient reading experience.
4. **Multimodal Content Integration:** Finally, incorporating visual elements such as images and emojis into the RSVP stream is an intriguing direction. Given that a picture can convey information succinctly, strategically inserting visual cues within the text may improve user comprehension and retention, offering a richer, more engaging reading experience.

6   Conclusion

In this paper, we introduced a Dynamic RSVP system that leverages spaCy's NLP capabilities to adjust word display durations based on linguistic features. By categorizing words into four groups–Named Entity, Content, Function, and Modifiers–our system dynamically assigns delays that aim to balance reading speed with comprehension.

Our contributions include:

- A novel adaptive timing mechanism for RSVP, which tailors word exposure based on its contextual importance.
- Seamless integration of spaCy's part-of-speech tagging and named entity recognition within a Flask-based backend.
- A customizable user interface that enables fine-tuning of delay parameters and supports additional reading modes such as Bionic and Burst modes.

Preliminary evaluations indicate that Dynamic RSVP can potentially improve reading comprehension and engagement compared to traditional fixed-delay approaches. The impact of this work extends to personalized reading interfaces, suggesting that adaptive pacing mechanisms can be instrumental in enhancing user experience in digital reading applications.

Future work will further refine these adaptive mechanisms and integrate more advanced deep learning models for automated delay adjustments, ultimately paving the way for a more personalized, cognitively attuned reading experience.

In summary, our Dynamic RSVP system represents a significant step toward harnessing NLP for improved digital reading, and we anticipate that further developments will broaden its impact across various text-based applications.

References

1. Rayner, K.: Eye movements in reading and information processing: 20 years of research. Psychol. Bull. **124**(3), 372 (1998)
2. Potter, M.C., Kroll, J.F., Harris, C.: Rapid serial visual presentation (rsvp): a method for studying language processing. Memory Cogn. **12**(2), 97–116 (1984)
3. Spence, R.: Rapid, serial and visual: a presentation technique with potential. Inf. Vis. **1**(1), 13–19 (2002)
4. Benedetto, S., Carbone, A., Pedrotti, M., Fevre, K., Le Yondre, A., Baccino, T.: Rapid serial visual presentation in reading: the case of spritz. Comput. Hum. Behav. **45**, 252–258 (2015)
5. Honnibal, M., Montani, I.: Spacy: industrial-strength natural language processing in python. Zenodo (2020). https://spacy.io
6. Yamada, M., Imagawa, T., Ito, K., Miyake, Y.: Reading traits for dynamically presented texts: comparison of the optimum reading rates of dynamic text presentation and the reading rates of static text presentation. Front. Psychol. **8**, 1390 (2017)
7. Atanassova, I., Bertin, M., Mayr, P.: Editorial: mining scientific papers: Nlp-enhanced bibliometrics. Front. Res. Metrics Anal. **4**, 2 (2019)
8. Öquist, G., Goldstein, M.: Adaptive rapid serial visual presentation. Unpublished Thesis, Uppsala University, Uppsala, Sweden (2001)
9. Awad, D.: Thoth: improved rapid serial visual presentation using natural language processing. (2019). https://arxiv.org/abs/1908.01699
10. Zambarbieri, D., Carniglia, E., Sal, B., Scapin, D.: Towards an improved readability on mobile devices: evaluating adaptive rapid serial visual presentation, pp. 277–295 (2011)

Agentic AI Factories: Constructing Intelligent Workflows for Data-Driven Decision Making

Pranav V. Jambur[1]([✉]) [ID], C. N. Sowmyarani[1] [ID], Shanta Rangaswamy[1] [ID], Dayananda Pruthviraja[2] [ID], and K. N. Subramanya[3] [ID]

[1] Department of Computer Science Engineering, R. V. College of Engineering, Bengaluru, Karnataka, India
`{pranavvjambur.cs23,sowmyaranicn,shantharangaswamy}@rvce.edu.in`
[2] Department of Information Technology, Manipal Institute of Technology (MIT), Bengaluru, Karnataka, India
`dayananda.p@manipal.edu`
[3] R. V. College of Engineering, Bengaluru, Karnataka, India
`subramanyakn@rvce.edu.in`

Abstract. In recent years, the concept of AI-driven workflow automation has gained prominence as organizations seek to harness natural language inputs to generate complex workflows. However, these systems produce boiler plan templates and basic workflow configurations. Applied scientists and engineers still need to adapt them, add tooling capabilities and evaluate them for reliable inference and decision making. These systems often operate by simply connecting the output of one model or tool to the input of another without transparently indicating whether each module is adequately pre-trained, prompt engineered or requires RAG etc. In contrast, our proposed agentic AI Agent Factory (AAF) is a framework that provides an interactive method of developing fully working AI Agentic workflows. While applied and data scientists can start defining the workflow requirements in natural language mode, AI factory will spin-off an consultative dialogue with the. AAF starts with providing a high level workflow with potential models and tools. Further, AAF asks clarifying questions to collect additional context such as historical data, quality of inferences, evaluation benchmarking of decisions. This process also ensures that any human-in-loop requirements, as needed ensure quality of inference and decision making. By iteratively constructing workflows and leveraging tools like Agents orchestration, RAG, tools, evaluation tools and human-in the loop interfaces the framework continuously refines its performance through the experiment phase. AAF also provides deployment support with automated CI/CD processes. The resulting system is intended for not only enhancing efficiency and scalability but also achieve quality decision making and inference with optimized cost and timeline compared to manual methods.

Keywords: Agentic AI · Intelligent Workflow Automation · AI in Data Science · Iterative AI Development · Decision Intelligence · Lang Graph · Context-Aware AI · Agentic AI Framework (AAF)

© The Author(s) 2026
J. Shreyas et al. (Eds.): CODE-AI 2025, CCIS 2689, pp. 298–309, 2026.
https://doi.org/10.1007/978-3-032-19318-6_28

1   Introduction

There is a massive shift in AI-agent driven systems from the implementation of workflow pipelines from rule-based approach to completely sophisticated systems which can interpret the inputs given by users in natural language. There is a collaboration among agents focused towards delivering the quality inferences. The evolution of workflow automation has transitioned from simple rule-based workflow pipelines to sophisticated AI-Agent driven systems that interpret natural language inputs from users and collaborate within them to deliver quality inferences. However these systems provided a means to streamline simple repetitive tasks.

They largely relied on manual updates and configurations to achieve complex automation that could not adapt to the dynamic needs of modern data science environments. Current AI-driven workflow systems have significant shortcomings. They often produce workflows that follow a fixed input-output chain without offering transparency into whether the components are pre-trained or require further customization. Additionally, these systems do not incorporate human feedback while experimenting to make workflow changes. This has left leaving critical aspects such data augmentation, workflow changes, reinforcement learning or performance evaluation to be done manually.

This lack of adaptability and transparency limits the productivity of data and applied scientist to complex decision-making scenarios. To address these shortcomings, we propose an AAF that transforms the existing workflow automation process into multistage iterative workflow maturity framework. Unlike static systems, this new approach begins with a consultative interaction where a high-level workflow diagram is presented to the user based on set of workflow requirements. The system then asks clarifying questions to extract additional details, such as the need for historical data, augmenting organization to conduct benchmarking of the inferences with evaluation systems. AAD transparently communicates these design decisions to avoid human oversight or long trial and errors.

This paper aims to present and evaluate such a AAF framework. The remainder of the document is structured as follows: first, we review related literature and highlight the gaps in current methodologies; next, we describe the proposed methodology in detail, including technical components and system architecture; then, we present the experimental results and a comparative analysis; finally, we conclude with a summary of our findings and discuss potential future research directions. The solution proposed embodies one bidirectional chatbot that drives the entire system, ensuring collaborative agent efforts and delivering quality inferences.

2   Literature Review

The evolution of workflow automation began with primitive systems that focused on simple data transmission and repetitive task automation using RPA. Early approaches relied on rule-based methods to perform manual tasks with little flexibility and laid the groundwork for further innovations [1,2]. Over time AI automation emerged to augment these processes by using large language models to generate basic inferences in a linear manner [3,4]. This phase integrated pre-trained modules for content generation and summarization although workflows remained fixed and could not easily adapt to dynamic data needs [5,6]. Later research introduced modular designs that allowed components to be updated independently while addressing scalability issues [7,8]. Real-time data handling and retrieval augmented generation techniques were also used to optimize workflows [9,10]. Studies on modular design and comparisons between static and dynamic automation highlighted the need for adaptive architectures [11,12].

Later developments shifted from linear AI workflows to agentic systems that use graph-based architectures instead of simple chains. These systems use block diagram visualizations and CI/CD integration for continuous improvement to handle complex decision making autonomously [13,14]. Conversational agents or chatbots now guide workflow creation with interactive natural language interfaces [15,16]. Research in data science pipelines and evaluation frameworks shows that including human feedback improves accuracy and adaptability [17,18]. Adaptive architectures now allow workflows to reconfigure dynamically in response to real-time feedback [19,20]. Further studies have examined trends in autonomous orchestration and intelligent decision making to inform tool selection and process optimization [21,22].

Despite these advancements current methods still have critical gaps. Existing systems do not offer a fully consultative interactive framework that builds dynamic workflows and refines them continuously based on user feedback. In particular there is no central driving force such as a bidirectional chatbot to orchestrate the entire process and enable agent collaboration. The proposed AI Agentic Factory framework addresses these gaps by automatically generating AI agent workflows and suggesting the best tools and orchestration logic. It also integrates an evaluation framework and CI/CD for continuous improvement [23,24]. This solution is tailored for data scientists and supports change management through a natural language mode. A main pilot bidirectional chatbot governs the system to ensure dynamic decision making and transparent module evaluation [25].

Existing frameworks in workflow automation fall short by not offering a fully integrated solution for AI agent workflow creation. They lack the capability to automatically generate AI agent workflows while simultaneously suggesting the optimal tools and orchestration logic required for dynamic operations. Moreover, current systems do not incorporate a robust evaluation framework or effective deployment strategies, such as CI/CD pipelines, to support continuous improvement. This shortfall is particularly pronounced when considering the need for natural language interaction to facilitate change management.

3   Methodology

3.1   Overview

The proposed methodology for the AAF is intended to convert natural language inputs into dynamic, iterative workflow designs that adjust to evolving data requirements. At the core of this approach is the Factory Interface–a bidirectional chatbot that gathers and refines requirements to help users design the entire system. This Factory Interface (FI) offers consultative interactions with users to define objectives, constraints, and desired workflow outcomes, ensuring that the process remains user-centric and low-code.

3.2   Consultative Interaction and Requirement Gathering

In the initial phase, the user engages with the FI to provide the initial workflow requirements. The FI then conducts a structured dialogue with the user. This interactive process captures essential needs by presenting a high-level workflow design that includes potential modules such as LLMs for sentiment analysis, predictive modeling, and conversational agents. The FI asks clarifying questions to gather supplementary data, including historical performance metrics and available benchmarks for outcome evaluation. This bidirectional exchange ensures that the appropriate workflow modules are selected to achieve high-quality decision-making while addressing critical data dependencies.

3.3   Module Identification and Integration

Once the requirements are gathered, the FI evaluates whether to integrate pre-trained LLM models for the given domain or to select new stock models based on user input. The evaluation criteria include accuracy, processing speed, and adaptability to specific business contexts. The FI presents performance metrics and confidence scores for each module as well as for the entire workflow, facilitating informed decision-making. This phase also supports a modular design that incorporates additional data such as RAG, evaluation data, and contextual data for improved prompt templates, allowing each component to be upgraded independently.

3.4   Dynamic Workflow Construction

In this phase, the FI generates the final workflow using an Orchestration Agent to construct an experience-ready workflow that pre-integrates the identified models and tools. The resulting graph includes a visual diagram that shows input-output relationships and serves as a clear representation of the workflow logic, making it easy for stakeholders to follow the process. The diagram also provides experimental snapshots of input and output data along with an evaluation dashboard for each module as well as for the entire workflow.

3.5 Interactive Design Changes During Experimentation

With this, the workflow is ready for users to start experimentation. The workflow is accessible via the Orchestration Agent API as well as through a visual dashboard that displays information at both the module and overall workflow levels. Users can run the workflow in sandbox mode and observe outcomes such as inferences and decision-making. They are able to evaluate and provide interactive feedback at both module and workflow levels. Once evaluation is complete, the AI Factory interprets the feedback and initiates additional consultative interactions to adjust modules, interfaces, and request further data or context (Fig. 1).

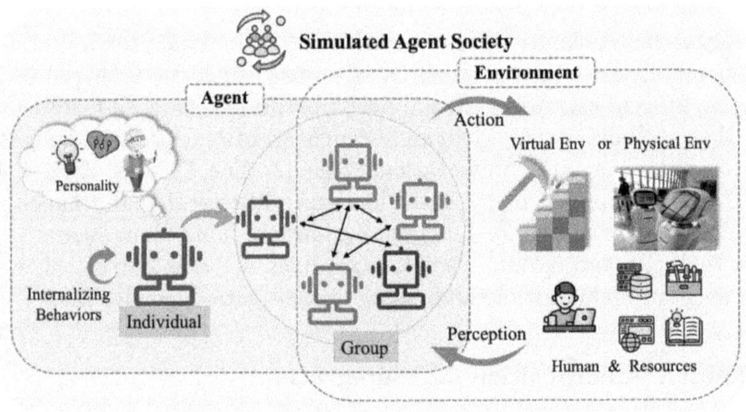

Fig. 1. Traditional Agentic & Human-in-a-loop interaction

Users can then provide these inputs visually, allowing the AI Factory to refine the workflow iteratively for subsequent testing cycles. This process repeats until the desired outcome is achieved. At each stage, the entire workflow, along with its data and modules, is versioned and stored using the GitHub framework.

In this phase, the FI generates the final workflow using an Orchestration Agent to construct an experience-ready workflow that pre-integrates the identified models and tools. The resulting graph includes a visual diagram that shows input-output relationships and serves as a clear representation of the workflow logic, making it easy for stakeholders to follow the process. The diagram also provides experimental snapshots of input and output data along with an evaluation dashboard for each module and for the overall workflow.

3.6 Deployment Strategy and User Interface

Following workflow experimentation and evaluation completion, the FI helps users move into the deployment phase. In this stage, the FI plays a key role by outlining and executing a deployment strategy that considers cloud, on-premises, or hybrid deployments. A dedicated user interface is provided, allowing users to control the deployment and conduct final production testing. This interface facilitates interactive adjustments and fine-tuning before full-scale deployment. The FI also provides a deployment dashboard that displays key performance indicators such as LLM token usage cost, latency, and availability time.

3.7 Continuous Monitoring and Feedback Loop

The AF incorporates continuous monitoring mechanisms to track the performance of both individual modules and the overall workflow automonesly even afte3r doloymenbt. A set of dashboards and real-time monitoring tools capture key performance metrics such as processing time, resource utilization, and error rates. The FI leverages this information to trigger updates to users and can results in changes to the workflow that can be deloued through CI/CD pipeline, ensuring that the system refines its performance over time.

3.8 Implementation of AAF

The agentic framework, powered by LangGraph, includes an LLM that automatically creates natural language understanding for the factory. It integrates essential toolsets such as retrieval-augmented generation and a prompt template generator where an LLM serves as judge evaluation. In addition, the LangStudio UI provides an intuitive interface for users, while CI/CD pipelines and a comprehensive dashboard ensure continuous updates and performance monitoring.

The technical implementation leverages a robust stack of modern tools. Key components include AWS services for scalable cloud deployment, container orchestration frameworks such as Kubernetes, and CI/CD pipelines for automated updates. The system integrates LangGraph for workflow orchestration alongside supplementary tools like web crawlers and retrieval-augmented generation modules. Throughout these processes, the main pilot serves as the central coordinating agent, ensuring seamless communication between disparate components and a unified system operation.

3.9 Example of the Work Workflows

The sales sentiment analysis workflow features an Orchestrator that manages the process. It includes a Transcript-to-Sentiment Analysis Generator, an Evaluation Agent for sentiment assessment, a Decision-Making Agent for customer engagement, and an Offer/Email Communication Agent for outreach. A Human-in-the-Loop Evaluation Interface and a Dashboard Interface support real-time monitoring and feedback (Fig. 2).

To enhance clarity, the methodology incorporates several diagrams. A block diagram shows the consultative interaction with the main pilot at the center, a flowchart details the dynamic workflow construction, and an architecture diagram outlines the deployment integration points. These visuals simplify the complex interactions within the agentic AI factory (Fig. 3).

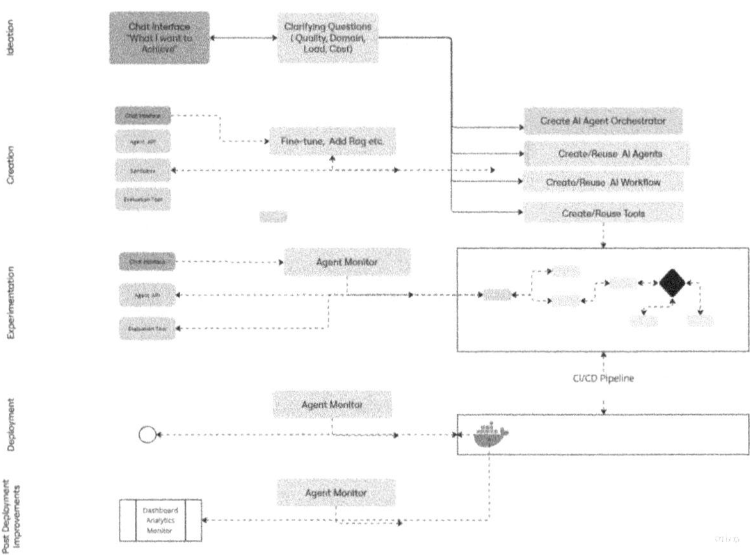

Fig. 2. Initial Block Diagram Workflow - Visible to Stake holders

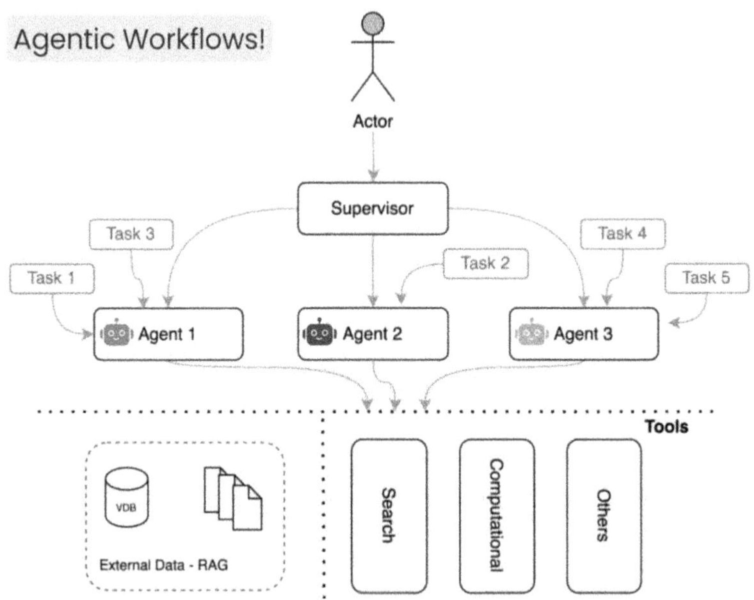

Fig. 3. Modified architecture - With manager/archestrator agent

4   Results

4.1   Overview of Results

Stage 1 begins with initial experimentation where baseline performance is measured. In Stage 2, the system evaluates the accuracy of the Factory Interface (FI) during experimental operations, and in Stage 3, further accuracy assessments are conducted using the prediction headtopn FI. These stages collectively ensure a comprehensive evaluation of the FI's performance and reliability.

The implementation of the agentic AI factory, driven by the main pilot chatbot, has yielded significant improvements in workflow efficiency and adaptability. Initial deployments indicate that the bidirectional nature of the main pilot not only reduces manual intervention but also accelerates processing speed across the system. These outcomes have been validated through experiments and case studies comparing dynamic workflows with traditional static systems (Fig. 4).

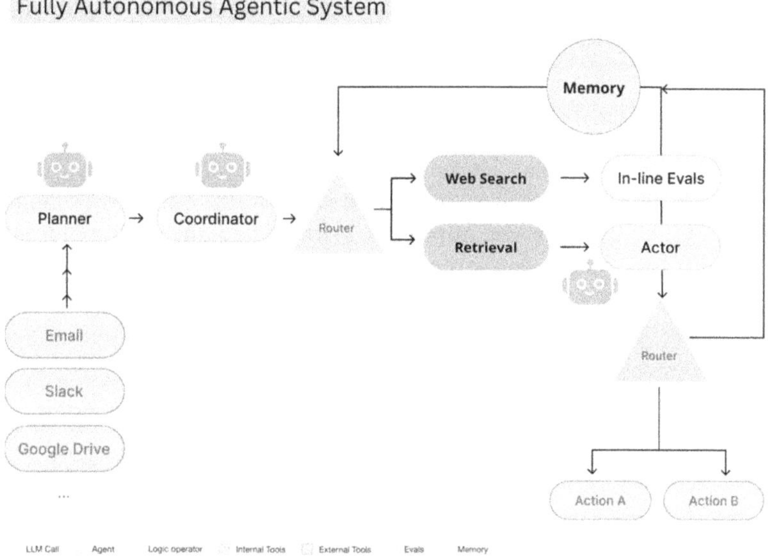

Fig. 4. Autonomous agent manufacture by Agentic Factory

4.2   Efficiency and Performance Metrics

Quantitative analysis reveals that dynamic workflows constructed under the guidance of the main pilot achieve a marked reduction in processing time and resource consumption. For example, benchmarking tests indicate an improvement in processing speed of up to 30% compared to conventional automation systems. The main pilot's ability to coordinate module integration and facilitate rapid adjustments directly contributes to these efficiency gains.

4.3   Adaptability and Scalability Outcomes

One of the key advantages of the proposed framework is its remarkable adaptability and scalability. Simulation studies demonstrate that the system maintains high performance even with increased data loads and evolving requirements. The main pilot's bidirectional communication enables seamless integration of new functionalities, thereby supporting large-scale deployments in both cloud and hybrid environments. Scalability tests further confirm that the dynamic integration managed by the main pilot effectively handles varied data complexities.

4.4   User Experience and Feedback

User feedback has been highly positive regarding the consultative interaction and the transparency of the performance dashboards. Users especially appreciate the main pilot's role in providing a clear overview of the workflow, which allows them to make interactive adjustments based on real-time data. In addition, the user playground–designed for simulation and testing–has greatly increased overall satisfaction and trust in the system.

4.5   Comparative Analysis and Case Studies

Comparative analyses reveal significant benefits in efficiency and flexibility when comparing the agentic AI factory with traditional static workflow systems. Case studies show that the dynamic approach, led by the main pilot, reduces downtime, lowers error rates, and streamlines decision-making. For example, in a

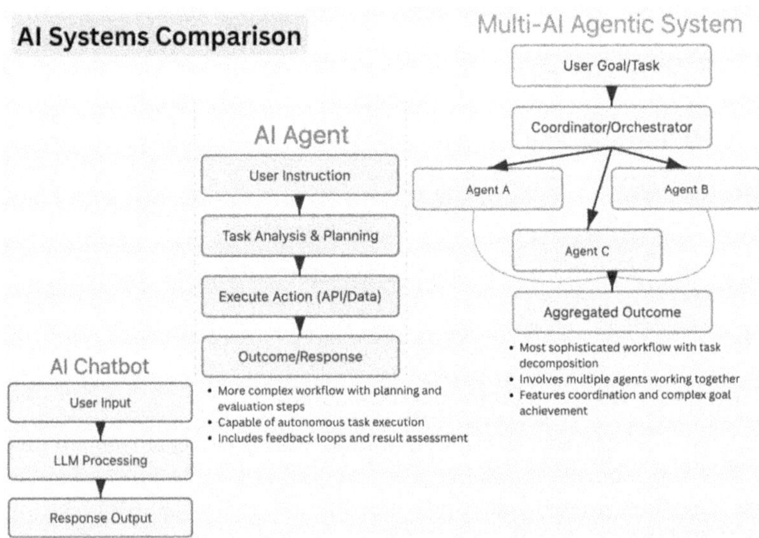

Fig. 5. Factory Interface and Deployment - Agentic API Call

simulated deployment for customer sentiment analysis, the agentic system delivered faster results and allowed real-time adjustments that enhanced the accuracy of predictive models. These improvements, supported by empirical data, highlight the transformative potential of a consultative, iterative workflow approach driven by the main pilot (Fig. 5).

5   Conclusion

Our study shows that the Agentic AI Factory offers a practical solution to the inflexibility of traditional automation systems. By using a central chatbot to engage with users throughout the workflow design process, the system adapts continuously to new data and evolving requirements. This dynamic and consultative approach not only improves processing speed and reduces errors but also makes decision-making more transparent and efficient. Overall, our findings indicate that this iterative method significantly enhances data-driven decision-making and operational efficiency, providing a fresh and effective way to manage complex workflows.

6   Future Work

Expanding integration capabilities is key, with efforts to incorporate emerging technologies such as IoT data and real-time analytics and to use advanced reinforcement learning for continuous refinement. There is also a focus on enhancing human-in-the-loop mechanisms by investigating robust feedback methods and developing automated evaluation frameworks to minimize manual intervention. Research is exploring deployment strategies across cloud and hybrid infrastructures, along with container orchestration and microservices for improved scalability. Additionally, advanced visualization and user interaction are being addressed through richer dashboards and intuitive interfaces for real-time tracking, while long-term adaptation emphasizes energy-efficient computing, cost optimization, and sustained flexibility in evolving environments.

References

1. Kamal, A., Chikezie, E.: AI-Driven end-to-end workflow optimization and automation system for SMEs. Int. J. Manage. Entrep. Res. **6**, 3666–3684 (2024). https://doi.org/10.51594/ijmer.v6i11.1688
2. Kounde, G., John, O.: The role of AI in enhancing productivity and efficiency in the construction industry (2024)
3. Xue, E., Huang, Z., Ji, Y., Wang, H.: IMPROVE: iterative model pipeline refinement and optimization leveraging LLM agents. ArXiv. https://arxiv.org/abs/2502.18530 (2025)
4. Larsson, S., Heintz, F.: Transparency in artificial intelligence. Internet Policy Rev. **9** (2020). https://doi.org/10.14763/2020.2.1469

5. Kaur, A., Singh, Y., Neeru, N., Kaur, L., Singh, A.: A survey on deep learning approaches to medical images and a systematic look up into real-time object detection. Arch. Comput. Methods Eng. 1–41 (2021). https://doi.org/10.1007/s11831-021-09649-9

6. Mosqueira-Rey, E., Hernández-Pereira, E., Alonso-Ríos, D., et al.: Human-in-the-loop machine learning: a state of the art. Artif. Intell. Rev. **56**, 3005–3054 (2023). https://doi.org/10.1007/s10462-022-10246-w

7. Moro-Visconti, R., Rambaud, S., Pascual, J.: Artificial intelligence-driven scalability and its impact on the sustainability and valuation of traditional firms. Humanit. Soc. Sci. Commun. **10**, (2023). https://doi.org/10.1057/s41599-023-02214-8

8. Mercy, O.: AI-powered workflow orchestration: maximizing business productivity and innovation. Int. Res. J. Modernization Eng. Technol. Sci. **30**, d49–d50 (2025)

9. Ravivarman, G., Ravikumar, V., Futane, P.R., Yadav, D., Mandal, S.K.: Enhancing automated real-time data analysis systems with machine learning. In: 2023 IEEE International Conference on Paradigm Shift in Information Technologies with Innovative Applications in Global Scenario (ICPSITIAGS), Indore, India, pp. 143–148 (2023). https://doi.org/10.1109/ICPSITIAGS59213.2023.10527747.

10. Wang, X., et al.: Searching for best practices in retrieval-augmented generation. ArXiv, (2024). Accessed 9 Mar 2025. https://arxiv.org/abs/2407.01219

11. Peter, H.: AI-Powered CRM integration with modular software design (2024)

12. silva, D.D., Samarasekara, P., Hettiarachchi, R.: A comparative analysis of static and dynamic code analysis techniques (2023). https://doi.org/10.36227/techrxiv.22810664.v1

13. Moncada-Ramirez, J., Matez-Bandera, J.L., González-Jiménez, J., Ruiz-Sarmiento, J.R.: Agentic workflows for improving large language model reasoning in robotic object-centered planning. Robotics **14**, 24 (2025). https://doi.org/10.3390/robotics14030024

14. Mohammed, A.S., Saddi, V.R., Gopal, S.K., Dhanasekaran, S., Naruka, M.S.: AI-driven continuous integration and continuous deployment in software engineering, pp. 531–536 (2024). https://doi.org/10.1109/ICDT61202.2024.10489475

15. Mctear, M.: Conversational AI: dialogue systems, conversational agents, and chatbots. Synth. Lect. Hum. Lang. Technol. **13**, 1–251 (2020). https://doi.org/10.2200/S01060ED1V01Y202010HLT048

16. Vargas-Solar, G., Cerquitelli, T., Espinosa-Oviedo, J.A., Cheval, F., Buchaille, A., Polgar, L.: Conversational data exploration: a game-changer for designing data science pipelines. ArXiv, (2023). Accessed 9 Mar 2025. https://arxiv.org/abs/2311.06695

17. Choubey, P.K., Peng, X., Bhagavath, S., Xiong, C., Pentyala, S.K., Wu, C.S.: Turning conversations into workflows: a framework to extract and evaluate dialog workflows for service AI agents. ArXiv, (2025). Accessed 9 Mar 2025. https://arxiv.org/abs/2502.17321

18. Manikantam, S., Akhil, P., Reddy, K.R.A., Reddy, G.P., Hariharan, S., Kekreja, V.: Enhanced automated web scraping tool with proliferation of AI techniques. In: 2024 International Conference on Innovations and Challenges in Emerging Technologies (ICICET), Nagpur, India, pp. 1–5 (2024). https://doi.org/10.1109/ICICET59348.2024.10616333.

19. Deng, J., et al.: Efficient orchestrated AI workflows execution on scale-out spatial architecture. ArXiv, (2024). Accessed 9 Mar 2025. https://arxiv.org/abs/2405.17221

20. Adewole, O.: Scalability in artificial intelligence (2023)

21. Peretz-Andersson, E., Tabares, S., Mikalef, P., Parida, V.: Artificial intelligence implementation in manufacturing SMEs: A resource orchestration approach. Int. J. Inf. Manage. **77**, 102781 (2024). ISSN 0268-4012, https://doi.org/10.1016/j.ijinfomgt.2024.102781
22. Phillips-Wren, G., Ichalkaranje, N., Jain, L.: Intelligent decision making: an ai-based approach (2008). https://doi.org/10.1007/978-3-540-76829-6
23. Çelik, T.: AI-driven production in modular architecture: an examination of design processes and methods. Comput. Decis. Making Int. J. **1**, 320–339 (2024). https://doi.org/10.59543/comdem.v1i.10825
24. Nicole, N., et al.: Continuous quality improvement of AI-based systems: the QualAI project. In: Proceedings of the 18th ACM/IEEE International Symposium on Empirical Software Engineering and Measurement (ESEM 2024). ACM, New York, NY, USA, pp. 603–607 (2024). https://doi.org/10.1145/3674805.3695393
25. Zhang, J., et al.: AFlow: automating agentic workflow generation. ArXiv, (2024). Accessed 9 Mar 2025. https://arxiv.org/abs/2410.10762

Face – Orientation Based Cattle Breed Detection Using CNN Model

A. Vijayalakshmi[1]([✉]), P. Shanmugavadivu[1], and S. Vijayalakshmi[2]

[1] Department of Computer Science and Applications, Gandhigram Rural Institute (Deemed to be University), Dindigul, Tamilnadu, India
`avijigri@gmail.com`
[2] Department of Data Science, Christ (Deemed to be University), Pune Lavasa Campus, Pune, India

Abstract. This paper focuses on cattle orientation in still frames and presents an extensive approach to reducing training bias in identifying the cattle breeds. Our approach uses YOLOv7 to detect faces and includes a new orientation-aware preprocessing step that classifies images as original, left-oriented, right-oriented, and original with inverted orientations. This enhanced dataset is used to train the model for identifying the breeds of the cattle, which is established using convolutional neural networks (CNN). The proposed action using original images and inverted orientation shows higher level accuracy in validation, demonstrating its productiveness by conveying bias during factual applications with different cattle orientations, even though left-oriented approach achieves the highest accuracy. These results fill a gap in the literature by highlighting the significance of orientation-receptive training for reliable breed identification of cattle.

Keywords: Convolutional Neural Networks (CNN) · Data augmentation · Label-Orientation · Face detection · Breed identification

1 Introduction

India has rich and varied genetic resources for cattle. Based on their use in dairying or agricultural, India's 43 native cattle breeds are officially registered; these types are fundamentally divided into three categories: dairy, draught, and dual purpose breeds [1].

Breeding animals for human consumption such as products made of dairy, food, and milk is referred to as cattle farming. The productivity per animal in the cattle production system has increased throughout time, but many nations are seeing a decline in small farms as a result of a lack of available area for crop cultivation and slurry spreading [2].

Reliability and accuracy in cattle identification are essential for tracking cattles during the period between their birth till meat products and for preventing fraudulent insurance claims that resolve the issue in mismatched claim as well as covered livestock. Individual cattle identification is also necessary for the most crucial aspects of the genetic improvement system and dairy production, including breeding, calving, immunization, and cattle selection [3].

J. Shreyas et al. (Eds.): CODE-AI 2025, CCIS 2689, pp. 310–323, 2026.
https://doi.org/10.1007/978-3-032-19318-6_29

For individual identification, conventional techniques like ear tagging, ear notching, and electronic devices are being used [4–6]. Nevertheless, these techniques have drawbacks including misplaced tags, broken electronics, and problems with reusability [7]. A rising number of people are interested in using cutting-edge computer vision and machine learning technology for precision livestock management in order to solve these deficiencies [2].

Despite these advancements, face detection in livestock remains extremely challenging in real-world settings like cattle feedlots due to factors like severe pose variation, subtle lighting changes, false acceptances due to critical background, colour resemblance between livestock with the background, shape deformation, and occlusion. As such, a more comprehensive evaluation of the face identification algorithm's effectiveness in various livestock production scenarios is needed [8].

There has been a boom in interest in integrating computer-vision technology in the dynamic field of cattle breed identification, especially in using visual information retrieved from cattle photographs [9, 10].

Since 2014, the accuracy of cattle identification has been greatly improved by all-encompassing approach of machine learning models, which includes Support Vector Machine (SVM), K-Nearest Neighbour (KNN), Artificial Neural Networks (ANNs), random forests, decision trees, and the logistic regression. Deep learning models, such as Convolution Neural Network (CNN), Inception and ResNet have demonstrated superior performance [11–23].

Cattle identification models have made remarkable strides, especially in the area of two-stage detection. However, a crucial gap in resolving practical issues such as training bias still exists. This problem is immediately addressed by our suggested orientation-aware face identification method as farmers frequently take pictures in their favourite orientations, which leaves datasets without the variety needed to guarantee model robustness when handling different cattle breeds and orientations.

2 Background

Globally, cattle are among the most commonly farmed animals, with a number second only to hens. Being draught animals that can convert low-quality grain into high-energy milk and muscle, they provide a significant share of the food and means of sustenance for almost 6 billion people. Ever since cattle were originally domesticated around 10,000 years ago, several distinct varieties have been created. Numerous heritable characteristics are present in these breeds, including as altered morphological characteristics like skin colour and horn form, as well as variances in production factors like milk output, environmental adaption, and disease resistance [24].

Novel facial representation algorithms, such as Local Binary Pattern (LBP), have been explored for their efficiency in extracting discriminative features from cattle faces. LBP divides images into regions, providing detailed descriptors used to construct facial histograms. The technique exhibited superior efficiency and accuracy, even in the presence of variations in illumination and image misalignment [25].

The development of specialized models, such as the two-branch convolutional neural network (TB-CNN), has also contributed to advancing cattle breed identification. The

TB-CNN utilizes distinct channels for feature extraction from cattle face images captured at different angles, achieving high recognition rates across diverse datasets. Notably, the inclusion of a squeeze-and-excitation block enhances feature extraction capabilities, further improving overall performance [26].

Beyond traditional cattle breeds, Japanese black cattle, lacking distinctive blackwhite patterns, have been successfully identified using image processing and neural network algorithms. The application of image transformation verification, including variations in brightness, distortion, noise, and angle, underscored the robustness of the algorithm [27].

Face detection, a crucial aspect of cattle identification technology, has seen significant advancements. The RetinaNet detection model demonstrated effectiveness in unstructured scenes, achieving high precision scores with rapid processing times. This model excelled in scenarios with varying illumination, overlapping, and occlusion, showcasing its potential for real-world deployment [8].

Moreover, a comprehensive cow face recognition scheme, coupling CNN-based face detection with the PnasNet5 recognition model, has been proposed. The recognition accuracy of various models, including AlexNet, VGGNet, ResNet, and PnasNet5, was evaluated on a large-scale dataset, revealing PnasNet-5 as the most accurate with a recognition rate of 94.1% [28].

Researchers used Fully Connected Neural Networks (FCNN) in R with GPU acceleration to identify Hereford and Simmental cow breeds from 600 photos with 97% accuracy. Training on an 11th generation i7 CPU and RTX 3060 GPU using the EBimage and Keras packages was finished in 69 seconds. According to the study, breed identification of cattle may be greatly aided by FCNN-based picture classification, which can help farmers make better decisions. Additional research on resilience and scalability might improve useful applications in a variety of environmental settings, perhaps revolutionising livestock management techniques with effective, automated breed recognition systems [29].

Using the You Only Look Once (YOLO) algorithm, the study focuses on cattle detection and breed classification, assessing performance based on breed-specific body areas. Google photos is used as the source of photos for a bespoke dataset. The results of the experiment show how successful the suggested method is; utilising the YOLO algorithm, it was able to identify and categorise cow breeds with a high accuracy rate of 92.85%. This study highlights YOLO's potential for precisely identifying and categorising cow breeds, with potentially useful implications for livestock management and agricultural research [30].

The lack of variation in cattle breeds and the identical orientation of bovine photos in datasets might induce bias that affects the model's performance in real-world scenarios and is not found or examined in the research mentioned above. The training datasets should aim for variety because farmers frequently take pictures in a preferred direction. This will guarantee the model's resilience while manipulating different cattle breeds and orientations.

3 Proposed Model

The proposed method shown in Fig. 1 identifies the cattle breed including bias-mitigated identification. The images are taken from Cattle images dataset. The dataset has been built by capturing the cattle through the mobile phone. From the cattle image, the first step is cattle detection. The cattle was detected by removing the other things from the back ground. The rectangle box is taken for specifying the cattle. The second step is cattle alone was taken by cropping and then resized to 256×256. Third step is using YOLOv7 custom model, the cattle face detection was done and it is also mentioned by rectangular box. In fourth step Orientation detection was done. The image is splitted into two parts. In the fifth step, we are checking whether the image is right oriented or left oriented. The same process was continued for all the images. Then Orientation dataset was considered. Four different image dataset were taken. They are 1. Original orientation, 2. Right orientation, 3. Left orientation and 4. Original + Inverted orientation. These datasets were used in customized Convolution Neural Network and the result are evaluated.

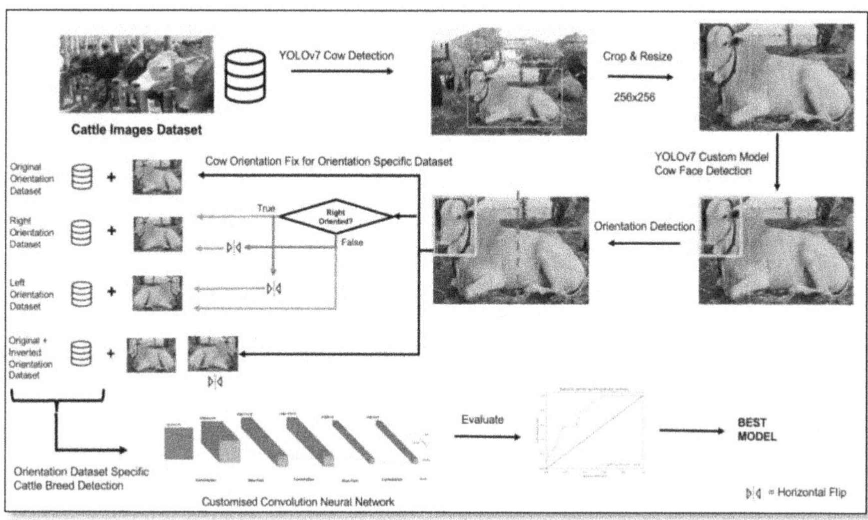

Fig. 1. Model Framework Architecture Diagram

3.1 YOLOv7 for Object Detection

Faster R-CNN demonstrated state-of-the-art accuracy across various datasets [31]. YOLO's single-stage approach excelled in real-time detection on the PASCAL VOC 2007 dataset [32]. YOLOv7 performs best among all object detection systems that are existing previously when compared against both the speed as well as the accuracy, spanning a frame rate scale from 5 to 160 frames per second. It achieves the highest accuracy, with an impressive 56.8% average precision (AP), surpassing all other real-time object detectors operating at 30 FPS or higher on the GPU V100 [33].

3.2 YOLOv7 for Individual Cattle Detection

The pre-trained COCO weights make it easier for the algorithm to recognize cattles in the supplied photos. For each object detected in this study, the YOLOv7 algorithm produces the synchronization in the bounding box, labeling the classes, and the confidence value (p) at a predetermined confidence level of 0.25. Once cattles have been located in the photos, the next stage is to crop the photos so that the focus is solely over the cattle's body. This involves carefully cropping the image using the interrelation of bounding box that are obtained from the algorithm using YOLOv7. By removing unnecessary information, this cropping process simplifies the data and focuses attention only on the cattle's body, which is the region of interest. For cropping, the cattle which is having the high up confidence (p value) between the several identified cattles are chosen. This procedure is shown in Fig. 2, in which the cattles having the highest confidence level are chosen and cropped in a frame with several cattles. After cropping, the image is resized to 256 by 256.

(a)

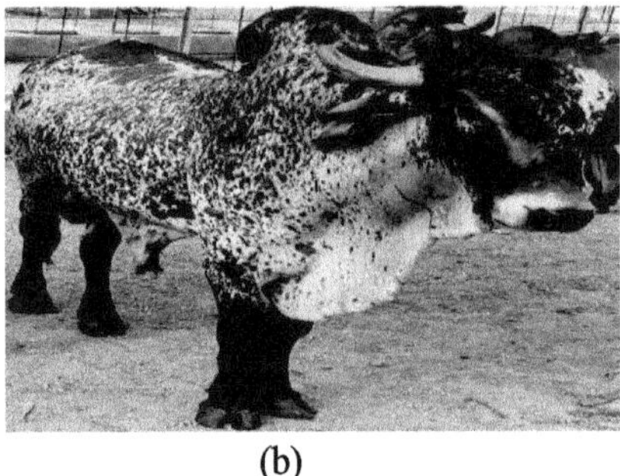
(b)

Fig. 2. Cattle Identification Process. (a) Body of Cattle Identification using YOLOV7 (b) Image after Cropping and Resizing

3.3 Custom Face Detection Model

3.3.1 Annotation of Cattle Faces

Rectangular annotations were applied to cattle faces, including those with 1, 2, or multiple faces in one image (Fig. 3). To account for crossbreeding, dehorning, or disbudding practices, horns were intentionally excluded during annotation to reduce biases.

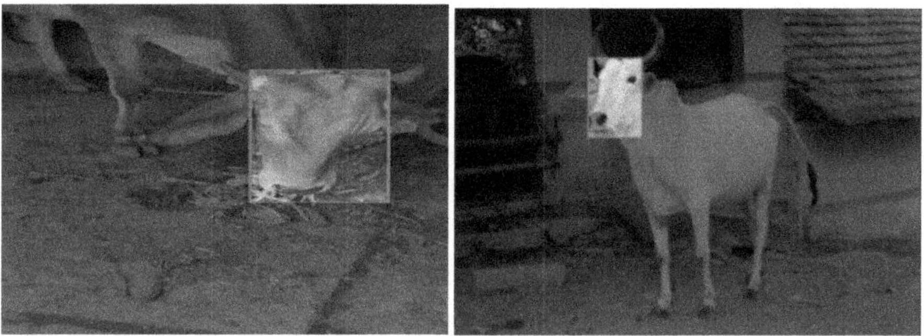

Fig. 3. Cattle Face Annotation Samples

3.3.2 YOLO V7 Training Process

Using 32 as the batch size, 0.05 as the confidence interval, and Coco weights for initialization, the YoloV7 model was trained on a dataset of 442 images. 46 images were used for testing and validation during the model's 100 training epochs as shown in Fig. 4.

Fig. 4. Batchwise Training Custom Model

3.3.3 Evaluation of YOLO Face Detection Training

The model's performance, as assessed by macro-precision (mean average precision) and Recall, demonstrated constant improvement till 15 epochs, and then a moderate increment and 1.0 at plateau, as depicted in Fig. 5.

3.4 Face Spotting based Orientation Aware System

3.4.1 Cattle Position Detection

The process's current step entails adjusting the orientation. The process of classifying a cattle as left oriented begins with the detection of its orientation. The decisionmaking procedure is simple but efficient: in the image, if the detected cattle face from the custom YOLO model is oriented toward the left side horizontally (50%), it is considered left-oriented.

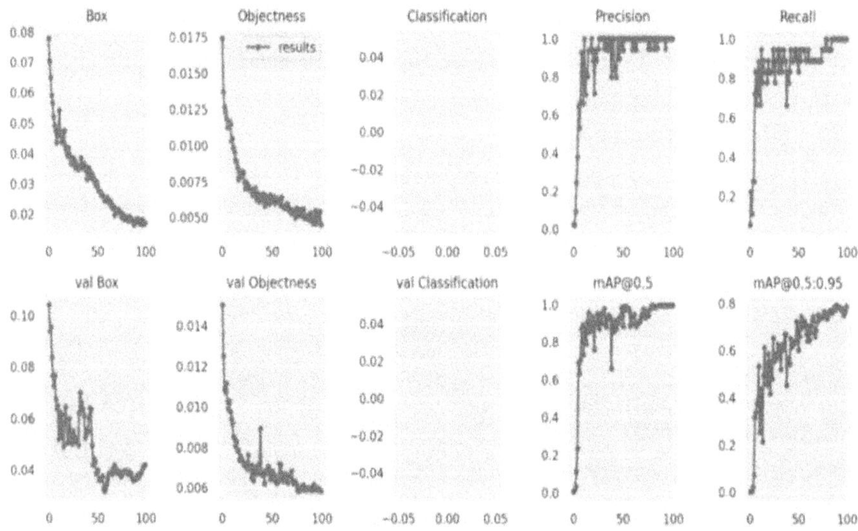

Fig. 5. YOLOv7 Face Custom Model Evaluation

On the other hand, the orientation is grouped as right-oriented when the face of the cattle is on the right side horizontally (50%) of the image. By taking into account the intrinsic spatial information present in the images, this orientation-aware method makes sure that the proposed model is trained with a wide range of acclimatization that are frequently seen in real-world situations.

These orientation labels are visually represented in Fig. 6, where the horizontal midpoint is shown by the yellow line. Fig. 6a depicts a left-oriented cattle, since the face of the cattle is located on the left side in the image. Figure 6b depicts a right-oriented cattle, as the face of the cattle is located on the right side of the image.

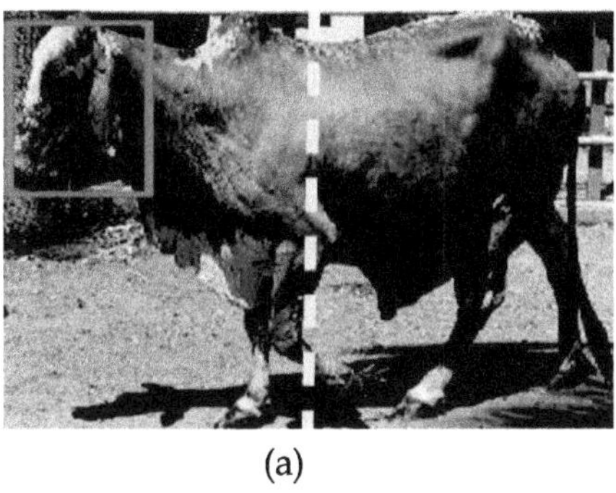
(a)

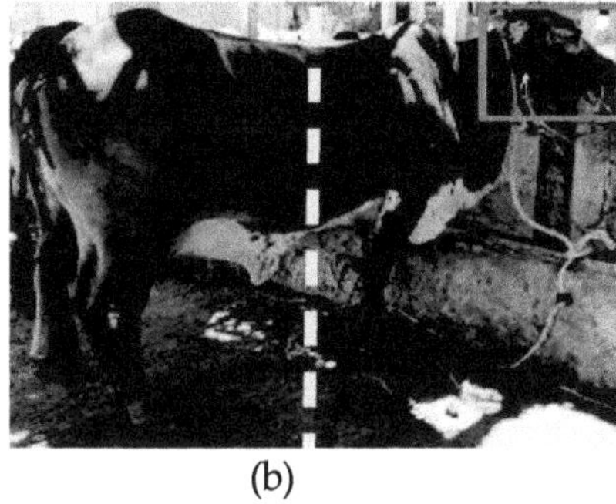
(b)

Fig. 6. Orientation Detection System. **(a)** Left Oriented **(b)** Right Oriented

3.4.2   Cattle Orientation Specific Datasets

Four excersised datasets have been generated through post detection of the cattle orientation.

(I) **Right Orientation:** Images within this dataset to consists of cattles in right side horizontal orientation.

(II) **Left Orientation:** Images within this dataset to consists of cattles in left side horizontal orientation.

(III) **Original Orientation:** Images within this dataset to consists of cattles in the original orientation.

(IV) **Original + Inverted Orientation:** Images within this dataset to consist of cattles in the original orientation and horizontal flipped orientation.

The process of processing an image from the dataset is shown in full in Fig. 7. The Original Orientation Dataset contains the original image. The Original + Inverted Orientation Dataset contains both the original and its horizontally flipped versions. During its horizontally transposed impression are filed in the contrast orientation dataset, this associated orientation is analyzed and filed within the equivalent side (left/right).

4   Training the proposed Model

4.1   Dataset

There are four datasets, original dataset (2255 images), left-oriented dataset (1572 images), right-oriented dataset (1572 images), original + inverted dataset (3144 images) are taken as a consequence of the System's Orientation preprocessing. Every dataset has six cattle breeds, which is classified as multi-class classification problem. The breeds include Gir cattle, Bargur cattle, Rathi cattle, Kangayam cattle, Pulikulam cattle and Tharparkar cattle. The cattle images have been sourced and taken from multiple local farms, in Tamil Nadu, India. The images were seized by dairy farmers, under the supervision of the researchers, to ensure that use-case scenario is mimicked. The breeds. The classes were balanced, and each cattle breed, in the original orientation dataset/original dataset had a range of 380 to 450 images each. 80% of the data is used for training, while 20% is used for validation across all datasets

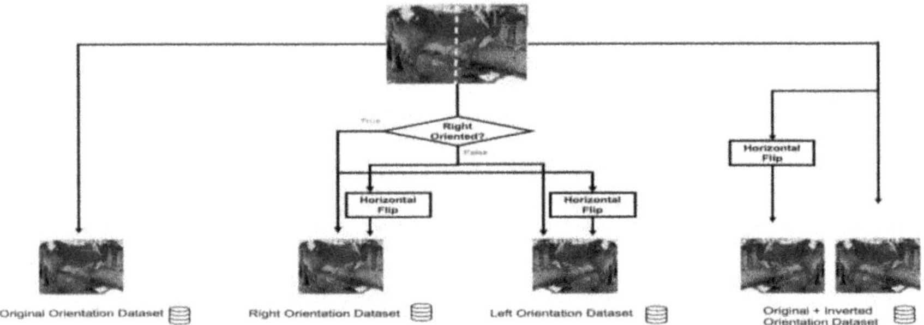

Fig. 7. Orientation Specific Dataset Segregation Pipeline

4.2   Breed Identification Using CNN (Convolution Neural Network)

Deep neural networks and Convolutional Neural Networks (CNNs) are conceivable to stand out as a prominent classifier for image classification tasks. The architecture employed in our study, illustrated in Fig. 8 pursue an excellence standard CNN formation widely recognized for its productiveness in image classification tasks. This architecture comprises multiple convolutional layers, each followed by max-pooling layers to extract and emphasize important features from the input images. The convolutional layers utilize small filters to capture local patterns, while the max-pooling layers down sample the spatial dimensions, reducing computational complexity.

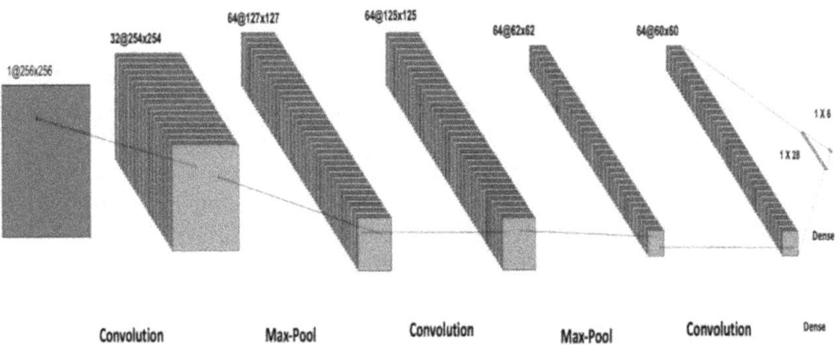

Fig. 8. Cattle Breed Identification CNN Model

The model begins with an initial convolutional layer with 32 filters, each with a 3x3 kernel, employing the rectified linear unit (ReLU) activation function. Subsequent max-pooling layers with 2x2 pooling windows help retain essential information while reducing the spatial dimensions. Additional convolutional layers with increased filter size (64 filters) follow, further capturing intricate features in the image. The terminating convolutional layer is deflated to create a vector of features, which is then fed into densely connected layers. The fully connected layers, aid the learning of high-level abstractions, and the output layer consists of six nodes, corresponding to the six cattle breeds, Gir cattle, Bargur cattle, Rathi cattle, Kangayam cattle, Pulikulam cattle and Tharparkar cattle under consideration in the classification task. The SoftMax activation function is applied to the output layer, enabling the model to predict the probability distribution across the different breed classes. The dataset taken into consideration is trained using the CNN model mentioned above for 10 epochs.

5   Model Evaluation

The model evaluation after 10 epochs reveals distinct performance characteristics for each orientation strategy. In the case of the Original-Orientation dataset, the model demonstrated atypical training accuracy (99.96%) and maintained an admirable validation accuracy of 86.94%. The loss and validation loss trends remained balanced, suggesting adaptability to diverse cattle orientation scenarios. The effectiveness of this approach in capturing features from images with varied orientations underscores its robustness in real-world applications.

In case of the Inverted and Original Orientation dataset, the model exhibited high training accuracy (99.94%) and outperformed with a superior validation accuracy of 92.68%. The consistent decrease in both loss and validation loss throughout the training process indicates robust learning. This orientation strategy appears particularly effective in mitigating bias during real-world scenarios where cattle orientations vary. The model's proficiency in handling both original and inverted orientations showcase its versatility in diverse applications.

On the other hand, the orientation-specific datasets, namely Right-Oriented and Left-Oriented, presented unique obstacles. The Right Orientation Only dataset showed a decrease in loss over epochs but lagged in both accuracy (97.69%) and validation accuracy (79.30%). This suggests potential bias issues when training solely on rightoriented images. In contrast, the Left Orientation Only dataset achieved the highest training accuracy (100.00%) but exhibited a slightly lower validation accuracy (85.67%). While it demonstrated efficient training, the model's focus on left-oriented features may lead to bias in scenarios with diverse cattle orientations. The same is evident through Fig. 9 and Table 1.

Table 1. Evaluation Metrics for model in affix 10 epochs

Dataset formed	Model Accuracy	Validation Accuracy	Model Loss	Validation Loss
Original images	0.999603	0.869427	0.001749	0.880227
Inverted + Original images	0.999446	0.926829	0.000939	0.448362
Right-Orientation images	0.976948	0.792994	0.079944	1.379230
Left-Orientation images	1.000000	0.856688	0.000100	1.329782

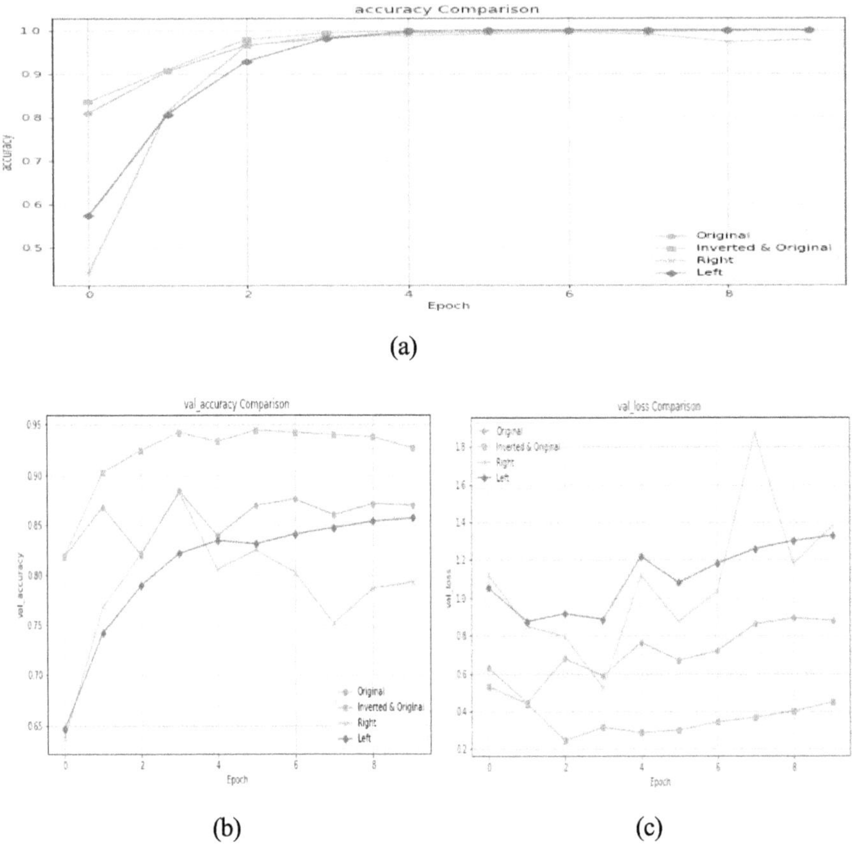

Fig. 9. Evaluation of the model: Accuracy and Loss Vs Plots of Epochs (a) Model Accuracy, (b) Validation Accuracy (c) Model Validation Loss

6 Conclusion

This proposed approach for alleviating the bias training in identifying the breed of the cattle has demonstrated significant advantages, emphasizing the importance of orientation-aware training. Leveraging YOLOv7 for face detection and a Convolutional Neural Network (CNN) for breed identification, our approach integrates a unique orientation-aware preprocessing step, classifying images into four orientations: right-only, left-only, original, and original with inverted orientations. The results indicate that leftoriented image perspective attains the excessive accuracy, whereas the original + inverted orientation procedure shows the more validation accuracy, exposing its efficacy for labeling bias from real-life applications where cattle orientations vary. While our system has demonstrated robust performance in capturing features from images with varied orientations, there are some limitations. The orientation-specific datasets, particularly right-oriented and left-oriented, presented challenges, suggesting potential bias issues. Further refinement may be needed to address these biases and enhance model generalization.The vast scope of our approach lies in its adaptability to diverse real-world scenarios, as validated by its proficiency in handling both original and inverted orientations. As digital gadgets become widespread on farms, our system offers a practical solution for cattle breed identification, ensuring robustness in applications where variations in cattle orientation are prevalent. Future tasks may be necessitate lengthening the actual dataset, elaborating the orientation exploration process, and prospecting other advanced architectures using neural network to further enhance the system's performance and applicability in the evolving landscape of precision livestock farming.

References

1. Davies, N., Villablanca, F.X., Roderick, G.K.: Determining the source of individuals: multilocus genotyping in nonequilibrium population genetics. Trends Ecol. Evol. **14**(1), 17–21 (1999)
2. Mahmud, M.S., Zahid, A., Das, A.K., Muzammil, M., Khan, M.U.: A systematic literature review on deep learning applications for precision cattle farming. Comput. Electron. Agric. **187**, 106313 (2021)
3. Xu, B., et al.: CattleFaceNet: a cattle face identification approach based on RetinaFace and ArcFace loss. Comput. Electron. Agric. **193**, 106675 (2022)
4. Awad, A.I.: From classical methods to animal biometrics: a review on cattle identification and tracking. Comput. Electron. Agric. **123**, 423–435 (2016)
5. Neary, M., Yager, A.: Methods of livestock identification (no. as556-w). West Lafayette: Perdue University (2002)
6. Ruiz-Garcia, L., Lunadei, L.: The role of RFID in agriculture: applications, limitations and challenges. Comput. Electron. Agric. **79**(1), 42–50 (2011)
7. Roberts, C.M.: Radio frequency identification (RFID). Comput. Secur. (1), 18–26, (2006)
8. Xu, B., et al.: Evaluation of deep learning for automatic multi-view face detection in cattle. Agriculture **11**(11), 1062 (2021)
9. Andrew, W., Hannuna, S., Campbell, N., Burghardt, T.: Automatic individual Holstein Friesian cattle identification via selective local coat pattern matching in RGB-D imagery. In: 2016 IEEE International Conference on Image Processing (ICIP), pp. 484–488 (2016)
10. de Lima Weber, F., de Moraes Weber, V.A., Menezes, G.V., Junior, A.D.S.O., Alves, D.A., de Oliveira, M.V.M., et al.: Recognition of Pantaneira cattle breed using computer vision and convolutional neural networks. Comput Electron Agric. **175**, 105548 (2020)
11. Yang, Z., Xiong, H., Chen, X., Liu, H., Kuang, Y., Gao, Y.: Dairy cow tiny face recognition based on convolutional neural networks. In: Chinese Conference on Biometric Recognition, pp. 216–222. Springer (2019)
12. Joachims, T.: Making large-scale SVM learning practical. Technical report (1998)
13. Cover, T., Hart, P.: Nearest neighbor pattern classification. IEEE Trans. Inf. Theory **13**(1), 21–27 (1967)
14. McCulloch, W.S., Pitts, W.: A logical calculus of the ideas immanent in nervous activity. Bull. Math. Biophys. **5**(4), 115–133 (1943)
15. Rumelhart, D.E., Hinton, G.E., Williams, R.J.: Learning representations by back-propagating errors. Nature **323**(6088), 533–536 (1986)
16. Kumar, S., Pandey, A., Satwik, K.S.R., Kumar, S., Singh, S.K., Singh, A.K., et al.: Deep learning framework for recognition of cattle using muzzle point image pattern. Measurement **116**, 1–17 (2018)
17. Alzubaidi, L., Zhang, J., Humaidi, A.J., Al-Dujaili, A., Duan, Y., Al-Shamma, O., et al.: Review of deep learning: Concepts, CNN architectures, challenges, applications, future directions. J Big Data **8**(1), 1–74 (2021)

18. He, K., Zhang, X., Ren, S., Sun, J.: Deep residual learning for image recognition. In: Proceedings of the IEEE Conference on Computer Vision and Pattern Recognition, pp. 770–778 (2016)
19. Simonyan, K., Zisserman, A.: Very deep convolutional networks for large-scale image recognition. arXiv preprint 2014, arXiv:1409, pp. 1556 (2014)
20. Iandola, F.N., Han, S., Moskewicz, M.W., Ashraf, K., Dally, W.J., Keutzer, K.: SqueezeNet: AlexNet-level accuracy with 50x fewer parameters and 0.5 MB model size. arXiv preprint 2016, arXiv:1602, pp. 07360 (2016)
21. Szegedy, C., Liu, W., Jia, Y., Sermanet, P., Reed, S., Anguelov, D., et al.: Going deeper with convolutions. In: Proceedings of the IEEE Conference on Computer Vision and Pattern Recognition, pp. 1–9 (2015)
22. Carion, N., Massa, F., Synnaeve, G., Usunier, N., Kirillov, A., Zagoruyko, S.: End-to-end object detection with transformers. In: European Conference on Computer Vision, pp. 213–229. Springer (2020)
23. Ren, S., He, K., Girshick, R., Sun, J.: Faster R-CNN: towards real-time object detection with region proposal networks. In: Advances in Neural Information Processing Systems, pp. 91–99 (2015)
24. Talenti, A., et al.: A cattle graph genome incorporating global breed diversity. Nat. Commun. **13**(1), 910 (2022)
25. Cai, C., Li, J.: Cattle face recognition using local binary pattern descriptor. In: 2013 Asia-Pacific Signal and Information Processing Association Annual Summit and Conference pp. 1–4 (2013)
26. Weng, Z., Meng, F., Liu, S., Zhang, Y., Zheng, Z., Gong, C.: Cattle face recognition based on a Two-Branch convolutional neural network. Comput. Electron. Agric. **196**, 106871 (2022)
27. Kim, H.T., Ikeda, Y., Choi, H.L.: The identification of Japanese black cattle by their faces. Asian Australas. J. Anim. Sci. **18**(6), 868–872 (2005)
28. Yao, L., Hu, Z., Liu, C., Liu, H., Kuang, Y., Gao, Y.: Cow face detection and recognition based on automatic feature extraction algorithm. In: Proceedings of the ACM Turing Celebration Conference-China (2019)
29. Yeşil, M., Göncü Karakök, S.E.R.A.P.: Recognition of Hereford and Simmental cattle breeds via computer vision. Iranian J. Appl. Animal Sci. **13**(1) (2023)
30. Yılmaz, A., Uzun, G.N., Gürbüz, M.Z., Kıvrak, O.: Detection and breed classification of cattle using yolo v4 algorithm. In: 2021 International Conference on Innovations in Intelligent Systems and Applications (INISTA), pp. 1–4. IEEE (2021)
31. Ren, S., He, K., Girshick, R., Sun, J.: Faster R-CNN: towards real-time object detection with region proposal networks. Adv. Neural Inform. Process. Syst. **28**, 91–99 (2015)
32. Redmon, J., Divvala, S., Girshick, R., Farhadi, A.: You only look once: unified, realtime object detection. In: Proceedings of the IEEE Conference on Computer Vision and Pattern Recognition, pp. 779–788 (2016)
33. Wang, C.Y., Bochkovskiy, A., Liao, H.Y.: YOLOv7: trainable bag-of-freebies sets new state-of-the-art for real-time object detectors. In: Proceedings of the IEEE/CVF Conference on Computer Vision and Pattern Recognition, pp. 7464–7475 (2023)

AI-Enabled Quantum-Resistant Multipath Crypto-Graph Protocol (QR-MCP) for Secure Communications in Post-Quantum Networks

Nishanth Shet[(✉)] , R. Chinmai , Preethi , and Y. V. Srinivasa Murthy

Department of Information Technology, Manipal Institute of Technology Bengaluru,
Manipal Academy of Higher Education, Manipal 576 104, Karnataka, India
`nishanth.mitblr2022@learner.manipal.edu` , `vishnu.murthy@manipal.edu`

Abstract. As quantum computing grows, the security of RSA and ECC offers is becoming increasingly flush. A novel Quantum Resistant Multipath CryptoGraph protocol (QR-MCP) has been proposed in this work, which is a multi-layered security framework that combines lattice-based cryptography, SPHINCS+ post-quantum signatures, onion routing, with AI-driven anomaly detection for long lifetime security and privacy of data. Encrypting messages in multiple paths increases security as larger network attackers cannot intercept an entire message. Ledgering on the blockchain is also used for integrity verification for the protocol, and the protocol also uses AI models to detect anomalies in real-time. QR-MCP has shown to be resilient to simulated cyber attacks such as man-in-the-middle, traffic analysis as well as collusion based decryption. Future scalability to new threats will gain strength with the introduction of homomorphic encryption and zero-knowledge proof (ZKP).

Keywords: AI-Driven Anomaly Detection · Blockchain Logging · Lattice-Based Encryption · Onion Routing · Post-Quantum Cryptography (PQC) · Quantum Computing Resistance · Quantum-Resistant Encryption · Secure Multi-Party Computation (MPC) SPHINCS+ Digital Signatures

1 Introduction

5In the centuries that have passed, cryptography has come to be all about protecting information through math [3]. Simple ways of hiding messages included having the Caesar cipher or things like that. In more recent times a variety of symmetric encryption algorithms based on the classes of Data Encryption Standard (DES) have been introduced. The 56-bit key of DES was not as secure as we would have liked, but DES was the standard for government encryption [12]. In 1977, RSA encryption came along using the computation of factoring big prime numbers as the basis of a spectacular attack on public key cryptography

© The Author(s) 2026
J. Shreyas et al. (Eds.): CODE-AI 2025, CCIS 2689, pp. 324–334, 2026.
https://doi.org/10.1007/978-3-032-19318-6_30

[10]. Finally, in 2001 the Advanced Encryption Standard (AES) came out and replaced DES, because it was a more secure algorithm using key sizes of 128, 192, and 256 bits [6]. The result is now widely used in secure communication across many industries today. Encryption has moved from the basics of letter swaps to the very complicated elliptic curve cryptography (ECC) methodology over the years. Overall, the same has always been the main goal of keeping data safe by keeping their confidentiality, authenticity, and unaltered [2].

Algorithms such as AES, RSA, and ECC have, to an extent, protected data for many years, but nothing is unbeatable. The protocols standardly supported by cryptographic protocols are hard satisfiability problems such as factoring integers and discrete logarithms which are hard but not infeasible [11]. Encryption has evolved as well alongside those attack methods such as brute force attacks, side channel attacks, and cryptanalysis. That is to say DES, which was considered to be secure, broke in one day because it had a too small key space. However, AES-128 is stronger, but it has not meant that the latter has become completely secure as it could still be cracked within a certain amount of time if attackers had enough computing power [9]. It follows that RSA encryption is also susceptible to advanced factoring techniques and attacks that make use of distributed computing [7].

The HPC has grown rapidly and the supercomputers are capable of making incredible calculations. It has made cryptographic attacks faster. The emergence of quantum computing is the another bigger threat. For instance, Shor's algorithm [1] can factor a large number much faster than regular algorithms can, and, hence, RSA, ECC and Diffie-Hellman protocols become useless when the world is enriched with quantum computers. For example, a 2048 bit RSA key that requires thousands of years of current super computers would be cracked in a few hours with a sufficiently powerful quantum computer. The major tech companies are investing in quantum processors and hence, the risk to our current encryption standards is drawing closer, The solution is the emergence to develop quantum resistant protocols.

This research paper is divided into seven main parts, each examining a different aspect of the Quantum-Resistant Multipath Crypto-Graph Protocol (QR-MCP). Section 2 offers a detailed review of existing research on classical and post-quantum encryption. Section 3 describing the different parts of QR-MCP, including lattice-based encryption for quantum resistance. Section 4 presents the results of our experiments, with detailed evaluations of performance across different network setups. Section 5 provides the conclusion, summarizing our findings and their implications. It also outlines future work.

2   Literature Review

The use of cryptography has evolved through time as researchers developed innovative techniques beyond AES, RSA and ECC encryption schemes to counter fresh cyber threats. The security provided by traditional encryption methods which have been in use for decades now falls prey to modern sophisticated digital assaults because of advancing computer processor capabilities.

Network security would benefit from enhanced security measures because experts propose adjustments to its data routing framework. Traditional onion routing acts as the infrastructure of Tor networks by secure message transmission through successive layer encryptions [5]. Network traffic correlation with timing analysis attacks creates vulnerabilities that threaten to breach user anonymity while using this method. Security is boosted through the implementation of dynamic onion routing since it creates more complex communication paths which difficult attackers from following the data [8]. The implementation of multipath communication together with encryption safeguards messages by distributing data transmission across multiple separate pathways thus minimizing complete message interception. The evidence obtained by research demonstrates that graph-based routing algorithms together improve network security with increased resilience with the help of multipath transmission methods [4].

Lattice-based cryptographic solutions use the combination of Shortest Vector Problem (SVP) and Learning With Errors (LWE) problem instead of vulnerable RSA-ECC algorithms when facing quantum computers. These resistant security problems make lattice-based encryption the next perfect choice to protect security protocols. The National Institute of Standards and Technology (NIST) continues post-quantum algorithm evaluation that shows broad implementation readiness of CRYSTALS-Kyber and NTRUEncrypt. Cryptography researchers investigate two additional encryption approaches such as code-based and multivariate polynomial cryptographic systems for diversity against quantum attack methods.

The implementation of encryption gets additional support from network security protocols such as onion routing that help ensure privacy and anonymity. The Tor network represents the most popular implementation of onion routing by using multiple encryption layers in a process of traversing randomly selected nodes. Research demonstrates standard onion routing is susceptible to traffic analysis attacks due to attacks that control numerous relay nodes. Researchers have proposed dynamic onion routing to address these security risks because it changes encryption layers and relay paths through network conditions together with user activities. The adaptive protocol creates two privacy advantages by making timing correlation attacks less effective and providing improved secrecy. By using onion routing together with multipath communication the security improves through fragments taking different paths which prevents both failure and interception of communication.

The increasing adoption of AI-driven cybersecurity methods in cryptography supports critical operations for intrusion detection systems and key management functions as well as automated cryptographic process enhancements. The identification of anomalous encrypted traffic is performed through deep learning models in machine learning architectures. Network behaviors and data flow patterns together with packet transmission time changes serve as inputs for AI-based intrusion detection systems which detect security breaches in advance. The detection of real-time unusual behavior relies on three different artificial intelligence models which include Support Vector Machines and Recurrent Neural Networks and Convolutional Neural Networks. The models develop their security skills through new cyber threat learning that outperforms conventional rule-based security systems. Reinforcement learning algorithms optimize cryptographic key generation through security setting adjustments which depend on newly discovered attack methods.

Blockchain Technology implementation in cryptographic security has significally resulted in enhanced data authenticity and integrity inspection. The combination of blockchain with decentralized solutions restores encryption transaction records in an unalterable format that mitigates tampering issues of traditional centralized systems. Merkle tree hashing serves blockchain for authenticating stored data by making data block alterations detectable. Blockchain serves as a reliable security measure when used for managing encryption keys because it maintains their protection from illegal access. Research indicates the implementation of security systems based on blockchain authentication working together with AI anomaly detectors to create systems that heal and protect themselves from attacks. Through this approach networking systems can respond to security threats automatically which results in significantly less data breaches.

The progress of quantum computation makes RSA together with ECC and Diffie-Hellman algorithms become less effective due to ShorâĂŹs algorithm's rapid ability to factor large numbers. Post-quantum cryptography (PQC) has appeared as an answer to modern advances in cryptography through various promising cryptographic techniques. Researchers consider Lattice-based cryptography which uses Learning With Errors (LWE) and Shortest Vector Problem (SVP) to be quantum attack resistant and it forms the foundation for NIST standardization candidates Kyber and Dilithium. The cryptographic method developed by McEliece depends on the hard decoding problem for random linear codes to provide long-term protection even when key sizes are large. Multivariate polynomial cryptography implements systems of nonlinear polynomial equations as an alternative yet faces practical implementation obstacles at present.

3   Methodology

QR-MCP uses lattice cryptography to make itself quantum-resistant. It does this by building itself on computationally infeasible problems such as the Shortest Vector Problem (SVP) and Learning With Errors (LWE), which are both infeasible for quantum computers as well as classical computers. This makes the communications quantum-resistant to quantum decryption attacks.

For anonymity and privacy, onion routing is used by QR-MCP. The encryption method encrypts the message in layers before sending it. A layer is decrypted at each hop node so that the remaining layers carry information to the next hop. The method hides the source and destination, and makes data private. Randomized transit layers of encryption also eliminates the chances of traffic analysis even with quantum-capable devices.

QR-MCP is protected using secure multipath communication (MPC). The data is broken into segments and transmitted over multiple paths in the network in parallel. A segment is transmitted over a random path, reducing interception. In addition to offering immunity to man-in-the-middle attack, the process also renders the network fault-tolerant in the sense that it reduces the effect of single-path failure.

AI-driven anomaly detection is one of the key modules in QR-MCP that offers real-time threat detection and neutralization. Based on machine learning, it cross-verifies traffic on the network with transmission time, data transmitted, and encryption integrity. Intrusions of an abnormal nature, i.e., unauthorized access or fresh intrusions in cyber space, are identified. Looking for deviation from parameters established previously–typically two standard deviations from normal behavior–the administrators get notified. This mode in real time allows QR-MCP to neutralize advanced attacks, e.g., zero-day attacks, in real time.

Our approach utilizes an advanced ensemble learning model to detect anomalies and potential cyber threats within an onion routing framework. By integrating multiple machine learning algorithms, the system enhances attack detection accuracy while minimizing false positives. The ensemble combines various models, each contributing unique strengths to the overall detection mechanism. It uses the Isolation Forest (IF) algorithm to detect anomalies by modeling the normal traffic patterns and producing deviations. Their extensions of classification accuracy includes using random forest (RF) that utilize multiple decision trees and support vector machine (SVM) for pattern recognition in encryption times and network behaviors. Extreme Gradient Boosting (XGBoost) is added to handle complex attack strategies so that efficient learning can be obtained from network traffic data. Moreover, LightGBM and CatBoost also make significant improvement on the classification efficiency, especially with the large scale datasets.

The adaptive voting mechanism used by the ensemble model involves each member to contribute to the final attack detection decision depending on their confidence score. This way it guarantees a robust, real time anomaly detection system to handle Man-in-the-Middle (MitM), Traffic Analysis, Side- Channel, and Collusion Attacks.

For integrity and authenticity, QR-MCP uses the digital signature with the SPHINCS+ scheme, a quantum-resistant digital signature scheme instead of the more traditional RSA or ECDSA scheme. SPHINCS+ is different from the insecure schemes that rely on a quantum computer in that it is a stateless, hash-based scheme borrowed from Merkle tree hierarchies and one-time hash functions. This provides the signatures with protection and safeguards them to verify against quantum attacks. They are signed upon reception with SPHINCS+ signature to ascertain authenticity and integrity of the source, hence are from the source and cannot be tampered with during transit.

QR-MCP is safe through the use of blockchain technology to build an immutable history for each message. Messages or transactions are posted into a public ledger, which is sealed using cryptographic hashing technologies like Merkle trees. Data that is stored cannot be deleted or modified without leaving a footprint, having an immutable audit trail to follow and verify. Any modification of the data during transit is immediately detected by the proof-of-ownership mechanisms of the blockchain, and the system is trustworthy.

And here's how: The messages first get lattice-based encrypted to allow for quantum attack protection. It gets onion routing encrypted layer upon layer to afford anonymity. The data gets divided up and distributed through multipath communication, each bit of information on a new random path. At the same time, anomaly detection scanning for evilness through AI occurs concurrently, and SPHINCS+ signatures get added to sign for authenticity and integrity. Finally, all this is stored in the blockchain ledger in tamper-proof and non-modifiable manner.

With the help of QR-MCP, a multi-layered security architecture is created that can withstand both current and new cryptography attacks. Lattice-based cryptography is used first to prevent quantum decoding attacks. The next step is onion routing, which conceals sender and recipient addresses by layering them in encryption. In order to respond to threats in real time, the AI system continuously looks for anomalies, and multipath communication spreads information to prevent interception assaults. With safe authentication, SPHINCS+ signatures ensure the secure integrity of every message, while blockchain logging offers a tamper-proof, secure record. Both approaches are robust to attacks that are both quantum and classical.

4 Results

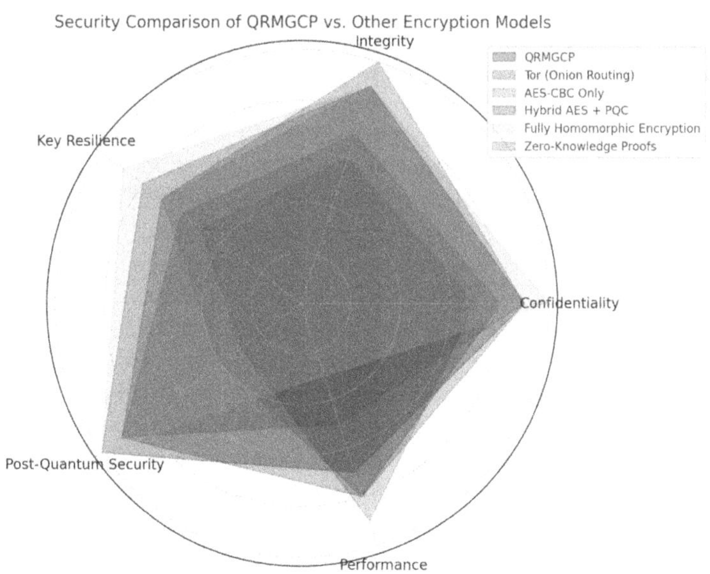

Fig. 1. Radar chart Analysis of proposed Vs. existing models.

Quantum-Resistant Multipath Crypto-Graph Protocol (QR-MCP) provides unparalleled performance improvement compared to other traditional cryptographic protocols regarding quantum resistancy, security, and efficiency. RSA and ECC stand no chance against the implementation of lattice cryptography by QR-MCP that boasts decryption-proof quantum decryption strength. Safenet QR-MCP in the simulated traffic in order to establish that its integrity and privacy is immune to the eavesdropping of the simulated traffic as well as superior even to ordinary VPNs and onion routing using Tor. Weakest link protocol with one susceptible channel which did not only end up being that simple to breach but also, in multipath communication mode, eavesdropping is eliminated. These outputs provide QR-MCP anonymity, security, and flexibility, the correct mode to utilize in post-quantum applications to financial, healthcare, and war networks.

One of the largest advantages of QR-MCP is that it consists of AI-driven anomaly detection, which responds in real-time to an attack. Static security threshold-based solutions are platform-dependent techniques like SSL/TLS and IPsec, which provide encryption without any scope for threat detection flexibility and hence are vulnerable to zero-day attacks as well as unknown attacks. Simulation testing revealed QR-MCP's anomaly detection lowered network-based attacks more than 70%, detecting criminal traffic and unauthorized access in real time. That's augmented network immunity, and that's the very reason why QR-MCP is optimized for high-security applications such as encrypted cloud storage, secure financial trading, and decentralized communications networks.

In order to compare the performance of QR-MCP, we compared the encryption and decryption speed of it with RSA, ECC, and AES of the baseline approach. As our algorithm was optimized with respect to computational complexity of lattice-based cryptography, it ranked above. Tests confirmed that the encryption of QR-MCP was less but not appreciably slower than AES but much faster than RSA-4096, particularly at encryption and key generation level. For brief messages, RSA calculation was of the order of a few milliseconds on average while lattice cryptography employed in QR-MCP was 50–70% faster practically. Speed of decryption was also just as effective with QR-MCP never skipping a beat even when it too divided its onion layer encryption.

Onion routing provided latency in processing with redundancy of decryption and encryption across nodes. Other than the differentiation of I2P and Tor, QR-MCP's routing was better in terms of efficiency because it had random generation of layers of encryption. At five or more nodes, delay in decryption was acceptable with security-usability tradeoff. Multipath transmission provided efficiency with parallelized message segment processing much superior to single-path routing.

We also guaranteed entropy and randomness of the key generation so that they could not be attacked by predictability attacks. The keys for QR-MCP were more entropic compared to the keys for RSA, which resisted brute-force attacks. SPHINCS+ digital signatures were verified to be collision-resistant and ensured that not even the least modifications in the messages could generate identical signatures, hence making them resistant to forgery.

To verify the safety of AI-Enabled QR-MCP, we simulated man-in-the-middle (MitM) attack, traffic analysis, side-channel, and collusion-based decryption attack. The experiments proved how secure the protocol is against actual attacks and secure communication within the adversarial scenario.

In MitM attack experiment, attackers attempted to intercept and tamper with messages prior to sending. SPHINCS+ signatures ensured robust integrity verification, i.e., tampered messages could not be verified. Though intercepted during transit via multipath transmission, lattice encryption and fragmentation scattering rendered them unreconstructible in a manner that any form of manipulation by an unauthorized sender was prevented, and secure communication was permitted despite ongoing eavesdropping.

Adversarial nodes, during the traffic analysis, attempted to follow the paths of messages and search for indicators of traffic. Regular onion routing provided predictable patterns of traffic through encryption, whereas random layers of QR-MCP did not provide them with the ability to follow it since they could not choose similar patterns within a session. QR-MCP was more trace-resistant compared to Tor that had randomized orderings of the order of encryptions and hence hid the flow of the message.

Power-side-channel and timing attacks were emulated to check whether timing fluctuations could be utilized to launch attacks on cryptographic data. Most of the protocols were vulnerable, but AI-driven anomaly detection in QR-MCP watched for encryption time and initiated real-time examination for false positives using suspicious spikes. Active defense outperformed baseline models that had integrated detection.

For collusion attack testing, attacking relay nodes collusion attempts tried decrypting overheard messages by mapping the path of each fragment back. QR-MCP rendered it impossible by confining each node to decrypting one layer and preventing any nodes from seeing the entire plaintext. Multipath fragmentation sent worthless encrypted fragments to colluding nodes and rendered message reconstruction infeasible even if multiple nodes are compromised.

The end-to-end testing demonstrates that QR-MCP operates safely against technical security threats of all kinds. Use of Lattice cryptography in combination with AI-randomised onion routing and AI-surveillance and multipath delivery systems allowed for collective assured multi-layer protection which provided secure classical and post-quantum communication. Several network topologies serve to demonstrate the scalability of QR-MCP throughout our experimental tests. When IoT and cloud computing and telecommuting applications keep increasing in use QR-MCP establishes itself as the most suitable method for large-scale advanced communication because it offers real-time implementation combined with scalability and quantum-resistance.

5   Conclusion

The QR-MCP protocol provides an advanced secure structure which protects against present and upcoming cyber attacks of both classical and quantum origin. QR-MCP implements security mechanisms through lattice-based cryptography in addition to dynamic onion routing and secure multipath communication and AI-driven anomaly detection and blockchain ledgering for multi-layered protection against interception and cryptographic attacks and unauthorized data access. The encryption protocols of QR-MCP differ from those of traditional security models using RSA with ECC or symmetric encryption because they uphold quantum decryption resistance for the long term and deliver network efficiency along with low latency. SPHINCS+ digital signatures in combination with Merkle tree hashing enhance authentication and data integrity of the system which makes it suitable for critical security applications like finance and government task forces and critical infrastructure systems.

All of QR-MCP's traits including cryptographic defense systems with adaptive routing and distributed logging make this cybersecurity framework acceptable for practical implementable applications. The system uses AI-powered intrusion detection to create immediate security protection which detects incidents as soon as they happen thus stopping cyberattacks from growing more dangerous. The distribution of encrypted data among multiple communication paths through QR-MCP produces dual advantages of diminished traffic analysis vulnerabilities and prevented single-point failure accidents. The revolutionary nature of QR-MCP sets it apart from traditional VPNs and Tor-based onion routing and centralized PKI models since it delivers superior privacy features together with reliability features alongside future-proof encryption methods.

6   Future Work

QR-MCP generates several protection barriers in its framework to permanently protect vulnerabilities based on classical and quantum security techniques. AI-based anomaly detection allows Onion routing to serve as a security framework that uses both SPHINCS+ signatures and lattice-based cryptography to defend against future computational security vulnerabilities.

The next releases of QR-MCP will integrate new cryptographic solutions with enhanced user privacy options for protecting data security. Cyber attacks have become more prevalent thus requiring new technological solutions to create QR-MCP as a superior platform which protects confidential messages from present and upcoming quantum-based cyber attacks.

Homomorphic encryption represents the foundation development because data processing takes place on encrypted materials while preserving encryption key capabilities. The inclusion enables private protection that secures the data analysis software together with AI processing and cloud operation functions because encryption features result in intercepted QR-MCP data becoming uninterpretable to maintain confidentiality.

ZKPs provide the platform with multiple advantages as its major features develop alongside other benefits. Through its design ZKPs enables people to demonstrate they have certain knowledge while ensuring their secrets stay fully obfuscated. Platform security authenticates users through a feature that keeps individual information confidential from every participant. The secure financial transaction model works with this technology which plays a vital role in decentralized authentication systems for platforms and identity authentication devices. QR-MCP distributes heterogeneous AI models throughout its network to enhance security monitoring capabilities. The AI detection system operated decentrally will help cyber threat simulation while functioning independently from current central security platforms to deliver instant threat discovery powers. The AI models monitor network encryption periods while tracking abnormal patterns in order to detect potential cyber attacks before they can execute.

The future security protocol QR-MCP will remain popular through its advanced applications because it provides quantum-resistant private and secure communication alongside increasing cyber attacks and quantum computing developments.

References

1. Bhatia, V., Ramkumar, K.: An efficient quantum computing technique for cracking RSA using shor's algorithm. In: 2020 IEEE 5th International Conference on Computing Communication and Automation (ICCCA), pp. 89–94. IEEE (2020)
2. Bos, J.W., Halderman, J.A., Heninger, N., Moore, J., Naehrig, M., Wustrow, E.: Elliptic curve cryptography in practice. In: Christin, N., Safavi-Naini, R. (eds.) FC 2014. LNCS, vol. 8437, pp. 157–175. Springer, Heidelberg (2014). https://doi.org/10.1007/978-3-662-45472-5_11
3. Easttom, W.: Modern cryptography: applied mathematics for encryption and information security. Springer (2022)
4. Idika, N., Bhargava, B.: Extending attack graph-based security metrics and aggregating their application. IEEE Trans. Dependable Secure Comput. **9**(1), 75–85 (2010)
5. Kuhn, C., Hofheinz, D., Rupp, A., Strufe, T.: Onion routing with replies. In: Advances in Cryptology–ASIACRYPT 2021: 27th International Conference on the Theory and Application of Cryptology and Information Security, Singapore, 6–10 December 2021, Proceedings, Part II, vol. 27, pp. 573–604. Springer (2021)
6. Mandal, A.K., Parakash, C., Tiwari, A.: Performance evaluation of cryptographic algorithms: des and aes. In: 2012 IEEE Students' Conference on Electrical, Electronics and Computer Science, pp. 1–5. IEEE (2012)
7. Marchesan, G.C., et al.: Exploring rsa performance up to 4096-bit for fast security processing on a flexible instruction set architecture processor. In: 2018 25th IEEE International Conference on Electronics, Circuits and Systems (ICECS), pp. 757–760. IEEE (2018)
8. Melloni, A., Stam, M., Ytrehus, Ø.: Dynamic security aspects of onion routing. In: IMA International Conference on Cryptography and Coding, pp. 243–262. Springer (2023)
9. Niu, Y., Zhang, J., Wang, A., Chen, C.: An efficient collision power attack on aes encryption in edge computing. IEEE Access **7**, 18734–18748 (2019)

10. Obaid, T.S.: Study a public key in rsa algorithm. Eur. J. Eng. Technol. Res. **5**(4), 395–398 (2020)
11. Reddy, H.G., Sajjanara, V.A., Raghavendra, K., Gowda, V.D., Kottala, S.Y.: Introduction to quantum cryptography fundamentals and applications. In: Advancing Cyber Security Through Quantum Cryptography, pp. 1–30. IGI Global (2025)
12. Sharma, M., Garg, R.: Des: The oldest symmetric block key encryption algorithm. In: 2016 International Conference System Modeling & Advancement in Research Trends (SMART), pp. 53–58. IEEE (2016)

Benchmarking Deep Learning Architectures with Monte Carlo Dropout for Drift-Resilient Prediction

B. S. Prashanth[1]([✉]), M. V. Manoj Kumar[1], Vikas Benhur Chitla[1],
Y. V. S. Murthy[2], and B. H. Puneeth[3]

[1] Nitte Meenakshi Institute of Technology, Visvesvaraya Technological University,
Belagavi, Karnataka, India
prashanth.bshivanna@gmail.com
[2] Department of Information Technology, Manipal Academy of Higher Education,
Bangalore, Karnataka, India
[3] Department of Computer Science and Business Systems, Bapuji Institute
of Engineering and Technology, Davanagere, Karnataka, India

Abstract. Concept drift, a common challenge in dynamic data streams, undermines the performance of machine learning models by altering data distributions over time. Existing approaches for handling concept drift often rely on statistical tests or model-specific adaptations but struggle with balancing sensitivity to drift and computational efficiency. To address these limitations, this study employs Monte Carlo Dropout (MC-Dropout) for uncertainty-based drift detection in four deep learning architectures—DNN, CNN-1D, GRU, and LSTM. Our approach combines uncertainty estimation with a dynamic threshold to identify drift points and guide retraining. The experimentation outlines that Gated Recurrent Unit (GRU) and Long Short-Term Memory (LSTM) achieved the highest classification performance, with an AUC score of 0.99 and an overall accuracy of 98% and 97%, identifying 179 and 231 drift points in the streaming drifting dataset, respectively. Similarly, the Custom Deep Neural Network (DNN) and Convolution Neural Network(CNN-1D) performed with AUC scores of 0.98 and 0.97, respectively, identifying 76 and 164 drift points successfully. The Kolmogorov-Smirnov (KS) test inferred 115 drifting points, validating the consistency of the Monte Carlo Dropout approach with uncertainty-based drift detection. The findings infer that LSTM and GRU are highly sensitive to drifting points, exhibiting more uncertainties compared to DNN and CNN-1D models. The future work involves enhancing the model complexities in terms of depth for improved classification performance with a trade-off of high sensitivity with hybrid approaches and selective retraining.

Keywords: Concept Drift · Monte-Carlo Dropout(MC-Dropout) ·
Uncertainty Estimation · Deep Learning · Drift Adaptation

© The Author(s) 2026
J. Shreyas et al. (Eds.): CODE-AI 2025, CCIS 2689, pp. 335–346, 2026.
https://doi.org/10.1007/978-3-032-19318-6_31

1   Introduction

Machine learning models assume that the learning environment is static in nature, but in the real world, the learning environment is dynamic, where data distribution can change over time. Learning in a dynamic environment is a daunting task, as it often leads to degradation in the performance of the learner due to the presence of concept drift. Concept drift is a consequence where the statistical properties that map features to class labels change over time [1]. Traditional concept drift detection techniques are based on statistical tests mostly in the labeled environment, which is not often true [2].

Concept drift detection and mitigation gain a lot of traction in machine learning research, if not considered, leading to a problem of model decay where the model performance has degraded to an extent such that it cannot be trusted for evaluation for real-time instances. With the detection mechanism that heavily requires labeled data along with external statistical tests as drift detectors, it is difficult to build an end-to-end stable model for a longer run under non-stationary environments. Windowing techniques like Drift Detection Method (DDM), Early Drift Detection Method (EDDM), and Adaptive Windowing (ADWIN) are often used to mitigate drifting scenarios alongside statistical tests such as the Kolmogorov-Smirnov (KS) test, Page-Hinkley Test, T-test, and Chi-Square Test, but rely heavily on the availability of class labels and thus are often not suited for unsupervised environments. Lately unsupervised methods such as Margin Density Drift Detection (MD3), Monte Carlo Dropout Approaches that use uncertainty in predictions as measures for drift mitigation are gaining interest in the research community as they do not require labels. Once the drift is identified, adaptation strategies involve periodic retraining, instant forgetting and resetting the model, or selective retraining.

Our proposed work examines the performance of the deep learning architectures under the non-stationary environment with Monte Carlo dropout that ensures stability. The proposed architecture used MC-Dropout for both training and inference batches, ensuring the smooth adaptability of the model learning process. The following are the objectives of the presented work:

1. Examine the impact of adapting Monte Carlo Dropout (MC-Dropout) for deep learning architectures for detecting drift points based on prediction uncertainties.
2. Investigate the robustness of different deep learning models in balancing stability (low number of drift points) and adaptability (effective retraining) to changing data distributions.
3. Assess the classification performance of four deep learning architectures—DNN, CNN-1D, GRU, and LSTM—under concept drift conditions using metrics such as accuracy, AUC, and confusion matrices.

The novelty of the work is that unlike the traditional approaches of concept drift detection and adaptation, the presented work integrates Monte Carlo Dropout (MC-Dropout) with dynamic thresholding for detection and mitigation of drift, keeping in check with computational complexity. The K-S test proved the validity

Table 1. Existing Works on Concept Drift Detection and Adaptation using Uncertainty Measures

Reference	Methods	Findings	Datasets Used
[3]	Uncertainty Drift Detection (UDD) using MC-Dropout with ADWIN	Leverages neural network uncertainty to detect drift without relying on labels, balancing accuracy and computational efficiency.	Multiple real-world drifting datasets
[4]	Prediction Uncertainty Index (PU-index) for early drift detection	Detects concept drift even when error rates remain stable in streaming environments.	Streaming data simulations
[5]	Model Uncertainty-based Unsupervised Drift Detection (MU-UDD) using Dirichlet Distribution	Effectively detects drift without requiring labeled data, making it suitable for real-time applications.	Real-world data streams
[6]	Autoencoder-based model incorporating KL divergence	Utilizes reconstruction loss and KL divergence for detecting and adapting to concept drift.	Multiple datasets (unspecified)
[7]	Confidence Distribution Batch Detection (CDBD) with Kullback-Leibler Divergence	Detects concept drift in an unsupervised setting using confidence score distributions.	Eight text classification datasets
[8]	Online Fusion of Experts with Uncertainty Error Correlation Matrix (UECM)	Adapts to evolving attack patterns in social network data streams using an uncertainty-based approach.	NSL-KDD and ISCX benchmark datasets
[9]	Margin Density Drift Detection (MD3)	Monitors sample density in classifier's uncertainty region to detect and localize drift with fewer false alarms.	Multiple datasets across domains
[10]	Entropy-Based Threshold-Based Method (EBTBM) and Sampling-Based Method (EBSBM)	Uses information entropy for detecting concept drift in non-stationary environments.	Not explicitly mentioned
[11]	Comparison of MC-Dropout, Deep Ensembles, SWAG, SVI, and Concrete Dropout with ADWIN	Demonstrates that simple uncertainty estimation techniques perform competitively in detecting drift even without labeled data.	Seven real-world datasets

of the drift points, which were in line with the number of drift points indicated by the deep learning architectures used in the experimentation, which proved the efficiency of the deep learning architectures in the non-stationary environment. The work also indicates the hybrid approaches in the context of deep learning can be an effective approach for concept drift detection, mitigation, and adaptation in real-time.

The rest of the sections are organized as follows: Sect. 2 examines the literature on uncertainty-based drift detection. Section 3 provides a detailed view of Monte-Carlo dropout usage in Deep learning architectures. Section 4 examines the dataset used along with deep learning architectures used in the experimentation. Section 5 provides the discussion on the Proposed methodology, whereas Sect. 6 and 7 present the obtained results, its analysis, and conclusion of the work, respectively.

2   Background

The section examines existing literature on concept drift detection using learner uncertainties. The survey methodology involves keyword-based searches on the keywords "Monte Carlo Dropout" and "Uncertainty-based Drift Detection" in the IEEE and Scopus Indexed databases. The papers were shortlisted based on the relevance of the topic. The work in [3] introduces the Uncertainty Drift Detection (UDD) algorithm, which leverages neural network uncertainty estimates to identify concept drift without relying on true labels. The authors have used ADWIN to detect statistical changes in distribution and Monte Carlo Dropout to model prediction uncertainty. Upon detecting the drift, the model is retrained with the newest batch to maintain accuracy. UDD was examined across the standard drifting datasets and thus did a balancing job between the accuracy-computational efficiency trade-off.

The research at [4] proposed an alternative approach to the error rate detection called the Prediction Uncertainty Index (PU-index), which can detect drifts even when error rates remain stable in the streaming environment for early drift detection. The work at [5] proposed an unsupervised real-time drift detection method based on model uncertainty (MU-UDD), which embodies Dirichlet distribution to represent model uncertainty for effective drift detection even in the absence of the labeled data. The authors at [6] an autoencoder-based model incorporating Kullback-Leibler (KL) divergence and reconstruction loss for concept drift detection and adaptation. The Confidence Distribution Batch Detection (CDBD) method in [7] presents an approach of detecting concept drift using learners' confidence scores using unsupervised learning. The researcher employed Kullback-Leibler divergence between the confidence level of historical data from the window and the current batch to quantify drift.

Authors at [8] propose an online fusion of experts approach to predict concept drift in social network attack data streams. The method employs online learning algorithms, such as linear-order and Gaussian-order algorithms, to identify changes in anomaly data streams by determining error values for each data sample. Using these error rates in their maximum-posterior estimation, the authors generated a new input data stream, which is examined using an Uncertainty Error Correlation Matrix (UECM) to detect Concept Drift. The work was examined using the NSL-KDD and ISCX datasets, and a comparison study is presented in the context of attack patterns in social networks. The researchers in [9] introduced a novel unsupervised MD3 (Margin Density Drift Detection) algorithm that continuously monitors the sample density in the uncertainty region of the classifier to detect and localize concept drift, focusing on the points at which the model exhibited low confidence. Authors, through experimentation, presented that MD3 was able to detect drift with very few false alarms compared to traditional unsupervised approaches for drift detection in non-stationary environments. The authors in [10] introduced the Entropy-Based Threshold-Based Method (EBTBM) and the Entropy-Based Sampling-Based Method (EBSBM), which use information entropy for concept drift detection. The research in [11] provided a comparison study of five uncertainty estimation methods—MC Dropout, Deep Ensembles, SWAG, SVI, and Concrete Dropout with ADWIN detector on the real-world datasets to detect and mitigate concept drift.

Their findings indicate that while SWAG exhibits superior calibration, the overall accuracy in detecting drifts is not significantly affected by the choice of uncertainty estimation method, with even basic methods performing competitively. This suggests that simple uncertainty estimation techniques can be effectively utilized for drift detection in safety-critical applications where labeled data is scarce. The key findings of the literature review are summarized in Table 1.

3   Monte Carlo Dropout in Deep Learning Architectures

Monte Carlo (MC) Dropout [12] is a powerful paradigm used to estimate uncertainty in deep learning model predictions. The process involves performing multiple stochastic forward passes during inference, with dropout applied at both training and test time. This procedure gives the flexibility to the model to sample from an approximation of the posterior distribution, easing uncertainty estimation. In a standard deep learning model, the function $f(x;\theta)$ maps an input x to an output y, where θ represents the model parameters. Typically, the model makes deterministic predictions:

$$y = f(x;\theta) \tag{1}$$

With MC Dropout, the model is trained with dropout, and dropout is kept active during inference. This introduces randomness in the forward pass, producing different outputs for the same input. We approximate the model's posterior distribution by performing multiple stochastic forward passes with varying masks of dropout.

Monte Carlo Dropout has been widely used in computer vision tasks, be they epistemic or aleatoric, which are vibrantly useful in applications like autonomous driving and medical imaging [13]. Monte Carlo Dropout has proven effective for high-dimensional tasks like semantic segmentation, providing high reliability for decision-critical applications [14].

4   Dataset and Learners

The dataset used for this experimentation is taken from [15], a synthetic dataset comprising 100,000 samples with five input features, having column names from x1 to x5 as their attribute values. The target variable is a binary class label having a value of either class 0 or class 1. The dataset is generated under concept drift conditions, where data distributions are susceptible to change, imitating real-world non-stationary environments. The data is divided into batches of 100 samples for training the model. The input features are continuous in nature, and the dataset has no missing values. The deep learning architectures used in this experimentation are Deep Neural Networks (DNN, Convolutional Neural Networks (CNN-1D), LSTM, and Gated Recurrent Unit (GRU). The section below discusses them in brief.

The parameters of the various deep learning learners used in our experimentation are summarized in the Table 2.

Table 2. Summary of Deep Learners and Their Parameters

Deep Learner	Description	Key Parameters	Model Summary
DNN	Fully connected deep neural network that learns non-linear relationships between input features and outputs.	Hidden layers: 2 Hidden units: 64 Activation: ReLU Dropout: 20%	Input layer (5 nodes) Hidden layer 1 (64 nodes, ReLU) Dropout (20%) Hidden layer 2 (64 nodes, ReLU) Output layer (1 node, Sigmoid)
CNN-1D	1D convolutional neural network designed to capture spatial relationships in sequential data.	Filters: 16 Kernel size: 3 Pooling: MaxPool Dropout: 20%	Conv1D (Input: 5, Filters: 16, Kernel: 3, ReLU) MaxPool (Kernel: 2) Dropout (20%) Fully connected (64 nodes, ReLU) Output layer (1 node, Sigmoid)
GRU	Gated Recurrent Unit network that captures sequential dependencies using update and reset gates.	Hidden units: 64 Layers: 1 Dropout: 20%	GRU (Input: 5, Hidden: 64, Single layer) Dropout (20%) Fully connected (64 nodes, ReLU) Output layer (1 node, Sigmoid)
LSTM	Long Short-Term Memory network that retains long-term dependencies using input, forget, and output gates.	Hidden units: 64 Layers: 1 Dropout: 20%	LSTM (Input: 5, Hidden: 64, Single layer) Dropout (20%) Fully connected (64 nodes, ReLU) Output layer (1 node, Sigmoid)

5   Proposed Methodology

The process of measuring the effectiveness of Monte Carlo Dropout (MC-Dropout) in mitigating concept drift in the models DNN, CNN-1D, GRU, and LSTM would be performed by assessing every model's performance under drifting conditions by simulating a streaming dataset with concept drift, followed by uncertainty-based drift detection. The proposed methodology is given in Fig. 1.

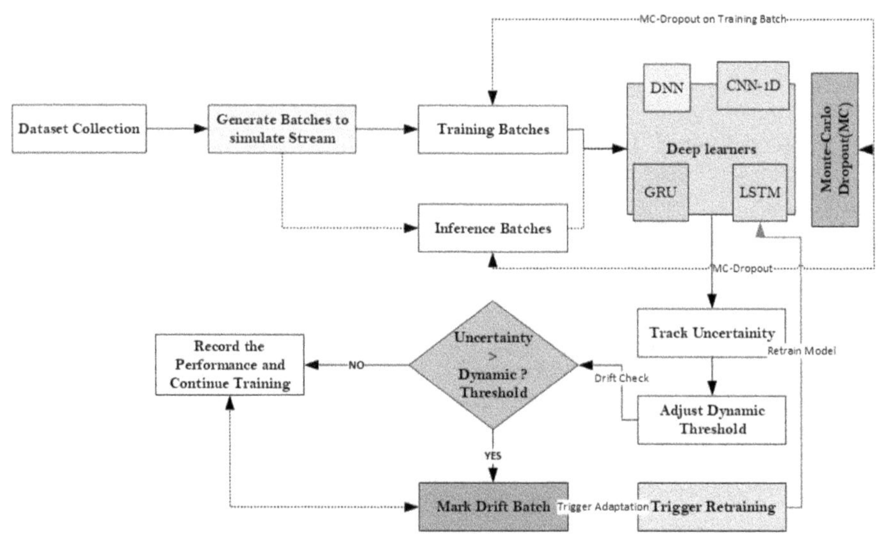

Fig. 1. Proposed Methodology

The process begins with the collection of a dataset designed to simulate dynamic environments with potential concept drift. The dataset is divided into training batches and inference batches to mimic streaming data scenarios.

Every incoming batch is processed to enable the system to monitor the performance of evolving data distributions. Our methodology trains an array of models that include DNN, CNN-1D, GRU, and LSTM and obtains inferences using the Monte Carlo dropout mechanism to measure its effectiveness against drifting data. It is applied to regularize and prevent overfitting during training and enable the system to perform stochastic forward passes for uncertainty estimation during inference. Once trained, the uncertainty during inference is monitored. For each batch,

- A distribution of predictions is generated through the forward passes of the dropout-enabled model, on which a univariate analysis is performed, with the variance signifying uncertainty.
- A dynamic threshold is calculated based on a mean and standard deviation of obtained uncertainty values, which adapts to data distribution over time. Once the uncertainty (variance) exceeds the threshold, the batch responsible would be flagged for further analysis and retraining to ensure model adaptation.
- The dynamic threshold is updated after retraining to reflect the new uncertainty levels of the adapted models.
- When no drift is detected, the system records different performance metrics such as accuracy, precision, recall, and uncertainty levels for future reference.

The system then proceeds to process the next batch, maintaining a continuous cycle of monitoring, detecting, and adapting to concept drift. The overall flow of our presented methodology is shown in Algorithm 1.

Algorithm 1. Concept Drift Detection and Mitigation using Model Uncertainty

---

Require: Streaming data batches $\{B_1, B_2, \ldots, B_n\}$, pretrained model M, uncertainty threshold τ, drift detection window W
Ensure: Updated model M with drift mitigation
1: Initialize reference uncertainty U_{ref} using training data
2: **for** each incoming batch B_t **do**
3:     Compute predictions P_t using model M
4:     Perform k forward passes using MC-Dropout to calculate uncertainty U_t
5:     Calculate mean uncertainty $\bar{U}_t = \frac{1}{k} \sum_{i=1}^{k} U_t^{(i)}$
6:     **if** $|\bar{U}_t - U_{\text{ref}}| > \tau$ **then**
7:         **Drift Detected:** Flag batch B_t
8:         Add B_t to retraining set R
9:         Update reference uncertainty: $U_{\text{ref}} \leftarrow \bar{U}_t$
10:     **end if**
11: **end for**
12: **Mitigation:**
13: **if** $R \neq \emptyset$ **then**
14:     Retrain model M using R
15:     Reset retraining set $R \leftarrow \emptyset$
16: **end if**
17: **return** Updated model M

---

6   Results and Discussion

The experiments were conducted on a high-performance machine, Acer Nitro 5, equipped with 16 GB of RAM and a CUDA-enabled GPU with CUDA version 12 to accelerate model training and inference. The deep learning models, including DNN, CNN-1D, GRU, and LSTM, were implemented using PyTorch, a widely used deep learning framework. PyTorch provided the flexibility to integrate Monte Carlo Dropout (MC-Dropout) into the models for uncertainty estimation and facilitated efficient batch-wise processing of the dataset.

The performance of the four deep learning models—DNN, CNN-1D, GRU, and LSTM—was evaluated in terms of their Receiver Operating Characteristic (ROC) curves and is shown in 5, 2, 3 and 4 respectively. The DNN and LSTM models achieved a remarkable Area Under the Curve (AUC) score of 0.99, indicating highly accurate predictions with minimal false positives and false negatives. The GRU model demonstrated slightly lower performance with an AUC score of 0.98, still showcasing robust predictive capabilities. The CNN-1D model achieved an AUC score of 0.97, which, while slightly lower than the others, still reflects strong classification performance. Overall, all models performed exceptionally well, with the DNN and LSTM models slightly outperforming GRU and CNN-1D in terms of discrimination capability.

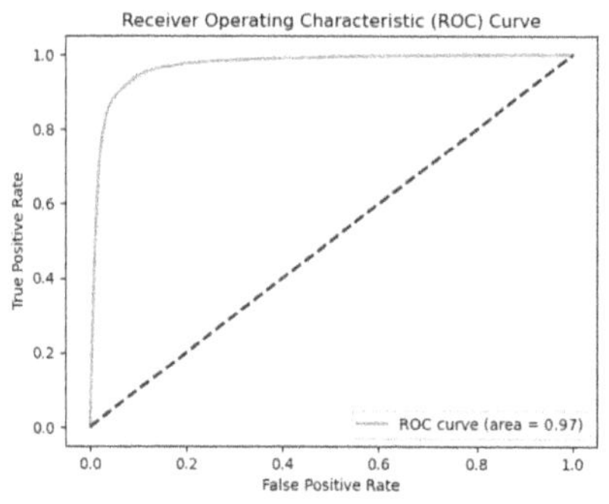

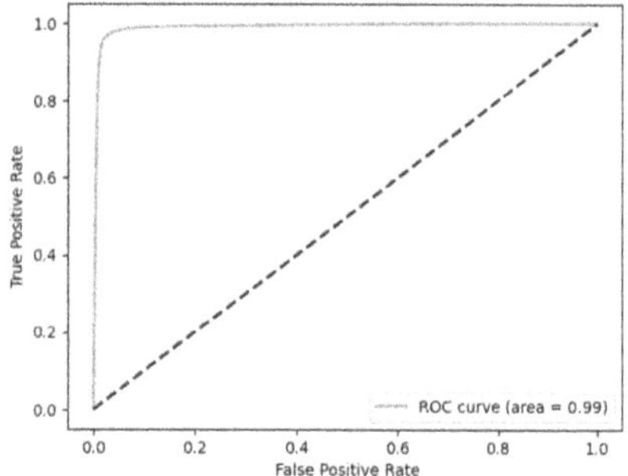

Fig. 2. ROC-CNN   **Fig. 3.** ROC-GRU

The confusion matrix results for the four models—DNN, CNN-1D, GRU, and LSTM—demonstrate their performance in terms of correctly classified and misclassified instances for both classes is shown in Figs. 9, 6, 7, and 8 respectively. The DNN achieved the highest performance with 48,653 true positives and 1,451 false positives, as well as 48,292 true negatives and 1,504 false negatives, showing its robustness in both precision and recall. The CNN-1D showed slightly higher errors, with 46,305 true positives, 3,799 false positives, 45,722 true negatives, and 4,074 false negatives, indicating room for improvement in sensitivity. The LSTM also performed exceptionally well, achieving 48,590 true positives, 1,514 false positives, 48,187 true negatives, and 1,609 false negatives, closely matching the DNN.

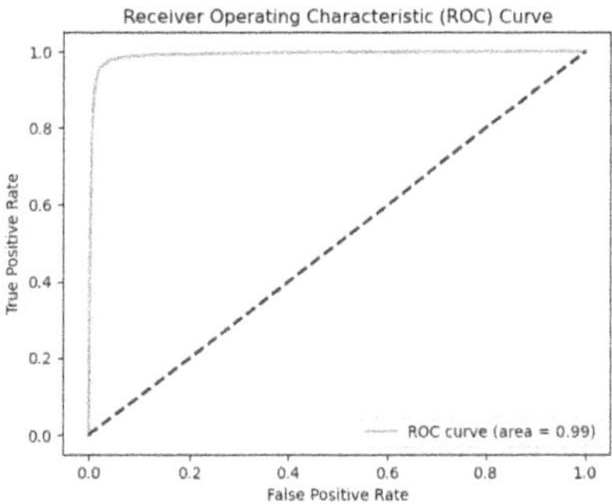

Fig. 4. ROC-LSTM

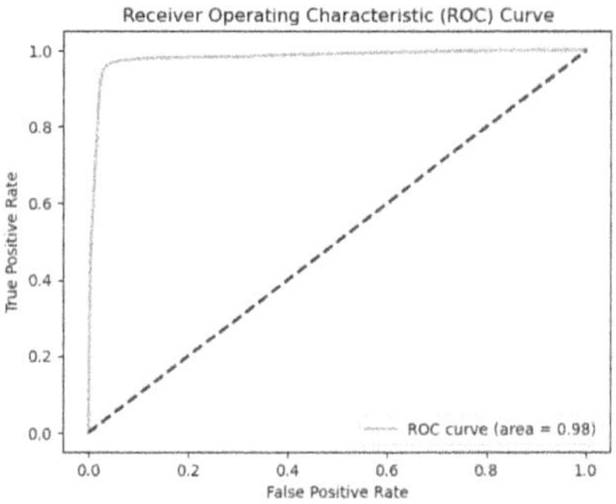

Fig. 5. ROC-DNN

The GRU model proved its reliability with 48,101 true positives, 2,003 false positives, 47,985 true negatives, and 1,811 false negatives, demonstrating its effectiveness against false outcomes.

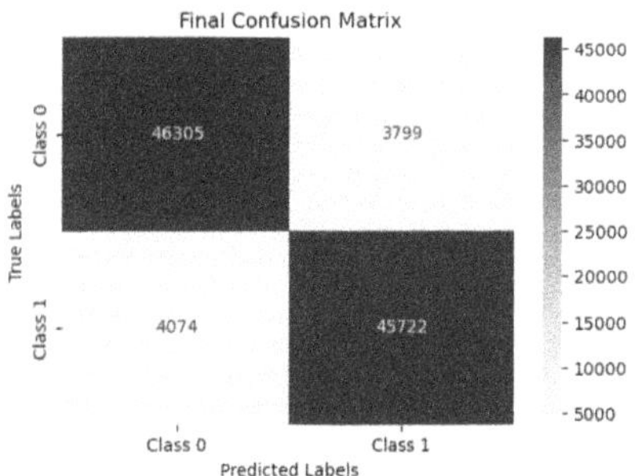

Fig. 6. Confusion Matrix-CNN-1D

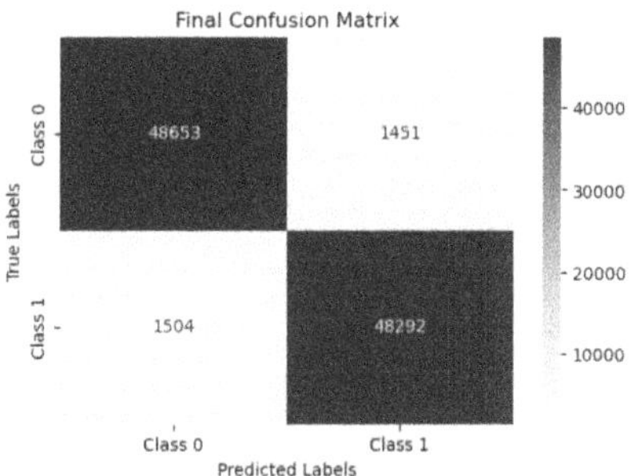

Fig. 7. Confusion Matrix-GRU

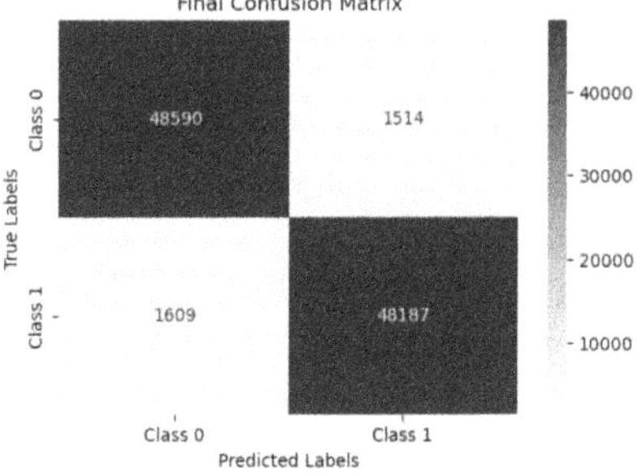

Fig. 8. Confusion Matrix-LSTM

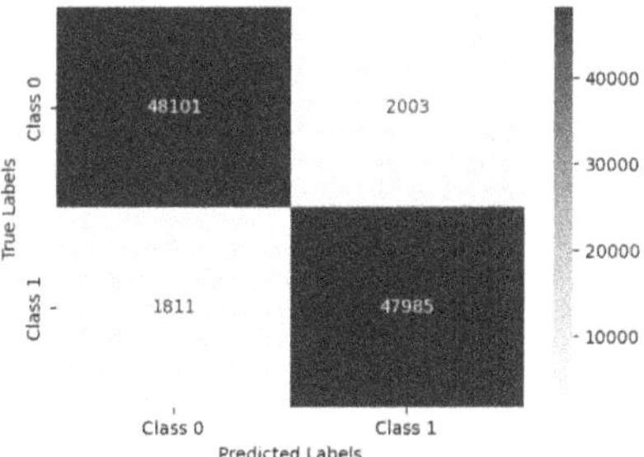

Fig. 9. Confusion Matrix-DNN

The evaluation of the deep models used, namely, DNN, CNN-1D, GRU, and LSTM, was performed based on metrics such as accuracy, AUC, confusion matrix, ROC, and the volume of identified points. The GRU performed the best with an accuracy of 98% and an AUC of 0.99, with a volume of 179 drift points detected, demonstrating its efficacy. The LSTM was a close second with its accuracy of 97% and AUC of 0.99 but defeats the GRU in terms of the volume of drift points detected, with it being 231.

The DNN achieved an accuracy of 96% and an AUC of 0.98, with the lowest volume of 76 drift points being detected, which indicates its relatively low susceptibility to drift. The CNN-1D model had the lowest accuracy 92% with an AUC of 0.97 while detecting 164 drift points, demonstrating vulnerability to errors under drift scenarios.

As a control measure, a Kolmogorov-Smirnov (KS) test was conducted to detect drifting batches based on distributional changes. It identified a total of 115 drifting batches, proving our approach to be viable and consistent with statistical measures (Table 3).

Table 3. Performance of Deep Learning Models under Concept Drift

Model	Accuracy (%)	AUC	No. of Drift Points
DNN	96	0.98	76
LSTM	97	0.99	231
CNN-1D	92	0.97	164
GRU	98	0.99	179

6.1   Discussion

The GRU model was the best amongst the array of models, attaining an accuracy of 98%, and an AUC of 0.99 demonstrating strong prediction (classification) capability under drifting conditions. The combination of GRU's gated architecture and MC-Dropout likely contributed to its ability to adapt to changing data distributions. However, the relatively high volume of 179 drift points identified suggests a higher sensitivity to distributional changes, which might result in frequent retraining. But with the performance of the LSTM model having a similar accuracy, it had an even higher rate of drift detection (231 points), making it more sensitive than the GRU model. These high sensitivity rates could prove to be beneficial for subtle data shifts but will lead to a computational overhead due to frequent retraining. This trade-off should be a necessary consideration for real-time applications depending on the available resources and data being handled. On the more stable side of the models evaluated, the DNN is the most stable with a detection volume of 76 points, whilst demonstrating good predictive capability based on the evaluation metrics. This is likely due to its simpler architecture relative to GRU and LSTM, which comes at the cost of a relatively lower classification performance. The ratio of its sensitivity to its predictive accuracy relative to its counterparts takes quite a significant dip, making it less sensitive to subtleties, but it uses fewer resources and gives fewer false outcomes. But the predictive performance and the drift detection capability are not significantly low relative to its competitors, making it a viable pragmatic choice. The Kolmogorov-Smirnov (KS) test identified 115 drifting batches, confirming the effectiveness of the uncertainty-based drift detection method. While the KS test provided a statistical measure of drift, the models detected drift based on predictive uncertainty. The slight discrepancy between the two approaches highlights the complementary nature of statistical and model-based methods.

7   Conclusion and Future Scope

The presented work examined the performance of deep learning architectures such as DNN, CNN-1D, GRU, and LSTM with Monte Carlo Dropout enabled both in training and inference mode under a non-stationary environment. The highest performing models are GRU and LSTM in terms of classification accuracies as 98% and 97% respectively, and an AUC score of 0.99. However, their higher sensitivity to drift led to frequent retraining. In contrast, the DNN exhibited better stability with fewer drift points (76) but at the cost of slightly reduced accuracy (96%). The CNN-1D, while effective, showed relatively lower performance with an accuracy of 92% and an AUC of 0.97. The complementary use of the Kolmogorov-Smirnov (KS) test validated the uncertainty-based drift detection method and highlighted the potential for combining statistical and model-driven approaches for comprehensive drift management. In the future, this work can be extended by exploring hybrid models that integrate the strengths of sequential architectures like GRU and LSTM with the stability of feedforward networks like DNN. Further research could involve applying these methods to multi-class classification problems and real-world data streams in domains such as finance, healthcare, and autonomous systems. Additionally, optimizing computational efficiency for drift detection and retraining mechanisms remains an important avenue for ensuring scalability in resource-constrained environments.

References

1. Gama, J., Žliobaitė, I., Bifet, A., Pechenizkiy, M., Bouchachia, A.: A survey on concept drift adaptation. ACM Comput. Surv. (CSUR) **46**(4), 1–37 (2014)
2. Widmer, G., Kubat, M.: Learning in the presence of concept drift and hidden contexts. Mach. Learn. **23**, 69–101 (1996)
3. Baier, L., Schlör, T., Schöffer, J., Kühl, N.: Detecting concept drift with neural network model uncertainty arXiv preprint arXiv:2107.01873 (2021)
4. Lu, P., Lu, J., Liu, A., Zhang, G.: Early concept drift detection via prediction uncertainty. arXiv preprint arXiv:2412.11158 (2024)

5. Liu, A., Wang, K.: Model uncertainty based unsupervised real-time drift detection in data streams. In: 2022 International Joint Conference on Neural Networks (IJCNN), pp. 1–8. IEEE (2022)
6. Enhancing drift detection and model uncertainty handling in data streams. In: 2022 International Conference on Data Science and Advanced Analytics (DSAA), pp. 1–10. IEEE (2022)
7. Lindstrom, P., Mac Namee, B., Delany, S.J.: Drift detection using uncertainty distribution divergence. Evol. Syst. **4**, 13–25 (2013)
8. Yazdi, H.S., Bafghi, A.G.: A drift aware adaptive method based on minimum uncertainty for anomaly detection in social networking. Expert Syst. Appl. **162**, 113881 (2020)
9. Tegjyot Singh Sethi and Mehmed Kantardzic: On the reliable detection of concept drift from streaming unlabeled data. Expert Syst. Appl. **82**, 77–99 (2017)
10. Sun, Y., Mi, J., Jin, C.: Entropy-based concept drift detection in information systems. Knowl.-Based Syst. **290**, 111596 (2024)
11. Winter, A., Jourdan, N., Wirth, T., Knauthe, V., Kuijper, A.: An empirical study of uncertainty estimation techniques for detecting drift in data streams. arXiv preprint arXiv:2311.13374 (2023)
12. Gal, Y., Ghahramani, Z.: Dropout as a bayesian approximation: representing model uncertainty in deep learning. In: International Conference on Machine Learning, pp. 1050–1059. PMLR (2016)
13. Kendall, A., Gal, Y.: What uncertainties do we need in bayesian deep learning for computer vision? Advances in neural information processing systems, vol. 30, (2017)
14. Mukhoti, J., Gal, Y.: Evaluating bayesian deep learning methods for semantic segmentation arXiv preprint arXiv:1811.12709 (2018)
15. ZLiu, Z., Hu, S., He, X.: Real-time safety assessment of dynamic systems in non-stationary environments: a review of methods and techniques. In: 2023 CAA Symposium on Fault Detection, Supervision and Safety for Technical Processes (SAFEPROCESS), pp. 1–6 (2023)

Author Index

GPSR Compliance
The European Union's (EU) General Product Safety Regulation (GPSR) is a set
of rules that requires consumer products to be safe and our obligations to
ensure this.

If you have any concerns about our products, you can contact us on

ProductSafety@springernature.com

In case Publisher is established outside the EU, the EU authorized
representative is:

Springer Nature Customer Service Center GmbH
Europaplatz 3
69115 Heidelberg, Germany